Sustainability through the Lens of Environmental Sociology

Special Issue Editor
Md Saidul Islam

MDPI • Basel • Beijing • Wuhan • Barcelona • Belgrade

MDPI

Special Issue Editor
Md Saidul Islam
Nanyang Technological University
Singapore

Editorial Office
MDPI AG
St. Alban-Anlage 66
Basel, Switzerland

This edition is a reprint of the Special Issue published online in the open access journal *Sustainability* (ISSN 2071-1050) in 2015–2017 (available at: http://www.mdpi.com/journal/sustainability/special_issues/EnvironmentalSociology).

For citation purposes, cite each article independently as indicated on the article page online and as indicated below:

Lastname, F.M.; Lastname, F.M. Article title. *Journal Name*. **Year**. *Article number*, page range.

First Edition 2018

ISBN 978-3-03842-660-8 (Pbk)
ISBN 978-3-03842-661-5 (PDF)

Table of Contents

About the Special Issue Editor

Md Saidul Islam is Associate Professor of Sociology and the Coordinator of the Environment and Sustainability Research Cluster, School of Social Sciences and Asian School of the Environment, Nanyang Technological University, Singapore. He also taught at York University in Canada and the College of William and Mary in the United States. Dr. Islam's key interest is environmental sustainability focusing on global agro-food system and climate change. He published five books, and over three dozen peer-reviewed journal articles and book chapters. In 2015, the Canadian Sociological Association (CSA) awarded him the Early Investigator Award/Prix jeune chercheur that translates into the top emerging sociologist of the year. He's currently working on, among other projects, Climate change and food security in the Asia-Pacific: Response and Resilience, a tier-1 project supported by a financial grant from the Ministry of Education, Singapore.

Preface to "Sustainability through the Lens of Environmental Sociology"

Twenty years after the Rio summit in 1992, world leaders met in Rio once again in 2012 to discuss the environmental challenges facing the humanity. It was a time for them to reflect on how successful and effective the international community has been over the past two decades in managing the major identified environmental problems. Are we still facing the same environmental problems? Have there been improvements in the situation or are we worse off? Environmental and social vulnerabilities will continue to exist twenty years from now and beyond. The question is what kind of steps can and should be taken to manage these vulnerabilities? Have they been taken? Do countries across the globe experience the same type and degree of vulnerabilities? Or is the distribution of these vulnerabilities uneven? How is the distribution of these vulnerabilities decided and by whom? What are the prospects for a sustainable planet? Given these critical questions of our time, there is an urgent need to explore and examine environmental sustainability from both local and global contexts. Environmental sociology provides a prowerful lens to understanding, managing and achieving a sustainable planet.

Our planet is undergoing radical environmental and social changes. Environmental sustainability has now been put into question by, for example, our consumption patterns, loss of biodiversity, depletion of resources, and imbalanced and exploitative power relations. On a global scale, every day humans:

Consume over 54% of the accessible runoff water on earth

Mine more materials from the earth than the natural erosion of all earths rivers

Add over 100 million tons of carbon to the atmosphere

Destroy over 180 square miles of tropical rain forest

Create over 60 square miles of desert

Eliminate at least 74 animal or plant species

Erode over 80 million tons of topsoil and

Add over 1400 tons of CFCs to the stratosphere.

With apparent ecological and social limits to globalization and development, the current levels of consumption are unsustainable, inequitable, and inaccessible to the majority of humans. Understanding the environmental sustainability is a crucial matter at a time when our planet is in peril - both environmentally and socially. This edited collection will hopefully show some possible pathways for a sustainable earth. Along with global environmental politics with an aim of a sustainable earth, we need to generate and inculcate new consciousness within the new social media generations about the environment and sustainability. We must develop a new understanding of the true purpose of our existence on this Earth, new models of behavior and a new set of values for the planet.

In closing, I would like to thank the editorial team of *Sustainability* journal for inviting me to guest-edit the special issue on Sustainability through the Lens of Environmental Sociology which has, with some additional chapters, subsequently been transformed into this edited collection. I also thank all the authors and reviewers for their crucial contributions.

Md Saidul Islam
Special Issue Editor

sustainability

MDPI

Editorial

Sustainability through the Lens of Environmental Sociology: An Introduction

Md Saidul Islam

Division of Sociology, Nanyang Technological University Singapore, 14 Nanyang Drive, Singapore 637332, Singapore; msaidul@ntu.edu.sg; Tel.: +65-6592-1519

Academic Editor: Marc A. Rosen
Received: 10 March 2017; Accepted: 15 March 2017; Published: 22 March 2017

Abstract: Our planet is undergoing radical environmental and social changes. Sustainability has now been put into question by, for example, our consumption patterns, loss of biodiversity, depletion of resources, and exploitative power relations. With apparent ecological and social limits to globalization and development, current levels of consumption are known to be unsustainable, inequitable, and inaccessible to the majority of humans. Understanding and achieving sustainability is a crucial matter at a time when our planet is in peril—environmentally, economically, socially, and politically. Since its official inception in the 1970s, environmental sociology has provided a powerful lens to understanding the challenges, possibilities, and modes of sustainability. This editorial, accompanying the Special Issue on "sustainability through the Lens of Environmental Sociology", first highlights the evolution of environmental sociology as a distinct field of inquiry, focusing on how it addresses the environmental challenges of our time. It then adumbrates the rich theoretical traditions of environmental sociology, and finally examines sustainability through the lens of environmental sociology, referring to various case studies and empirical analyses.

Keywords: environmentalism; environmental sociology; ecological modernization; treadmill of production; the earth day; green movement; environmental certification; global agro-food system

1. Introduction: Environmental Sociology as a Field of Inquiry

Environmental sociology is the study of how social and ecological systems interact with one another. Both social and ecological systems are very complex and vast on their own, and together the complexity grows. The interaction between social and ecological systems might seem clear when we think about the way our society is built. However, due to the complexity of the interaction, the development of separate disciplines, such as 'sociology' focusing on social relationships alone and 'ecology' based on environmental relationships without relating to society, bifurcation in intuitions and disciplinary bias, the profound relationships between society and the environment were hardly mentioned for a long period of time. Over time, as Gould and Lewis stated, "The increasing urgency of the negative impacts of social systems on ecosystems created both the social space and social need for the emergence of environmental sociology" [1] (p. 3). Environmental sociology is a subfield in sociology despite the fact that it also has roots in ecology. The roots are not equally split between sociology and ecology and environmental sociologists are not required to know the natural sciences in ecology. Gould and Lewis describe the core of environmental sociology: its "special focus is on how social systems are organized and change in response to the natural world, just as the changes they produce in the natural world force them to further respond and change" [1] (p. 3). This editorial, accompanying the Special Issue "sustainability through the Lens of Environmental Sociology", will first provide a brief sketch on the social and institutional trajectories in which environmental sociology emerged as a distinct field of inquiry.

Environmental sociology became an officially recognized subfield in sociology as late as 1976 [1]. In the late 1800s, environmental sociology was not a part of sociology at all, but in the early 1900s there were two sociologists who started to talk about the relationship between humans and nature, Henry Thomas Buckle and Ellsworth Huntington. According to Buckle, human society is a product of natural forces and his theory of social change made quite an impact on the intellectual circles in the 19th century [2]. Huntington, on the other hand, tried to establish a connection between climate and health, energy, and metal processes such as intelligence, genius and willpower. He used his theories to try to explain the rise and fall of ancient societies such as Rome, connecting the fall of the kingdom to changes in the climate. This has been questioned by, for example, Sorokin, who says that the correlations are fictive and that he overestimated the role of the geographical environment. However, he agrees that the geography has an impact on every social constellation. During this time, many sociologists applied Darwinism and the concept of "evolution" and "natural selection" to the human context, and the most prominent social Darwinist was an English social philosopher, Herbert Spencer. He opposed any suggestion that society could be transformed through education or social reform. Instead he believed that it is better to leave it alone and we will change as time goes on. Spencer also had a disciple, Sumner, who thought that we do not only fight with other species in nature, but also in society; however, these theories were largely rejected later on [1,2].

Between the years of 1955 and 1975, it was more evident that the sociological literature became more and more modern, and there are sociologists in particular that stand out during these two decades, David H. Smith, Alex Inkeles and Daniel Lerner. According to Smith and Inkeles, many individual members of certain communities were physiologically trapped in the past and they had problems doing what modern citizens could do, such as keeping to a fixed schedule, observing abstract rules, adopting multiple roles, and empathizing with others; this resulted in many developing nations failing to be a part of the modern world in the 1960s. Humans are not born knowing all these qualities, but through life experience and education we can obtain them. In his book *The Passing of Traditional Society*, Lerner discussed that the key to modernity is the media; they have the power to establish a physiological openness among the population [3]. One reason why the environmental aspect of sociology did not take off was an apprehension that it would take the focus away from the debate that many sociologists thought was more important—class. Even when no one could close their eyes from the environmental issues they were facing, they still said that it was class-related problems that were the cause of the issues, instead of using environmental reasons [2].

There are three important founders of classical sociological theories: Émile Durkheim, Max Weber and Karl Marx. Émile Durkheim is least likely to be recognized as an environmental sociologist. Émile Durkheim stated that social facts are more important than physical and environmental facts. He put very little effort in the environmental part of sociology and insisted that social facts "are consequently the proper field of sociology" [2] (p. 6). Max Weber, on the other hand, took environmental sociology into account. He connected economy, science, government and industries with geographical attributes. The third one, Karl Marx, was the one who has provoked the most widespread response from present-day environmental commentators. Marx has affected modern-day environmental sociology the most. He only touched the subject in his work, but many of his ideas later became the starting point for modern theories of the environment. Marx and his colleague, Friedrich Engels, believed that the class conflict did not profit any side of the conflict; instead it alienates people from their work and from nature. This was obvious in the industrial revolution when it was more profitable to use the land for industries rather than agriculture, which forced rural workers to give up their lands and move to cities that were polluted and crowded. Marx and Engels were convinced that capitalism was to blame for these events that eventually led to a bad state for the whole society. They wanted to make the gap between nature and people smaller and reinstate the bond between them, but did not know how to establish it. Marx talked about the "humanization of nature" which he said will give humans a better understanding of nature and how we can co-exist in a way that benefits both the environment and us. He even pointed out specific environmental issues and saw the importance of ecological sustainability. Both Marx and Engels were pro-organic agriculture since they saw the danger in taking away all the nutrients in the soil and using chemicals to get more crops. They suggested that to stop pollution from fertilizers, farmers could use recycled human waste from the city instead [2].

None of these three founders—Durkheim, Weber and Marx—spoke directly about environmental sociology, but they were all talking about it indirectly, as they were talking about humans and nature. It was not a single discovery that made environmental sociology a field of inquiry; it was more like a movement driven by political reasons for social reform. Various publications of books and reports during the 1970s drew more intellectual and public attention towards environmental issues and problems. When sociologists first started discussing environmental issues and problems in the 1970s, they applied social theories to the environmental issues; soon a distinct field of study began to emerge. They made a distinction between two parts, one of which studied the interaction between the society and the environment, and the other which dealt more with environmental issues. This separation today is not very clear as both parts often go under the umbrella of environmental sociology [2].

The term "environmental sociology" was first used in 1971 by Samuel Klausner in his book *On Man in His Environment* [4]. Riley Dunlap came across Klausner´s book and the term several years later and he is considered one of the founders of this field. He focused mostly on the relationship between modern industrial societies and the physical environments they inhabit. According to many, Earth Day in 1970 was the debut of the modern environmental movement. It all started as a small proposal for national awareness on the environment, but soon, it had grown into a much larger event, with many participants around the world. "Earth Day 1970" symbolizes "Day 1" of the new environmentalism and it is widely used in the American mass media [2]. It was during this time that environmental sociology became an officially recognized subfield in sociology and the environmental issues started getting more attention on a political level. Sociologists started to incorporate formal niches for environmental sociology, as the Rural Sociological Society´s Natural Resource Research Group was formed in 1960s, the Society for the Social Problem started a group in 1973 and the American Sociological Association´s Environment and Technology group was formed in 1976. Also, among the population, it became a bigger and bigger topic and due to several environmental crises, such as the energy crises in the early 1970s and the Love Canal incident in 1978, and people became more aware of environmental issues and problems. Political actions were also made, both on national and international levels; notable among them was the United Nations Conference on the Human Environment held in Stockholm in 1972 [1]. Later on, the Global Environmental Change Programme was established in the UK and the Kyoto protocol was signed in 1997. Further, in the 1990s, environmental sociology was being taught all over the world [2].

Today's world is facing a lot of different problems related to economic growth and environmental protection, and environmental sociology provides key tools to understand them. The "21 issues for the 21st century" is a list made by the United Nations Environmental program (UNEP) that proposes a sustainable earth through keeping the global environment under review and bringing emerging issues to the attention of governments and the international community for action. The problems are divided into five different main issues: cross-cutting issues, land issues, water issues, climate-change issues and development issues. All of the issues are ranked by the UNEP Foresight panel which includes 22 distinguished members of the scientific community from 16 developing and industrialized countries. The issue ranked number one is the cross-cutting issue: Aligning Governance to the Challenges of Global Sustainability. Many governments do not have the capacity to support environmental actions on a global level, but without governments' support it is going to be hard for the world to solve the environmental issues that lie ahead. Some other issues posing sustainability challenges on the list include: new challenges for ensuring food safety and food security for nine billion people, new challenges for climate change mitigation and adaptation, managing the unintended consequences and changing the face of waste, solving the impending scarcity of strategic minerals and avoiding electronic waste. All of the issues listed have one thing in common: they have become issues because of the way humans use the natural environment [5].

One challenge we find in today's society is the correlation between social and environmental vulnerability. This gives different countries different capabilities to cope with environmental disasters depending on economic and political factors. In history, it is mostly the developed countries that contributed to the anthropogenic environmental problems and issues since they were the first to build large-scale factories and their inhabitants had more money to purchase cars, indulge in mass consumption, and lead lifestyles that harmed the environment. The environmental problems caused by these anthropogenic factors are not only affecting these developed countries but rather the contrary: developing countries are often more exposed to disasters derived largely from human impacts. Developing countries are often more environmentally vulnerable. They are not only afflicted and affected by disasters from human activity, but also by natural disasters such as earthquakes, typhoons, tsunamis and extreme dry periods. Their lack of infrastructure, poor governmental establishment and tight economy make it harder for them to cope with these problems; rather, these factors damage these countries even more, making it even harder to recover from future disasters, and they end up in a vicious circle. The consumption patterns in the world lead to an increasing and unending demand. Developed countries, where the demand often comes from, put pressure on the developing countries to drain their natural resources. This can ruin their environment and lead to a massive loss of biodiversity, but not all developing countries will prioritize environmental protection over economic growth. The resources move from the developing countries to more affluent ones, leading to a core-periphery dynamics where the assets move from countries in the periphery into the core, while political pressure moves in the reverse direction [6]. The developed countries take advantage of more environmentally vulnerable countries. It not only forces the poor countries to drain their natural resources but also leads to higher emissions of greenhouse gases, thus speeding up global warming. For this to stop, demand needs to decrease. However, to decrease demand is hard since it is often correlated with economic development, something all countries strive for. Countries need to find a way to achieve development without necessarily having to increase demand.

There are many problems regarding the human impact on the environment such as the dilemma and tension between the economy and the environment, increasing demand and environmental vulnerability. Environmental sociology is a tool we can use to understand the complexity of the problems and find solutions, thus making sustainable development a reality and not just a dream. This is necessary if we are going to continue living on this earth and live together with other species in a harmonious manner.

2. Theoretical Traditions in Environmental Sociology

Environmental sociology is largely oriented towards the reciprocal relationship between the environment and the larger society. This relationship is problematized and conceptualized, and hence needs a reestablishment of social theories in order to better understand these issues. As a sub-field of sociology, environmental sociology employs various theoretical traditions in order to analyze and fathom the concerns raised in this discipline. Some of these traditions originate from the Marxist perspective, which are discussed and reintroduced as neo-Marxist theories. Similarly, neoliberalism theories in environmental sociology attempt to understand problems without contesting the economic and political structure. Symbolic interactionism theories discuss how meanings received from social interactions can influence and interpret the relationships between human society and the environment. There are several theories under each tradition that are discussed in further detail below.

2.1. Neo-Marxist Theories

2.1.1. Metabolic Rift

Metabolic rift is an important neo-Marxist theory as explained by John B. Foster and Karl Marx. It describes how society and ecology should not be classified as two different entities. Instead, they should be seen as one metabolism as one cannot function without the other. The theory explains that man started to view society and ecology as two separate entities with the rise of the capitalist system, creating a "rift" between humans and earth [7]. Marx discusses how capitalism has disrupted the "metabolic interaction between man and the earth, i.e., it prevents the return to the soil of its constituent elements consumed by man in the form of food and clothing" (p. 379). Through our growing patterns of consumption, man starts to only care about the product and forgets that nature is the root to all the resources going into making the products that we consume. The capitalist system places much emphasis on the process of production, rather than the source of factors of production. Man begins to lose touch with nature, and sees no problem in further exploiting the ecological system for natural resources in order to produce what society demands for.

Marx, as cited in Foster [7] (p. 380), also discusses how the "long-distance trade in food and clothing made the problem of the alienation of the constituent elements of the soil that much more of an 'irreparable rift'". In today's globalized economy, a large proportion of food and clothing are being produced in peripheral countries such as Brazil and China, before being shipped to core countries such as the USA for consumption. This distancing between the location of production and consumption further alienates humans' relationship with the goods (p. 380). Human beings do not take into consideration the factors of production, such as the process of it, as well as the extraction of natural resources during their consumption of goods. This further enables humans to lose "touch" and their relationship with nature, resulting in the "irreparable rift" between society and the ecological system.

2.1.2. Treadmill of Production

The treadmill of production theory, propounded by Allan Schnaiberg in his book *The Environment: From Surplus to Scarcity* [8], suggests a never-ending cycle of production is the central characteristic of the capitalistic mode of production [9]. Human societies are dependent on flows of energy from nature, and hence Schnaiberg argues that this energy can only be transformed, and that each transformation is in itself a degradation [10]. The treadmill of production depicts the two dialectic processes of societal-environmental interaction: withdrawals and additions. Withdrawals from the environment are raw materials that are extracted from nature in order to transform them into economic goods, while additions are pollutive or toxic waste that is returned to the natural world. Capitalism generates profit through consumerism, with advertisements and marketing constantly driving human consumption to a grotesque level to generate profit [1,10]. Basic supply and demand predict that with increased demand, supply will increase to match that level, which results in the rising rate of the production process. That production process results in the exponential rate of both withdrawals and additions which rapidly depletes raw materials from nature and dumps toxic waste back into it. Hence, treadmill theory predicts environmental destruction with the current trend of human activities. This treadmill is unsustainable as the carrying capacity and resources on this planet are finite, while humans' wants are infinite [6].

On top of ecological destruction, there is also the innate contradiction of capitalism as highlighted by Marx: exploitations of labor. However, social institutions are rooted and embedded in capitalism [1,6,10], and hence they seek only to strengthen and reinforce the capitalist ideology. For example, labor unions are in favor of the treadmill as it creates jobs for the workers, and governments prioritize economic growth over environmental protection [6]. Hence, without a radical change to the economic and political structure of the world system, the treadmill of production will not cease. Critics of this theory are convinced that capitalism will be able to adapt to consider environmental problems without radical changes, and that this theory is based too heavily on a materialist approach [1].

Others critics argued that treadmill of production is too pessimistic and it will discourage individuals from environmental efforts.

2.1.3. World Systems Theory

Immanuel Wallerstein's world systems theory was built on Marxist foundations, and emerged as a reaction and counter to modernization theory, by arguing that colonialism is one of the main causes of underdevelopment in the third world. Like Karl Marx, Wallerstein believed that capitalism is based on a system of exploitation, to achieve maximization of profit. However, it is seen as neo-Marxist as the world systems theory rejects the notion that capitalism only occurs in nation-states, and argues instead that it encompasses the entire world into an economy. Wallerstein presents three positions inside this world system: the core, the periphery and the semi-periphery. Labor and surplus are commoditized and extracted from the periphery to the core, which creates an unequal exchange of resources. The semi-periphery stands in the middle, being an exploiter as well as an exploited country [11]. This ultimately creates a global stratified system that works based on rationalization and commoditization as the driving forces behind this exploitative relationship. The environment can be subject to commoditization, with land being one example. A further theory of unequal ecological exchange builds on Wallerstein's world systems theory, as it contends that more developed countries and their high consumption-based environmental costs are redistributed to less developed countries, which results in environmental degradation and disasters [12].

Through extensive agricultural practices that benefit the cores, environmental damage can occur in the periphery. One example is the Hamburger Connection, where deforestation happens in order to give rise to cattle ranching. Much of the Amazon rainforest in Brazil has been cleared for pastures, and this has been a practice for decades. Brazil was seen as the world's top exporter of beef in 2003, and it was exported to developed countries for their consumption in fast food restaurants [13]. The land is hence seen as a commodity for exchange value, in order to raise cattle for the growing consumption of beef in developed countries. This creates an unequal ecological exchange because the core reaps the benefits of the beef product, while Brazil as a semi-periphery is exploited for the land that they own. Extensive environmental damage is the cost suffered by the periphery for the interest of the core countries, and this exemplifies how it occurs in the modern capitalist world system as proposed by Wallerstein.

2.2. Neoliberalism Theories

2.2.1. Risk Society

According to Beck, as cited in Adam, Beck and Van Loon [14] (p. 5), a risk society can be understood as "a particular mode of organization as a response to new challenges enforced upon the world by technologies and practices". Present society is said to be fraught with risks as a result of modernization where there has been a rapid increase in the advancement and employment of new technologies. While such technologies have brought about increased convenience, productivity and benefits, they are not without risks. Risks are defined as "a systematic way of dealing with hazards and insecurities induced in and introduced by modernization itself" [15] (p. 21). Beck, as cited in Matten [16], argues that while potential threats have always existed in society, such as natural disasters, the difference between threats and risks is that risks are a result of human decisions. Giddens has coined such risks as "manufactured risks" and believes that people today tend to focus their concerns on manufactured risks as a result of their actions over external risks. In particular, there have been rising anxieties and debate on the kind of environmental issues and problems brought about by the risk society due to new technologies in the field of nuclear, chemical and genetic industries, which have generated environmental hazards that creates risks in modern society [17,18].

An example of a manufactured risk in society is the use of nuclear energy. Nuclear energy is a welcomed alternative source of energy as it is deemed to be the most environmentally friendly in terms of its lower greenhouse gas emissions as compared to other forms, such as coal or electricity. However, there are environmental risks that accompany it. Environmental impacts and nuclear accidents such as radioactive waste produced from nuclear energy have impactful consequences on the environment and the health of individuals [19]. The Chernobyl nuclear accident in 1986 shows how manufactured risks translated into actual environmental hazards and damage. It is evident that nuclear energy is a risk because no one is able to fully understand the kind of far-reaching consequences it can have in the event of an accident. Based on the Chernobyl nuclear accident alone, over 300,000 people had to be evacuated and resettled after the accident as the area surrounding the nuclear power plant was deemed to be unsuitable for living. A study has shown that the increase in reported cases of thyroid cancer was linked to the high levels of radioiodine exposure during the Chernobyl accident [20]. In terms of environmental damage, over 200,000 square kilometers of Europe was contaminated from the release of radionuclides [21]. The danger here is that as Beck has claimed, there is no form of insurance against the kind of risks that emerge out of risk societies, yet societies continue to take deliberate risks in the name of modernization. Another clear example would be the Fukushima nuclear disaster in Japan. Although past experiences such as the nuclear accident in Chernobyl have shown the far-reaching consequences of nuclear accidents, society continues to create human-induced risks by investing in nuclear technology. Modern societies continue to take risks in the name of modernization and profit-making, despite the fact that that political institutions lack the competence to manage risks that accompany new technologies [16].

2.2.2. Ecological Modernization

Mol and Spaargaren's ecological modernization aims to resolve environmental issues without forsaking modernization, through the use of new technologies from more modernization instead of drastically reforming society [2]. Hence, this theory attempts to take a middle stance between pessimistic environmentalists pushing for de-industrialization and capitalists who would rather ignore the issue of the environmental crisis altogether. Ecological modernization assumes that capitalism is flexible enough to self-regulate, craft solutions and evolve towards "sustainable capitalism" [1,6]. Huber posits that the development of an industrial society occurs in three stages: the industrial breakthrough, the construction of the industrial society and superindustrialization. Superindustrialization refers to the final stage where an "ecological switchover of the industrial system" happens, where environmentally friendly technologies are developed [22].

Another key point of ecological modernization theory is reflexive modernization, whereby a capitalist society re-examines its circumstances and develops a heightened awareness of environmental problems. Thus, consumers will push institutions such as governments and corporations for change towards a green society [6]. However, critics claim that any improvements made from pressure are not real, and are attained through misreporting [2]. In addition, ecological modernization theorists are criticized for being over-optimistic about the potential of technology, forgetting that 'clean' technologies such as nuclear power were once lauded until their more undesirable features and risks were discovered. Ecological modernization theory also disregards the political-economic dimension characterizing environmental processes, assuming that social and political forces will align for the sake of environmental conservation [1,2,6].

2.3. Symbolic Interactionism Theories

Naturework

The idea of naturework was first explored in Gary Alan Fine's book *Morel Tales: The Culture of Mushrooming* [23]. In this book, Fine studies how Americans assign meaning to the natural world that they live in. "Nature" always has been in unquestionable existence, but Fine argues that the meanings we give to the natural environment are culturally grounded. This implies that there can be no nature if we detach it from culture and he terms the cultural construction and interpretation "naturework". Fine illustrates his point by dabbling in the field of mushrooming. Fine examined present-day concerns of nature, environment and culture and how we give meaning to the first two. Mushroomers practice naturework by giving names to their fungi, assigning different values to different types of mushrooms and giving a gender identification to the mushrooms. In his opinion, the mushroomers adopt the "humanist" view of nature by making use of nature to meet the needs of humans. The other two views are the "protectionist" view that nature should be left untouched and the "organic" view that humans have no need to control themselves in exploiting nature compared to any other living thing as humans are also a part of nature.

The concept of "naturework" talks about how human beings adjust their behaviors and attitudes around socially constructed symbols [6,23]. Humans structure and revolve their actions and relationship with nature based on meanings that they assign to nature themselves. For example, human beings have constructed a social meaning for farm animals (e.g., pigs and cows) as animals raised for food. Thus, society does not see a problem with raising animals for food and the use of inhumane methods on animals in order to produce sufficient livestock for human consumption. In comparison, when human beings claim that they need a break from work and escape the city for a holiday by joining farm stays, they claim that they are being "in nature" and would never treat the farm animals in an exploitative manner. Through this example, it shows how human beings renegotiate the meanings of nature, and perceive nature as a source of relaxation instead of a natural resource for human consumption. This showcases the concept of naturework and how human beings entail the capacity to shape their behaviors and relationships with nature around social meanings [6]. In addition, by "enframing" nature as a separate entity to society (human beings), it creates the perception of distance and separation between society and nature. Thus, society does not develop a sense of accountability towards nature, allowing society to develop a power relation over nature to result in the exploitation of nature through technology [24].

3. Sustainability through the Lens of Environmental Sociology

The powerful lens provided by environmental sociology is important not only to understand the current environmental problems and challenges, but also to devise solutions for a sustainable earth. This Special Issue of *Sustainability* provides an environmental sociology approach to understanding and achieving the widely used notion of "sustainability", focusing on, among other topics, the inherent discursive formations of environmental sociology, conceptual tools and paradoxes, competing theories and practices, and their complex implications on our society at large.

Some papers in this Special Issue have solid conceptual and theoretical contributions to the study of sustainability. Longo and his colleagues, for example, problematized the prevailing notion of sustainability and sustainable development as mired in a "pre-analytic vision" that naturalizes capitalist social relations and closes off important questions regarding economic growth. To overcome this problem with the sustainability discourse, the authors highlight how several environmental sociology perspectives—such as human ecology, the treadmill of production, and metabolic analysis—can serve as the basis for a more integrative "socio-ecological conception" and can help advance the field of sustainability science [25]. To better understand and theorize sustainability in a post-natural age, Arias-Maldonado, on the other hand, suggested that environmental sociology should incorporate and reconsider the "anthropocenic turn" in its fold for a realistic understanding of sustainability. The anthropocene, as he explains, is a scientific notion, grounded in geology and Earth-system science, which plausibly suggests that human beings have colonized nature to a degree that has irreversibly altered the functioning of planetary systems, and, consequently, social and natural systems have become "coupled". Elucidating the consequences of the "anthropocenic turn" for sustainability studies, his paper explores the related notions of hybridity and relational agency as key aspects of a renewed view of nature [26].

Other papers applied various tools of environmental sociology in addressing various environmental issues and problems affecting societies and communities around the world. Islam and his colleagues, for example, applied the treadmill of production theory and environmental governance to understand the causes and consequences of trans-boundary haze pollution in Southeast Asia and proposed sustainability through a plural coexistence framework [27]; Hui-Ting Tang and Yuh-Ming Ling, assembling disparate information across time, space and discipline in their paper, aim to build a clear and concise synthesis of sustainable urban development not only to serve as an essential reference for decision- and policy-makers, but also to encourage more strategically organized sustainability efforts [28]. Sustainability with "academic ecohealth" literature, focusing on existing engagements and future prospects [29]; certified organic farming, posing a "metabolic rift" similar to conventional agriculture [30]; hybrid arrangements and governance as a form of "ecological modernization" in understanding the complexity of climate governance and energy efficiency in US cities [31]; and the extent to which forms of certification in global agro-food value chains guarantee sustainability [32] are among the key case studies in this Special Issue that advance our understanding of sustainability through the lens of environmental sociology.

Two papers clearly signal towards methodological innovations within environmental sociology in understanding and addressing today's sustainability challenges. Mark Brown made a large-scale textual and discourse analysis to show how multinational corporations manage and naturalize "nature-business" through developing a vocabulary and a "grammar" which enables them to manage natural spaces in the same way that they are able to manage their own far-flung business operations [33]. Sing Chew and Daniel Sarabia, on the other hand, suggest a robust historical analysis of nature-culture relations, focusing on early globalization dating back 5000 years, climate change and system crisis. They believe a long-term tracing of the socioeconomic and political processes of the making of the modern world will allow us to have a more incisive understanding of the current trajectory of world development and transformations [34].

Papers published in this issue thus focus on how sustainable development has been understood through different theoretical lenses in environmental sociology, such as ecological modernization, policy/reformist sustainable development, and critical structural approaches (such as the treadmill of production, ecological Marxism, metabolic rift theory, etc.). Also, review papers and original manuscripts draw on how sustainable development has been practiced in, or by, various stakeholders, such as states, corporations, and local communities, for various ends, through the use of specific case studies, showing, for example, the discursive shifts, dynamic formations, and diverse contours of sustainable development. The lens of environmental sociology on sustainability in this Special Issue has therefore been expressed through conceptual and theoretical contributions, methodological innovations, and critical analyses of various cases around the world.

Acknowledgments: The author thanks the editorial team of the *Sustainability* journal for inviting him to guest-edit the Special Issue of "Sustainability through the Lens of Environmental Sociology". The author acknowledges the contribution of his students of the Environmental Sociology course in finding materials, generating debates and helping during the writing phase.

Conflicts of Interest: The author declares no conflicts of interest.

References

1. Gould, K.A.; Lewis, T.L. (Eds.) *Twenty Lessons in Environmental Sociology*; Oxford University Press: New York, NY, USA; Oxford, UK, 2009.
2. Hannigan, J. *Environmental Sociology: A Social Constructivist Perspective*, 2nd ed.; Routledge: New York, NY, USA, 2006.
3. Lerner, D. *The Passing of Traditional Society*, 1st ed.; Free Press: New York, NY, USA, 1964.
4. Klausner, S. *On Man in His Environment*, 1st ed.; Jossey-Bass: San Francisco, CA, USA, 1971.
5. United Nations Environmental Program (UNEP). *21 Issues for the 21st Century*; United Nations Environmental Program: Nairobi, Kenya, 2012.

6. Islam, M.S. *Development, Power and the Environment: Neoliberal Paradox in the Age of Vulnerability*; Routledge: New York, NY, USA; London, UK, 2013.

7. Foster, J.B. Marx's Theory of Metabolic Rift: Classical Foundations for Environmental Sociology. *Am. J. Sociol.* **1999**, *105*, 366–405. [CrossRef]

8. Allan, S. *The Environment: From Surplus to Scarcity*; Oxford University Press: Oxford, UK, 1980.

9. Islam, M.S.; Hossain, M.I. *Social Justice in the Globalization of Production Labor, Gender, and the Environment Nexus*; Palgrave Macmillan: London, UK, 2015.

10. Barbosa, L.C. Theories of Environmental Sociology. In *Twenty Lessons in Environmental Sociology*; Kenneth, A.G., Tammy, L.L., Eds.; Oxford University Press: Oxford, UK, 2009.

11. Wallerstein, I. *The modern World System I: Capitalist Agriculture and the Origins of the European World-Economy in the Sixteenth Century*; Academic Press: New York, NY, USA, 1974.

12. Jorgenson, A.K. Unequal Ecological Exchange and Environmental Degradation: A Theoretical Proposition and Cross-National Study of Deforestation, 1990–2000. *Rural Sociol.* **2006**, *71*, 685–712. [CrossRef]

13. Kaimowitz, D.; Benoit, M.; Sven, W.; Pablo, P. *Hamburger Connection Fuels Amazon Destruction*; Center for International Forest Research: Bangor, Indonesia, 2004.

14. Adam, B.; Ulrich, B.; Van Loon, J. (Eds.) *The Risk Society and Beyond: Critical Issues for Social Theory*; Sage Publications: London, UK, 2000.

15. Beck, U. *Risk Society: Towards a New Modernity*; Sage Publications: London, UK, 1992.

16. Matten, D. The impact of the risk society thesis on environmental politics and management in a globalizing economy—Principles, proficiency, perspectives. *J. Risk Res.* **2004**, *7*, 377–398. [CrossRef]

17. McGrail, S. Anthony Giddens on the Rise of Futures Thinking and Risk Management. Desperately Seeking Sustainability. Available online: http://www.facilitatingsustainability.net/?p=2620 (accessed on 4 October 2016).

18. Dunlap, R.E. *Sociological Theory and the Environment: Classical Foundations, Contemporary Insights*; Rowman & Littlefield Publishers: Lanham, MD, USA, 2002.

19. Pros and Cons of Nuclear Energy. Available online: http://www.conserve-energy-future.com/pros-and-cons-of-nuclear-energy.php (accessed on 4 October 2016).

20. Nuclear Energy Institute. Fact Sheets. Chernobyl Accident and Its Consequences. 2015. Available online: https://www.nei.org/master-document-folder/backgrounders/fact-sheets/chernobyl-accident-and-its-consequences (accessed on 4 October 2016).

21. World Health Organization. Chernobyl: The True Scale of the Accident. Available online: http://www.who.int/mediacentre/news/releases/2005/pr38/en/index1.html (accessed on 4 October 2016).

22. Huber, J. *Die Verlorene Unschuld der Ökologie. Neue Technologien und Superindustrielle Entwicklung*; Fisher: Frankfurt/Main, Germany, 1982.

23. Fine, G.A. *Morel Tales: The Culture of Mushrooming*; Harvard University Press: Cambridge, MA, USA, 1998.

24. Heidegger, M. *The Question Concerning Technology, and Other Essays*; Harper Torchbooks: New York, NY, USA, 1977.

25. Longo, S.B.; Clark, B.; Shriver, T.E.; Clausen, R. Sustainability and Environmental Sociology: Putting the Economy in its Place and Moving Toward an Integrative Socio-Ecology. *Sustainability* **2016**, *8*, 437. [CrossRef]

26. Arias-Maldonado, M. The Anthropocenic Turn: Theorizing Sustainability in a Postnatural Age. *Sustainability* **2016**, *8*, 10. [CrossRef]

27. Islam, M.S.; Yap, H.P.; Shrutika, M. Trans-Boundary Haze Pollution in Southeast Asia: Sustainability through Plural Environmental Governance. *Sustainability* **2016**, *8*, 499. [CrossRef]

28. Tang, H.-T.; Lee, Y.-M. The Making of Sustainable Urban Development: A Synthesis Framework. *Sustainability* **2016**, *8*, 492. [CrossRef]

29. Aryn, L.; Gregor, W. Sustainability within the Academic EcoHealth Literature: Existing Engagement and Future Prospects. *Sustainability* **2016**, *8*, 202. [CrossRef]

30. McGee, J.A.; Camila, A. Sustaining without Changing: The Metabolic Rift of Certified Organic Farming. *Sustainability* **2016**, *8*, 115. [CrossRef]

31. Galli, A.M.; Dana, R.F. Hybrid Arrangements as a Form of Ecological Modernization: The Case of the US Energy Efficiency Conservation Block Grants. *Sustainability* **2016**, *8*, 88. [CrossRef]
32. Mol, A.P.J.; Peter, O. Certification of Markets, Markets of Certificates: Tracing Sustainability in Global Agro-Food Value Chains. *Sustainability* **2015**, *7*, 12258–12278. [CrossRef]
33. Brown, M. Managing Nature–Business as Usual: Resource Extraction Companies and Their Representations of Natural Landscapes. *Sustainability* **2015**, *7*, 15900–15922. [CrossRef]
34. Chew, S.C.; Daniel, S. Nature–Culture Relations: Early Globalization, Climate Changes, and System Crisis. *Sustainability* **2016**, *8*, 78. [CrossRef]

Article

Sustainability and Environmental Sociology: Putting the Economy in its Place and Moving Toward an Integrative Socio-Ecology

Stefano B. Longo [1,*], Brett Clark [2], Thomas E. Shriver [1] and Rebecca Clausen [3]

[1]   Department of Sociology and Anthropology, North Carolina State University, Raleigh, NC 27695, USA; tom_shriver@ncsu.edu

[2]   Department of Sociology, University of Utah, Salt Lake City, UT 84112, USA; brett.clark@soc.utah.edu

[3]   Department of Sociology, Fort Lewis College, Durango, CO 81301, USA; clausen_r@fortlewis.edu

*   Correspondence: sblongo@ncsu.edu; Tel.: +1-919-515-2491

Academic Editors: Md Saidul Islam and Marc A. Rosen

Received: 22 February 2016; Accepted: 27 April 2016; Published: 3 May 2016

Abstract: The vague, yet undoubtedly desirable, notion of sustainability has been discussed and debated by many natural and social scientists. We argue that mainstream conceptions of sustainability, and the related concept of sustainable development, are mired in a "pre-analytic vision" that naturalizes capitalist social relations, closes off important questions regarding economic growth, and thus limits the potential for an integrative socio-ecological analysis. Theoretical and empirical research within environmental sociology provides key insights to overcome the aforementioned problems, whereby the social, historical, and environmental relationships associated with the tendencies and qualities of the dominant economic system are analyzed. We highlight how several environmental sociology perspectives—such as human ecology, the treadmill of production, and metabolic analysis—can serve as the basis for a more integrative socio-ecological conception and can help advance the field of sustainability science.

Keywords: economic development; growth; social theory; human ecology; treadmill of production; metabolic rift

---

1. Introduction

In environmental scholarship, the influential concept of sustainability has been discussed and debated by many natural and social scientists. Extraordinary efforts have been made to systematize sustainability and to set environmental goals over the last several decades. For example, since the 1970s, scholars have held national and international conferences to discuss the relationship between economic growth, natural limits, and environmental sustainability. At the same time, environmental problems have continued to worsen—such as the acceleration of global climate change, decrease in biodiversity, increase in water pollution, and depletion of fisheries, to name a few. Despite the ongoing effort of the sustainable development project, environmental problems have magnified [1–3].

The scientific literature on sustainability is quite vast. Part of this research employs a diversity of scientific assessments and indicators of sustainability and sustainable development, including the Ecological Footprint [4], Environmental Sustainability Index [5], Global Scenarios Group [6,7], and Genuine Progress Indicator [8]. These tools and measures are just a few of the well-known efforts aimed at increasing systematic knowledge for advancing sustainability goals and sustainability science. The distinct assessments highlight the diversity of ecological conditions and the complexity of interactions within and between social and natural systems. The prevalence of different research programs stems from myriad theoretical assumptions, scientific conceptions, and questions regarding the social implications of sustainability [9,10]. Additionally, the discourse around sustainability and sustainable development has generated debates and contrasting meanings and conceptions [11,12].

We argue that many mainstream conceptions of sustainability—and the related concept of sustainable development—are mired in a "pre-analytic vision" that naturalizes capitalist social relations, closes off important questions regarding economic growth, and hinders socio-ecological analysis. A pre-analytic vision provides the initial conceptual categories and base assumptions for analyzing a particular phenomenon [13]. We offer a critique of the extant pre-analytic vision found in many approaches to sustainability in order to present a critical inquiry of sustainability and sustainable development. We draw on critical social theorists, particularly Karl Polanyi and Karl Marx, to reveal how this pre-analytic vision, which readily privileges economic growth, developed. In doing so, we illustrate the importance of putting the economy in its place—namely, within the larger social and ecological systems. We address how theoretical and empirical research within critical perspectives of U.S. environmental sociology provides key insights to help overcome the aforementioned problems, whereby the social, historical, and environmental relationships associated with the dominant economic system are analyzed. Finally, we highlight how several environmental sociology perspectives—such as human ecology, the treadmill of production, and metabolic analysis—can serve as the basis for a more integrative socio-ecological conception of sustainability and contribute to the emerging field of sustainability science.

2. Development and Sustainability

Modern theories of sustainability and sustainable development appear in the post-Second World War era [1,14]. This particular period in world history influenced the institutional framework, meaning, and application of these concepts, especially in relation to increasing concern regarding the vast inequality between nations [11,12,15,16]. Specifically, the United Nations and other global institutions, such as the World Bank, helped construct what was meant by sustainable development in major debates and discussions regarding economic development. Within universities, scholars of development studies and development economics incorporated the concepts into their evaluation of the global political-economic system. Both modernization theory and development theory became the leading social science approaches for understanding and addressing the problems of the "Third World" [1]. These theoretical perspectives were rooted predominantly in neoclassical economic theory, which had several implications for the policies they informed and for the definitions of development and underdevelopment [14].

Some major tenets of neoclassical theory are that economic growth (or the expansion of market-based economic activity with a resulting increase in gross domestic product) will have beneficial effects on all sectors of society, that markets are self-regulating (*i.e.*, market equilibrium will produce optimal utility), and that rational actors make cost-benefit decisions that will maximize utility. Development theorists operating in a neoclassical economics paradigm argued that what was essentially needed for social progress in the formerly colonized societies was an unleashing of capital in the parts of the world where capital had not yet fully made its mark. This would increase the potential for expansion and economic efficiencies. They maintained that policies encouraging such actions would have the desired effects of propelling these areas into new, grander "stages" of economic growth that would result in progress toward "mature" societies [17].

The United Nations and the Bretton Woods institutions (the World Bank and the International Monetary Fund) have been central institutions promoting industrial capitalist development. These organizations provided much of the original planning and financing for development projects throughout the world, and such funding continues to this day [1]. Environmental problems gained more social attention in the latter part of the twentieth century, many of which could be directly tied to the global expansion of industrial capitalism. Some institutions, such as the United Nations, began to consider that environmental issues might need to be addressed within development models and funding plans. As a result, the mainstreaming of the concept of sustainable development is borne largely out of U.N. projects [18].

In 1972, the United Nations Conference on the Human Environment was held in Stockholm, Sweden. This was the initial conference in a series organized by the United Nations on development and the environment that took place over the next 40 years. These conferences were commissioned to examine the escalating environmental impacts occurring throughout the world and to work toward developing new global legal frameworks that could better address the growing environmental and social concerns associated with capitalist development. The initial Stockholm conference resulted in the creation of the United Nations Environment Program (UNEP) whose mission is "to provide leadership and encourage partnership in caring for the environment by inspiring, informing, and enabling nations and peoples to improve their quality of life without compromising that of future generations" [19]. This statement contained a preview of the popular conceptions of sustainable development.

Over the next 40 years, the United Nations hosted a series of conferences, including the well-known meeting in Rio de Janeiro, Brazil, in 1992 (*i.e.*, Rio Summit or Earth Summit), where "sustainable development" was the central theme. In 1992, the Commission on Sustainable Development was established out of proposals in Agenda 21 [20,21]. The best-known definition of sustainable development was a product of the U.N. World Commission on Environment and Development (WCED), which was formed in 1983 and is also known as the Brundtland Commission. The Report of the WCED, "Our Common Future", was published in 1987, creating a pleasingly formulaic definition of sustainable development as "development that meets the need of the present without compromising the ability of future generations to meet their own needs" [22].

Thus, for more than four decades the United Nations has been promoting a vision of development that has included a conception of sustainability. This process can be regarded as the greening of development theory, where the goals of economic development began to take ecological concerns into account [23]. While the new development model attempted to address physical realities associated with environmental degradation, the focus on economic growth did not change significantly, if at all. A common critique of the U.N. approach to sustainability is that it merely tacks on the term "sustainable" to the traditional economic development model in order to advance an era of neo-liberalism [15,18]. While the Brundlandt Report [22] does begin to address fundamental ecological and social concerns, and can be commended for some of its inclusive language and creative vision, it has been argued that the sustainability programs and initiatives created under the auspices of the United Nations have been nothing more than hollow efforts and platitudes for addressing ecological concerns [11,23,24]. Critics of the sustainable development approach have suggested that it fails to integrate ecological realties and the interdependence of humans with the rest of nature [25].

As discussed, the development project emerged from a set of historical circumstances that resulted in a definition of the concept that is largely a plan to expand the scope and scale of global capitalism. At the United Nations, resolutions and humanitarian goals, which are at the core of their mission, were often placed in a context of expanding industrialization, and ultimately global economic growth and modernization. Consequently, many view U.N. summits as key mechanisms through which transnational corporations have become principal contributors to the strategy, goals, and practices set forth for achieving sustainable development [11,24].

3. Neoclassical Economics, Sustainable Development, and the Pre-Analytic Vision

Following the 1950s, the "Great Acceleration" in the human disruption of Earth systems led to much more research regarding the types and range of environmental degradation [26]. In the 1960s and 1970s, the environmental movement gained traction, demanding fundamental changes in society. Books such as *Silent Spring* [27], *The Closing Circle* [28], *Limits to Growth* [29], and *Blueprint for Survival* [30] presented analyses that depicted how social processes, including economic growth, resulted in environmental problems. In an effort to diminish extreme forms of degradation, many wealthy nations passed environmental regulations and laws due to increasing social pressure.

As discussed, by the 1980s, the private sector, particularly large corporations, became more involved in environmental policy conversations, in large part to protect their economic interests. At the 1984 meetings of the Organization for Economic Co-operation and Development (OECD), the position that the economy and environment are "mutually reinforcing" was established as a focal point for sustainable development [31]. This proposed compromise between economic development and environmental protection served as the basis of the Brundtland Report. Along with the previously mentioned definition, this report stated that "The concept of sustainable development does imply limits—not absolute limits but limitations imposed by the present state of technology and social organization on environmental resources and by the ability of the biosphere to absorb the effects of human activities. However, technology and social organization can be both managed and improved to make way for a new era of economic growth" [22]. Effectively, this definition couples economic growth and sustainability. Further, in some prominent interpretations, economic development is even assumed to provide the basis for sustainability. These positions reveal the taken-for-granted epistemic presuppositions that are found in the mainstream neoclassical economics tradition, which influence many conceptions of sustainability and sustainable development.

For example, environmental economist David Pearce indicates that "most economists" define sustainability entirely in terms of economic growth, monetary wealth, and consumption, without any direct reference to the environment. From this perspective, sustainable development is really sustainable economic development, which necessitates "continuously rising, or at least non-declining, consumption per capita, or GNP" [32]. Indeed, economic development becomes the central feature of sustainable development, and nature becomes a secondary consideration, at best. Consequently, economic models predicated on the growth imperative are central to modern economics and dominate policy discourse. Ecological economist Herman Daly points out that the ecological fact that the earth is essentially a closed and limited system, in which there are absolute limits as determined by natural science, runs contrary to the dominant economic paradigm [33].

Mainstream economics and policy operating largely within the neoclassical economic paradigm generally conceive of nature as a subsystem of the economy. In this view, the macro-economy becomes the primary point of analysis, which subordinates ecosystems. Everything, including biophysical nature, falls within the dynamic of the macro-economy [33]. This perspective has had a long history in economic thought, going back to the classical economists who regarded nature as providing "free gifts", and up until the present period where some modern economists argue that everything in nature can be substituted with the help of technology. For example, Robert Solow, a Nobel laureate in economics, argued that "if it is very easy to substitute other factors for natural resources, then there is in principle no 'problem'. The world can, in effect, get along without natural resources, so exhaustion is just an event, not a catastrophe" [34]. Such notions of endless substitutability and rejection of ecological limits are characterized as "weak sustainability", at best.

The Brundtland Report, environmental organizations, and mainstream sustainability approaches embody these orienting assumptions, whereby sustainable development is, underneath it all, about maintaining endless economic growth and technological solutions. It is important to note that the discipline of economics has many subfields, which develop varying approaches to scholarship. While mainstream neoclassical economics has been challenged for its lack of ecological awareness, other heterodox approaches, such as ecological economics, have developed theories and methods that are consistent with our critique. In fact, we draw directly from heterodox economists, including Herman Daly and others, who argue that neoclassical economics omits essential conceptual categories for understanding the relationship between economic and ecological systems.

Daly—following Joseph Schumpeter—argues that a growth-oriented approach serves as an important part of a "pre-analytic vision" that guides academic and policy groups, omitting essential relationships and categories from evaluation [13,33,35,36]. This pre-analytic vision is at the core of neoclassical economic models, but also provides the underpinnings for a common conceptual framework of sustainability or sustainable development throughout contemporary environmental discourse, scholarship, analytical examinations, and policies. This vision has frequently been formulated into heuristic devices, and is often used to conceptualize sustainability using metaphors like the "three legged stool", "three pillars of sustainability", and a "sustainable development triangle" [37–40]. Each of these conceptions proposes a similar vision, revolving around environmental, economic, and social factors that converge on sustainability. These are regarded as practicable and reasonable approaches for evaluating and achieving this important socio-ecological goal.

The business community has adopted this perspective, labeling it the "triple bottom line", consisting of people, planet, and profits [41,42]. The triple bottom line view has received significant attention as a sensible approach toward sustainability [43]. Government agencies, international organizations, businesses, and universities alike employ these heuristic devices and depictions to plan and communicate their sustainable philosophies and practices [37]. For example, the OECD states: "All of the economic, social and environmental systems must be simultaneously sustainable in and of themselves. Satisfying any one of these three sustainability systems without also satisfying the others is deemed insufficient" [44]. In other words, environmental policies are deemed acceptable so long as they create opportunities for economic growth.

The triple bottom line/three pillars description of sustainability emphasizes the importance of environmental, economic, and social concerns, and all three are considered central to the broader definition of sustainability [42]. In fact, it is argued that these three pillars serve as the foundation upon which sustainable development must be built, providing the outline for achieving what Andres Edwards calls a "sustainability revolution" [38]. Numerous researchers have employed variants of these models that sometimes include other factors (e.g., climate, fresh water, fisheries) depending on the focus and scope of the particular questions and problems [45,46].

While these three pillars are obviously important, there are critical flaws in the pre-analytic vision garnered from neoclassical economics. One central problem is that what constitutes the economy—within the three pillars conception—is limited to a growth-oriented market system. As we will elaborate in the next section, neoclassical economic assumptions are based on a pre-analytical vision in which the economic order of capitalist growth is naturalized. Therefore, its central relationship as a driver of environmental degradation falls outside of the analysis. Economist John Kenneth Galbraith proposed that this "innocent fraud" avoids the importance of conceiving of capitalism as a historical system, preventing an adequate analysis and understanding of the forces shaping the world [47].

4. Why It Is Necessary to Put the Economy in Its Place

Ecological economists and environmental sociologists argue that a weak sustainability position as outlined above is completely insufficient for addressing the ecological challenges we confront. Instead, a strong sustainability perspective must serve as a starting point. We suggest that this can add depth to the meaningful contributions coming from the scholarly work in the field of sustainability science. A sociological conception of history, economic relations, and ecology provides a more systematic understanding of these matters. From this perspective, the existence of ecological limits and planetary boundaries are cornerstones to analyses of sustainability. The capacity to substitute—by technological means—for environmental resources, as suggested under the weak sustainability approach, is inadequate. Maintaining the conditions that support life is of utmost importance. As ecological economist Richard Norgaard suggests, the organization and operation of the economic system must be critically examined, in order to offer a more comprehensive understanding of how interactions between human society and the larger physical world influence each other [48]. These concerns establish a broader conception of sustainability, and present important questions such as: do we prioritize sustaining the economy or sustaining the environment? In much of the sustainable development literature, it is commonly assumed that there should be a focus on both sustaining the environment and the economy. We argue this seemingly sensible and balanced approach has been largely a pretext for furthering business as usual, since the economy is conceived of as always consistent with capitalist preconditions.

It is important to establish a foundation that allows alternative viable conceptions of sustainability and society to emerge, challenging the pre-analytic vision that informs many conventional economically oriented approaches, particularly those in the tradition of neoclassical economics, on which we elaborate below. A first step in creating a more integrated socio-ecological analysis, which can inform discussions of the complex interactions and relationships of sustainability, involves recognizing that economic systems are embedded within the biosphere. This position stands in contrast to standard models such as the triple bottom line and its variants, which regard economic concerns as an independent realm (*i.e.*, one leg of the stool), existing in their own right, as if they are separate from the larger biophysical world. Furthermore, it is just as important to analyze the dominant economic system as a socio-historical system, rather than naturalizing its social relations. In doing this, the distinct and general characteristics of the economic order can be examined, especially as far as they act as social drivers of environmental degradation. This analysis allows for the potential for social transformation to be recognized as a positive force. Here, we offer a brief critique of the naturalization of the modern economic system that is an essential part of the pre-analytic vision discussed above. We highlight the ecological and social contradictions that arose with the ascent of capitalism, and the necessity to re-embed the economic system within the socio-ecological order.

The naturalization of capitalist social relations is not a new analytical presupposition. Both Karl Marx and Karl Polanyi indicated that this was fundamental to the work of some of the most famous founders of political-economic thought, particularly William Thompson, Adam Smith, Jeremy Bentham, David Ricardo, and Thomas Malthus [49,50]. Elaborating on this issue in *The Great Transformation*, Polanyi asserted: "the drive for a competitive market system acquired the irresistible impetus of a process of Nature. For the self-regulating market was now believed to follow from the inexorable laws of Nature, and the unshackling of the market to be an ineluctable necessity" [50]. Consequently, in a vigorous political battle to renounce and ultimately eliminate the English Poor Laws, leading economic thinkers with clear class interests were ideologically compelled to assume that the capitalist economy is guided by laws comparable to those of governing nature [49–51].

Economist E.K. Hunt explains that this move to naturalize capital and the system was crucial to the new science of economics that took shape in the late nineteenth century, under the theoretical guidance of those who ushered in what is commonly referred to as "the marginalist revolution" [52]. In this neoclassical theory, economics was modeled after the natural sciences, specifically physics, which attempted to develop a dispassionate, value-free study that could interpret the so-called laws of the market [53]. In this view, the modern global economy, based in commodity production and exchange value, is universalized and theorized as a natural system, akin to biophysical systems. Accordingly, "a new abstract universal, namely 'the economy'" was objectified, which reified modern social relations as relations between things [54].

Paradoxically, while neoclassical economic theory tended to naturalize the existing economic order, the material basis of economic development was torn from its ecological foundations. As Marx explained, part of this is due to the inherent characteristics of capital as an economic system based on generalized commodity production. Economist Paul Sweezy elaborates, "it is this obsession with capital accumulation that distinguishes capitalism from the simple system for satisfying human needs [as] it is portrayed in mainstream [neoclassical] economic theory. And a system driven by capital accumulation is one that never stands still, one that is forever changing, adopting new and discarding old methods of production and distribution, opening up new territories, subjecting to its purposes societies too weak to protect themselves. Caught up in this process of restless innovation and expansion, the system rides roughshod over even its own beneficiaries if they get in its way or fall by the roadside. As far as the natural environment is concerned, capitalism perceives it not as something to be cherished and enjoyed but as a means to the paramount ends of profit-making and still more capital accumulation" [55].

As Sweezy clearly describes, capitalism is a dynamic system, with a general character rooted in endless accumulation. As a grow-or-die system, capitalist development must expand exchange value, which is purely seen as a quantitative measure [56,57]. Qualitative relations, such as the conditions of life, are not a primary part of capitalist accounting. This foundational tendency towards expansion or "development" pushes the economic system onward, increasing the scale and breadth of its impacts upon the biophysical world [49,58,59]. Thus, ecological limits are easily explained away within this neoclassical conception.

Polanyi provides further insight into this matter when describing the social transformations that arose due to capitalist social relations. Prior to this development, Polanyi argues, economic systems of production and consumption were clearly embedded within the institutions and cultural practices of societies. Labor and distributional activities were influenced by principles of behavior including householding, reciprocity, and redistribution, rather than limitless economic gain [50]. In other words, people, customs, and social institutions set limits and regulated economic productive activities, directing them to serve particular ends, such as human needs. Max Weber advanced a similar argument in his most famous work *The Protestant Ethic and the Spirit of Capitalism*. That is, he argued that a variety of social conditions, particularly religious institutions, limited the expansion of capitalism around the world until the specific influence of the Protestant religions (namely particular beliefs and actions of Calvinists) of Western Europe provided the impetus for its full development [60]. What is common in these analyses is the recognition that human economies and its organizations are embedded within society and, we emphasize, the larger ecological complex that support life.

Polanyi, similar to Marx, explains that under a capitalist market economy, all aspects of social life become subordinated to the requirements of the economic realm. He maintains that "all transactions are turned into money transactions"—in order to meet the needs of capital [49,50]. The emergence of an all-encompassing self-regulating market "disembedded"—in terms of coming to dominate and alienate social relations—practical human activity from its foundation in the broader sociocultural and natural conditions. As a result, market activity directed by commodity production "acquired the irresistible impetus of a process of nature". Accordingly, the organization of social production and consumption activities is fundamentally transformed from an emphasis on the exchange of qualities into the exchange of quantities. Alienation from each other and from nature increases, as qualitative relations are subsumed under the quantitative growth imperative of capital and a culture of quantity [61]. Polanyi explains that during the transformation toward capitalist social relations "it was necessary to liquidate organic society". This "divorcedness of a separate economic motive", which is unique to capitalism, and therefore relatively new in human history, became commonplace [50].

Drawing from these social theorists it is clear that conceptualizing the economic sphere in a way that detaches it from society and/or nature results in a flawed understanding of social relations and ecology. Consequently, the economic system can be easily formulated as an autonomous self-governing force, and meeting economic needs can become separated or "divorced" from social and ecological concerns. The divorcing of ecology from the economy is consistent with neoclassical economic theory and political ideology that prioritizes specific economic (class) interests and universalizes capitalist social relations, turning comprehension of issues such as sustainability inside out. That is, the economic system is prioritized. Transferred into the modern sustainability models and policies, putting the economy on an ahistorical platform that independently erects a pillar of sustainability becomes a plausible step.

From a sociological perspective, it is clear that economic institutions and relations arise through socio-historical processes. They cannot be analyzed outside their socio-historical environment any less than they can be extracted from the natural environment. Clearly, it would be inappropriate to exclude other social institutions from our conception of society, such as, for example, religious life and its related institutions. The various institutions that make up our modern economy are no less social than past economic arrangements, or other non-economic social institutions [60]. The emphasis granted to modern economic institutions in environmental policies, for example prominence given to "the market", is a reflection of ideological priorities and commitments, not material conditions. Thus, we must understand how social institutions interact with each other and with ecosystems when considering how to develop an integrated socio-ecological analysis that informs sustainability.

The economy obviously plays a major role in modern social life, but it should not be extracted from its place within the social sphere. Doing so mis-prioritizes historically specific and unique economic arrangements, giving them undo primacy and inevitability. It also results in approaches toward sustainability in which economic indicators, such as gross domestic product, become fundamental guides for addressing ecological and human welfare. This is the case with the sustainable development project, which essentially uses the template of the traditional development project whereby economic growth and free trade are the foremost goals, reflecting the priorities of global capitalism [1,18].

The larger contextual setting in a socio-ecological approach is the greater Earth system. When social and economic systems are properly conceptualized as existing within the Earth system, we can analyze economic systems as part of socio-historical processes, situated within relationships of power, that are dependent on ecological systems [16,26]. Economies by necessity involve social relations. Economic action is social action. Economies are comprised of social institutions and networks, and societies manage economies. As Polanyi made clear, the "economy, as a rule, is submerged in social relationships" and "the economic system is, in effect, a mere function of social organization" [50]. Thus, what the purpose of an economy is and how it should be organized are important questions to consider for a sustainable society.

While capitalist social relations have distinct properties, they arose through socio-historical conditions and processes. In this, they are not inevitable, and can be changed. There are alternative economic arrangements and systems that have existed and can be created. At the same time, it is necessary to understand the operations of the capital system, its inner characteristics, especially if and how the endless drive to accumulate generates environmental degradation and social inequalities. Sweezy declares that it is important "to ask whether there is anything about capitalism as it has developed over recent centuries to cause us to believe that the system could curb its destructive drive and at the same time transform its creative drive into a benign environmental force". He argues, "unfortunately, there is absolutely nothing in the historic record to encourage such a belief" [55]. The array of environmental problems we confront is in part a consequence of this economic system and its inherent drive to constantly increase the accumulation of capital. Its business-as-usual operations are contributing to escalating concentrations of greenhouse gases, the loss of freshwater, and decreasing biodiversity [62–66]. Any discussion of sustainability must account for the role that the modern economic system plays in influencing the human dimensions of environmental change. This requires calling into question the pre-analytic vision that supports maintaining the capitalist market as a cornerstone, or further, a precondition of sustainability.

If we continue to rely on conceptions of sustainability that exclude critical questions related to the role of the modern economy, models that are ecologically and socially problematic will continue to be produced. Importantly, it will become increasingly difficult to develop proper indicators to measure the improvements and/or deficiencies of sustainability projects. Excluding important data, such as the historically specific growth dynamics of the economy, ignores the elephant in the room and generates scientifically imprecise models that ignore the inner logic of capital and its operations. Tyndall Centre climate scientists Kevin Anderson and Alice Bows indicate that this is exactly what is happening in regard to discussions concerning climate change. There is an enormous "discontinuity" between climate science findings and mainstream neoclassical economic emphases on exponential growth and unregulated markets [62]. Rather than taking the implications of climate data seriously, assurances about future technical fixes and the feasibility of adaptation and resilience are offered as solutions. The importance of dramatic social change is sidestepped. Furthermore, contesting charges of being political, biased, and alarmist for contravening prevailing economic beliefs and interests of corporate and public policy elites, Anderson and Bows contend, climate scientists "repeatedly and severely underplay implications of their analyses" [62]. As a result, it is all the more important to recognize that if the theoretical assumptions of neoclassical economics are fatally flawed, ensuing policies cannot work toward achieving the essential goals.

Approaches to sustainability and sustainable development influenced by neoclassical economic models and its underlying pre-analytic vision have produced little, if any, progress toward addressing global ecological challenges. Indeed, upon examining the scientific research, it becomes clear that there has been decline of ecological and social systems since these matters were taken up during the last half-century or so [2,4,26]. Important work by natural scientists suggests that business as usual cannot solve the myriad ecological problems facing the world today. Addressing sustainability requires putting the economy back in its place, as a subsystem within the larger social and biophysical systems [13,67,68]. This re-conceptualization and re-embedding will benefit from employing an integrated socio-ecological approach, informed by insights from both the natural and social sciences. Specific perspectives, which, for example, have developed within U.S. environmental sociology, can help advance such work and add new dimensions to the analyses being provided by sustainability science scholars.

5. Towards an Integrated Socio-Ecological Approach

Environmental sociology provides necessary insights for developing a rich understanding of sustainability. We elaborate on specific approaches that emerged in the United States. The sub-discipline arose as a field of study in the 1970s, proposing that the biophysical world must also be a realm of social inquiry. Early scholars argued that the presuppositions of contemporary sociology, similar to the pre-analytical vision within neoclassical economic thought, were "human exemptionalist", and implied that society existed outside and/or largely independent from relationships with the biophysical world [69,70]. Since then, environmental sociology has become a leading field of study, analyzing the interrelationships between human systems and natural systems. Several distinct research traditions have been created that provide useful analytical tools for considering issues related to socio-ecological sustainability, which overcome several of the problems discussed earlier.

From an environmental sociological perspective, traditional sustainability approaches like the triple bottom line/three pillars or the United Nation's sustainable development agendas are inadequate. For environmental sociologists, human systems are embedded within the larger Earth system. Additionally, economic systems are seen as socio-historical products that arose through specific social relationships within the larger biophysical world. Thus, the capitalist economic order is not naturalized, avoiding the pre-analytical vision that plagues so much of sustainability studies and sustainable development policies. As a result, environmental sociologists advance a research program that begins to assess the importance of various social institutions that are crucial for understanding sustainability. Many of the perspectives within environmental sociology are focused on examining the dynamic relationships in socio-ecological systems. We will briefly discuss three prominent theoretical approaches—human ecology, treadmill of production, and social metabolic analysis—and emphasize the contributions that environmental sociology can make to sustainability science.

The human ecology tradition provides useful conceptions and tools that offer a basis for an integrated socio-ecological approach to sustainability, avoiding tendencies within social and economic analyses toward human exemptionalism. Counter to the prevailing approaches in sociology of the period, theorists and researchers within human ecology argue that society is embedded within the larger ecological complex, comprised of reciprocal relationships between population, organizations, environment, and technology [71–74]. Culture mediates human relationships with the natural environment, but does not exempt humans from ecological and biophysical limits [69,75–78]. They analyze how growth dynamics—especially population and economic—influence the Earth system, altering material conditions and the availability of resources. These scholars incorporate feedback loops into their studies, emphasizing that the larger biophysical environment also constrains and influences social conditions. Human ecologists argue that all organisms and societies require energy expenditure to be sustained and that social organization shapes material resource flows and energy consumption levels. The scale and intensity of energy consumption, in relation to natural cycles, such as the carbon cycle, and the availability of carbon sinks, influence whether or not these social actions are sustainable. Further, variable population characteristics (e.g., size, growth, age structure, and migration), combined with other factors in the ecological complex (e.g., economic inequality), generate distinct patterns of ecological impact.

Human ecologists offer multifaceted assessments of interactions within the ecological complex and account for historical change. They contend that preindustrial societies generally relied on limited supplies of biomass for energy, which restricted population growth and development of complex social organizations [79–81]. By employing steam engines and greatly increasing coal consumption and energy output, human ecologists argue, industrial societies enormously increased productive capacity, sociocultural complexity, and population growth, stimulating demand for more energy [80]. These scholars propose that modernity's exceptional rates of population and economic growth have generated unparalleled resource demands and waste and that it is a primary social driver of environmental degradation and the accumulation of carbon dioxide in the atmosphere, contributing to ecological overshoot [75,82]. Thus, the overall demands are expanding beyond ecological limits, and that, when controlling for population size, developed nations consume the bulk of resources and produce most of the pollution and carbon dioxide [82–87].

Treadmill of production scholars also situate the economic system within the larger Earth system. They primarily provide a political-economic analysis of the historical development of modern industrial society, particularly the demands that the capitalist economic order places on the environment, and the consequences of its operations. Avoiding the pitfalls of the pre-analytic vision of neoclassical economics and human exemptionalism discussed above, they do not proceed from capitalist social relations as a givens. Instead, the dominant economic system is created through historically distinct human and institutional interests and must constantly recreate itself. Given the inner dynamics of the economy, profits must constantly increase, which are reinvested to enlarge and intensify the scale of production. On this treadmill, accumulation takes precedence and drives a cycle of growth that necessitates ever-increasing production [88]. Treadmill theorists highlight how this growth imperative heavily influences the organization of production, and drives culture-nature relations. They argue that private capital, the state, and labor depend on economic growth for profits, taxes, and wages, creating a type of path dependency with an array of social and ecological consequences.

The constant pursuit of profit and expansion has "direct implications for natural resource extraction", pollution generation, and overall environmental conditions [89]. Treadmill theorists explain that each expansion in the production process to sustain economic operations on a larger, more intensive scale generates higher natural resource demand, often at rates that exceed ecosystem regenerative capacity and that contribute to an increased disorganization in nature [67,88–92]. Moreover, they contend that energy-intensive materials, such as plastics and chemicals, which are incorporated into manufacturing, generate widespread waste and pollution that producers externalize [91–94]. As a result, ceaseless economic growth generates environmental degradation. The treadmill of production approach suggests that the modern capitalist economic order is generally incompatible with sustainability.

Similar to treadmill of production scholars, theorists and researchers in the social metabolic tradition analyze capitalism as a historically specific regime of accumulation that drives the growth imperative. They view this social order as a distinct social metabolic system that operates in accord with a particular logic, reducing labor and nature to means of further capital accumulation. This system fundamentally shapes material exchanges with the environment. Relentless growth increases demands on ecosystems and the larger environment. Like human ecologists, social metabolic scholars incorporate the operation of natural cycles and systems into their analysis in order to better assess the interpenetration and exchanges between society and its biophysical bases. They indicate that capitalism's social metabolism exceeds natural limits, producing "metabolic rifts" in various cycles and processes, which are necessary for ecosystem maintenance and regeneration [57,58,95–97]. Metabolic theorists specify that capitalist social relations tend to generate ecological rifts in specific ecological cycles through the intensification of the social metabolism. For example, the capitalist growth imperative locks in dependence on burning massive quantities of coal, natural gas, and oil [98,99]. This process has resulted in breaking the solar-income budget, releasing enormous quantities of carbon that had been sequestered. At the same time, consequent growth-driven, ecological degradation (e.g., deforestation) substantially reduces carbon sinks, further contributing to the accumulation of atmospheric carbon dioxide, resulting in a carbon rift that drives climate change. Similarly, social metabolic studies have also been applied to help elaborate on the ways in which modern capitalist development has fundamentally altered marine systems [100–103] Capital accumulation processes have been demonstrated to play a primary role in the structure and function of, for example, the seafood industry, which has guided fishing activities on a global scale. This analysis has illuminated how economic forces lead to fish being harvested at a rate faster than they can reproduce, thus contributing to the "fishing down the food web" process of capturing species of a lower trophic level and potentially to the collapse of fisheries [104].

These three theoretical traditions in environmental sociology challenge the tendency within mainstream sustainability studies to treat environmental issues as largely technical problems. The techno-optimist position of many sustainable development models helps maintain the pre-analytic vision, ignoring the role of the economy in producing environmental problems. Promoting technological solutions as the answer, such as simply improving the energy efficiency of production, maintains the status quo and limits the potential for social change. A more sophisticated understanding of technology and socio-ecological relationships is necessary to advance a more comprehensive conception of sustainability.

Treadmill and metabolic theorists propose that technological innovation plays a crucial role in capitalist development, rationalizing labor processes and generating cost reductions via automated production. They hold that new technologies often make energy and raw material usage more efficient, but, contra neoclassical environmental economists, contend that innovation does not necessarily dematerialize society or contribute to an absolute decoupling of development from energy and resources. They suggest that more efficient resource usage can often increase aggregate consumption of that particular resource—creating a socioeconomic dynamic known as the Jevons paradox, named after the nineteenth-century economist William Stanley Jevons [105–108]. In *The Coal Question*, Jevons noted this paradoxical relationship, whereby increased consumption outstrips gains made in energy efficiency [106]. He, however, did not provide a full explanation for why this occurred. Metabolic theorists, drawing upon insights from Marxist political economy, explain that efficient operations produce savings that expand investment in production and thereby promote increased production and consumption, and accordingly total energy consumed, raw materials used, and carbon dioxide produced [67,109,110]. To understand why this paradox arises, it is necessary to consider how the growth imperative of capital and processes of accumulation influence these dynamics. These critical environmental sociological perspectives reveal that technological rationalization must be situated within the global economy's overall social relations and operations. It is inappropriate to assume that technological innovations, such as improvements in energy efficiency, automatically lead to less environmental impacts. The most efficient nations are often found to be the largest consumers of natural resources [111].

Human ecology, the treadmill of production, and social metabolic analysis offer more nuanced understandings of the interactions between social, economic, and ecological realms. They raise critical questions regarding how the economy is traditionally conceived within sustainability studies. They indicate that it is important to embed the economy within society and the larger Earth system. These perspectives also suggest that an economy should be organized to meet social needs in an ecologically sound way, rather than for the sake of endless growth and accumulation. The social metabolic approach offers important insights in regard to how the independent and relational dynamics of socioeconomic and ecological systems can be examined in a fruitful way to inform sustainability research and create an integrated socio-ecological analysis.

We contend that environmental sociology has great potential to address the social and ecological challenges associated with sustainability. A very important aspect of social metabolic analysis is the role it can play in better linking the natural and social sciences on sustainability. Sustainability science is an area that has gained increasing notability, which can benefit from further contributions from environmental sociology.

Some within the natural sciences have recognized the necessity of developing an environmental science that integrates "societal and political processes that were shaping the sustainable development agenda" [112]. Several established approaches aiming to accomplish these goals can be placed under the umbrella of sustainability science. As a growing area of interdisciplinary environmental research, sustainability science "seeks to understand the fundamental character of interactions between nature and society" [112]. Here we briefly mention two prominent approaches: Coupled human-natural systems (CHANS) and resiliency. These overlapping multidimensional approaches have done much to advance sustainability science.

CHANS and resiliency scholars analyze interconnected complex systems [113]. Based in a systems theory model of scientific analysis, these approaches recognize that reducing ecological systems to isolated parts, common in modern environmental management schemes, is fundamentally flawed. Unlike neoclassical economics, that models linear changes in simple cause-and-effect relationships, these approaches highlight the complexity of interacting systems, that systems can be subsumed in other systems, and multi-scalular effects. CHANS scholars emphasize that changes can occur at different levels of a system and can cascade up or down [114]. Further, these systems-based approaches stress that coupled human and natural systems or "social-ecological systems" are highly dynamic and heterogeneous [115,116].

Resilient systems are understood as systems that can maintain their structure and capacity for long-term renewal, even in the face of various impacts, shocks, or disturbances [115–119]. Resilient socio-ecological systems tend to be more robust with regard to diversity and health. In contrast, when socio-ecological systems are overstressed due to resource exploitation or overwhelming waste inputs, these systems are weakened, less likely to maintain regenerative capacity, and less resistant to perturbations. As thresholds are breached, "ecological discontinuity" can occur with few, if any, immediate warning signs, drastically changing the system conditions [119–121]. Such changes affect both social and ecological outcomes in complex and often surprising ways, which can have serious implications for associated human communities. Underlying CHANS and resiliency research is the conceptualization of human systems as embedded in ecological systems, and that social and ecological systems are in constant interaction. An emphasis is placed the unity of what the resiliency literature calls "social-ecological" systems, highlighting the integrated nature of conditions.

These two approaches are very important, yet they do not adequately address the general and specific characteristics of the socio-economic system, especially its inner driving force. Thus, sustainability science must better integrate critical political-economic insights, which environmental sociology can offer, such as the dynamics associated with the growth imperative of capital, the social and ecological contradictions that arise from commodity production systems, technological innovation and the Jevons paradox, power and inequality, and the institutional conditions that produce social tendencies toward particular ecological outcomes [96,103]. We contend that further incorporation of environmental sociological approaches, such as human ecology, treadmill of production, and social metabolism, will advance sustainability science toward an improved analysis of coupled human and natural systems. It is essential that the social dimensions of integrated systems are not simply folded into functionalist ecological models [122]. The growth of sustainability science approaches, such as CHANS and resiliency research, have produced important steps in moving away from the traditional sustainability paradigm and flawed pre-analytic visions, and would benefit from a deeper integration of environmental sociological approaches.

6. Conclusions

Conventional views of sustainable development and sustainability are rooted in problematic presuppositions that lack important political-economic insights. That is, historically specific social relations and institutions cannot be transposed into trans-historic, natural systems that exist alongside nature, or even conceived of as largely separate from the biophysical world. A central theme in both Marx's and Polanyi's work is that it is absolutely necessary to describe the social nature of the modern capitalist economic system. They also stressed the political and ecological dangers of mis-conceptualizing the economy as a natural system.

Sustainable development paradigms based on the ill-conceived triple bottom line/three pillars models are fundamentally flawed in that they do not recognize the socio-historical nature of economic conditions and that all social conditions exist in a material reality, namely within the larger Earth system. They suffer from a pre-analytic vision, drawn from neoclassical economics, which assumes that the economic pillar is independent and simply means the ongoing pursuit of endless growth. Here the role of capitalist social relations in the creation of environmental degradation is sidestepped. Critical questions are ignored and technological solutions are repeatedly proposed, neglecting a serious discussion of social transformations. Fortunately, several theoretical traditions within environmental sociology provide insights for better articulating the socio-ecological circumstances, and the roots of the socio-ecological concerns of modernity.

We suggest that the emergence and development of sustainability science has come a long way toward addressing the gaping holes in the more traditional sustainable development approaches. In particular, sustainability science that emphasizes that human and natural systems are coupled are promising, given the emphasis that these systems must be studied together, in interaction. While making headway, sustainability science has not sufficiently drawn upon the power of critical approaches within environmental sociology in developing analyses, and thus has not appropriately engaged social dynamics and has sometimes been susceptible to functionalism. Human ecology, treadmill of production, and social metabolism can greatly enhance the studies of sustainability, and particularly contribute to insights into matters of political-economic power, environmental degradation, inequality, and social justice.

Sustainable socio-ecological systems must not only be resilient, but also socially just. Environmental sociological theories can clarify the connections and interactions between institutional dynamics and their ecological outcomes. They call into question mechanical, functionalist analyses that are rooted in a pre-analytic vision that naturalizes the economic system. Societies that are more equitable have greater potential for socio-ecological sustainability and resiliency [117]. Under these conditions, individuals and communities participate in a social form and manner that prioritizes sustainability and equity, the purpose of which is to facilitate social/human development and to enhance human welfare and dignity. Thus, the larger community must be invested in and benefit from these processes of change. Under such a system, accumulation for accumulation's sake is not the focus. Instead, the economy is deeply embedded within a society and must recognize natural limits, which can allow for a radical qualitative and quantitative shift in humanity's relationship to the Earth system [90]. The goals are to better address environmental change, lessen the human demands placed on ecosystems, and promote socio-ecological sustainability.

Author Contributions: All authors contributed to this manuscript. The order of authorship signifies the proportion of the contribution to the manuscript. All authors read and approved the final manuscript.

Conflicts of Interest: The authors declare no conflict of interest.

References

1. McMichael, P. *Development and Social Change: A Global Perspective*; SAGE Publications: Washington, DC, USA, 2012.
2. Ponting, C. *A New Green History of the World*; Penguin Books: New York, NY, USA, 2007.

3.   United Nations. *Millennium Ecosystem Assessment Synthesis Report*; United Nations: New York, NY, USA, 2005.
4.   Wackernagel, M.; Rees, W.E. *Our Ecological Footprint*; New Society Publishers: Philadelphia, PA, USA, 1998.
5.   Esty, D.C.; Levy, M.; Srebotnjak, T.; de Sherbinin, A. *Environmental Sustainability Index: Benchmarking National Environmental Stewardship*; Yale Center for Environmental Law & Policy: New Haven, CT, USA, 2005.
6.   Raskin, P. The great transition today: A postscript. In *GTI Paper Series, Frontiers of a Great Transition*; Tellus Institute: Boston, MA, USA, 2006.
7.   Raskin, P.; Banuri, T.; Gallopin, G.; Gutman, P.; Hammond, A.; Kates, R.; Swart, R. *Great Transition: The Promise and Lure of the Times Ahead*; Stockholm Environment Institute: Boston, MA, USA, 2002.
8.   Talberth, J.; Cobb, C.; Slattery, N. *The Genuine Progress Indicator 2006: A Tool for Sustainable Development*; Redefining Progress: Oakland, CA, USA, 2006.
9.   Parris, T.M.; Kates, R.W. Characterizing a sustainability transition: Goals, targets, trends, and driving forces. *Proc. Natl. Acad. Sci. USA* **2003**, *100*, 8068–8073. [CrossRef] [PubMed]
10.   Parris, T.M.; Kates, R.W. Characterizing and measuring sustainable development. *Annu. Rev. Environ. Resour.* **2003**, *28*, 559–586. [CrossRef]
11.   Banerjee, S.B. Who sustains whose development? Sustainable development and the reinvention of nature. *Organ. Stud.* **2003**, *24*, 143–180. [CrossRef]
12.   Mebratu, D. Sustainability and sustainable development: historical and conceptual review. *Environ. Impact Assess. Rev.* **1998**, *18*, 493–520. [CrossRef]
13.   Daly, H.E.; Farley, J. *Ecological Economics: Principles and applications*; Island Press: Washington, DC, USA, 2004.
14.   Kanth, R.K. *Paradigms in Economic Development: Classic Perspectives, Critiques, and Reflections*; ME Sharpe Inc.: Armonk, NY, USA, 1993.
15.   Lele, S.M. Sustainable development: a critical review. *World Dev.* **1991**, *19*, 607–621. [CrossRef]
16.   Seghezzo, L. The five dimensions of sustainability. *Environ. Politics* **2009**, *18*, 539–556. [CrossRef]
17.   Rostow, W.W. *The Stages of Economic Growth: A Non-Communist Manifesto*; Cambridge University Press: Cambridge, UK, 1993.
18.   Redclift, M. Sustainable development (1987–2005): An oxymoron comes of age. *Sustain. Dev.* **2005**, *13*, 212–227. [CrossRef]
19.   About UNEP. United Nations Environment Program. Available online: http://www.unep.org/Documents. Multilingual/Default.asp?DocumentID=43 (accessed on 30 March 2016).
20.   UNCED. UN Conference on Environment and Development. Available online: http://www.un.org/geninfo/bp/enviro.html (accessed on 30 March 2016).
21.   Agenda 21. UN Division for Sustainable Development. Available online: http://www.un.org/esa/dsd/agenda21/ (accessed on 30 March 2016).
22.   WCED. Our Common Future. Available online: http://www.un-documents.net/wced-ocf.htm (accessed on 1 April 2016).
23.   Goldman, M. *Imperial Nature: The World Bank and Struggles for Social Justice in the Age of Globalization*; Yale University Press: New Haven, CT, USA, 2005.
24.   Chatterjee, P.; Finger, M. *The Earth Brokers: Power, Politics, and World Development*; Routledge: New York, NY, USA, 1994.
25.   Imran, S.; Alam, K.; Beaumont, N. Reinterpreting the definition of sustainable development for a more ecocentric reorientation. *Sustain. Dev.* **2014**, *22*, 134–144. [CrossRef]
26.   Steffen, W.; Broadgate, W.; Deutsch, L.; Gaffney, O.; Ludwig, C. The Trajectory of the Anthropocene: The Great Acceleration. *Anthr. Rev.* **2015**, *2*, 81–98. [CrossRef]
27.   Carson, R. *Silent Spring*; Houghton Mifflin: Boston, MA, USA, 1962.
28.   Commoner, B. *The Closing Circle*; Knopf: New York, NY, USA, 1971.
29.   Meadows, D.H.; Meadows, D.L.; Randers, J.; Behrens, W.W., III. *The Limits to Growth: A Report for the Club of Rome's Project on the Predicament of Mankind*; Universe Books: New York, NY, USA, 1972.
30.   Goldsmith, E.; Allen, R.; Allaby, M.; Davoll, J.; Lawrence, S. *Blueprint for Survival*; Pengiun Books: New York, NY, USA, 1973.
31.   Bernstein, S. Ideas, social structure and the compromise of liberal environmentalism. *Eur. J. Int. Relat.* **2000**, *6*, 464–512. [CrossRef]
32.   Pearce, D. *Blueprint 3: Measuring Sustainable Development*; Earthscan: London, UK, 1993.
33.   Daly, H.E. Steady state economics: A new paradigm. *New Lit. Hist.* **1993**, *24*, 811–816. [CrossRef]

34. Solow, R.M. The economics of resources or the resources of economics. *Am. Econ. Rev.* **1974**, *64*, 1–14. [CrossRef]

35. Daly, H.E.; Czech, B.; Trauger, D.L.; Rees, W.E.; Grover, M.; Dobson, T.; Trombulak, S.C. Are we consuming too much—For what? *Conserv. Biol.* **2007**, *21*, 1359–1362. [CrossRef] [PubMed]

36. Jackson, T. *Prosperity without Growth*; Earthscan: London, UK, 2009.

37. Dawe, N.K.; Ryan, K.L. The faulty three-legged-stool model of sustainable development. *Conserv. Biol.* **2003**, *17*, 1458–1460. [CrossRef]

38. Edwards, A.R. *The Sustainability Revolution: Portrait of a Paradigm Shif*; New Society Publishers: Gabriola Island, BC, Canada, 2005.

39. Munasinghe, M. *Sustainable Development in Practice: Sustainomics Methodology and Applications*; Cambridge University Press: New York, NY, USA, 2009.

40. Muschett, D.F. *Principles of Sustainable Development*; St. Lucie Press: Delray Beach, FL, USA, 1997.

41. Elkington, J. *Cannibals with Forks: The Triple Bottom Line of 21st Century Business*; New Society Publishers: Stony Creek, CT, USA, 1998.

42. Elkington, J. Partnerships from cannibals with forks: The triple bottom line of 21st-century business. *Environ. Qual. Manag.* **2007**, *8*, 37–51. [CrossRef]

43. Savitz, A.W.; Weber, K. *Triple Bottom Line: How Today's Best-Run Companies Are Achieving Economic, Social, and Environmental Success-and How You Can Too*; Jossey-Bass: San Francisco, CA, USA, 2006.

44. OECD. Glossary of Statistical Terms. Available online: https://stats.oecd.org/glossary/detail.asp?ID=6591 (accessed on 1 February 2016).

45. Anderson, J.L.; Anderson, C.M.; Chu, J.; Meredith, J.; Asche, F.; Sylvia, G.; Smith, M.D.; Anggraeni, D.; Robert, A.; Guttormsen, A.; *et al.* The fishery performance indicators: A management tool for triple bottom line outcomes. *PLoS ONE* **2015**, *10*, e0122809. [CrossRef] [PubMed]

46. Grafton, R.Q.; Kompas, T.; Ha, P.V. The economic payoffs from marine reserves: Resource rents in a stochastic environment. *Econ. Rec.* **2006**, *82*, 469–480. [CrossRef]

47. Galbraith, J.K. *The Economics of Innocent Fraud*; Houghton Mifflin: Boston, MA, USA, 2004.

48. Norgaard, R.B. *Development Betrayed*; Routledge: London, UK, 1994.

49. Marx, K. *Capital*; Vintage Books: New York, NY, USA, 1977; Volume 1.

50. Polanyi, K. *The Great Transformation*; Beacon Press: Boston, MA, USA, 1957.

51. Ricardo, D. *On the Principles of Political Economy and Taxation*; John Murray: London, UK, 1817.

52. Hunt, E.K. *History of Economic thought: A Critical Perspective*; Wadsworth Publishing Company: Belmont, CA, USA, 1979.

53. Yates, M. *Naming the System*; Monthly Review Press: New York, NY, USA, 2003.

54. Linebaugh, P. *The London Hanged: Crime and Civil Society in the Eighteenth Century*; Verso: New York, NY, USA, 2003.

55. Sweezy, P. Capitalism and the Environment. *Mon. Rev.* **2004**, *56*, 86–93. [CrossRef]

56. Burkett, P. *Marx and Nature*; St. Martin's Press: New York, NY, USA, 1999.

57. Mészáros, I. *Beyond Capital*; Monthly Review Press: New York, NY, USA, 1995.

58. Foster, J.B. *Marx's Ecology*; Monthly Review Press: New York, NY, USA, 2000.

59. Li, M. Capitalism, climate change, and the transition to sustainability: Alternative scenarios for the U.S., China, and the world. *Dev. Chang.* **2009**, *40*, 1039–1062. [CrossRef]

60. Weber, M. *The Protestant Ethic and the Spirit of Capitalism*; Routledge: New York, NY, USA, 2001.

61. Marx, K. *Early Writings*; Vintage Books: New York, NY, USA, 1975.

62. Anderson, K.; Bows, A. A new paradigm for climate change. *Nat. Clim. Chang.* **2012**, *2*, 639–640. [CrossRef]

63. Millennium Ecosystem Assessment, Ecosystems and Human Well-Being. 2005. Available online: http://millenniumasssessment.org/en/Index.aspx (accessed on 2 April 2016).

64. Rockström, J.; Steffen, W.; Noone, K.; Persson, Å.; Chapin, F.S., III; Lambin, E.; Lenton, T.M.; Scheffer, M.; Folke, C.; Schellnhuber, H.J. Planetary boundaries: exploring the safe operating space for humanity. *Ecol. Soc.* **2009**, *14*, 32. Available online: http://www.ecologyandsociety.org/vol14/iss2/art32/ (accessed on 30 April 2016).

65. Stocker, T.S.; Qin, D. Climate Change 2013 the Physical Science Basis: Summery for Policy Makers (WG1). IPCC Fifth Assessment Report. Available online: http://www.climatechange2013.org/images/uploads/WGI_AR5_SPM_brochure.pdf (accessed on 24 November 2013).

66. World Wildlife Fund, Living Planet Report 2008. Available online: http://wwf.panda.org (accessed on 2 April 2016).
67. Foster, J.B.; Clark, B.; York, R. *The Ecological Rift*; Monthly Review Press: New York, NY, USA, 2010.
68. Odum, H.T.; Odum, E.C. *A Prosperous Way Down*; University Press of Colorado: Boulder, CO, USA, 2001.
69. Catton, W.R.; Dunlap, R.E. Environmental sociology: A new paradigm. *Am. Sociol.* **1978**, *13*, 41–49.
70. Dunlap, R.E.; Catton, W. Environmental sociology. *Annu. Rev. Sociol.* **1979**, *5*, 243–273. [CrossRef]
71. Cottrell, F. *Energy and Society*; McGraw-Hill: New York, NY, USA, 1955.
72. Duncan, O.D. From social system to ecosystem. *Sociol. Inq.* **1961**, *31*, 140–149. [CrossRef]
73. Duncan, O.D. Social organization and the ecosystem. In *Handbook of Modern Sociology*; Faris, R.E.L., Ed.; Rand McNally: Chicago, IL, USA, 1964; pp. 36–82.
74. Duncan, O.D.; Schnore, L.F. Cultural, behavioral, and ecological perspectives in the study of social organization. *Am. J. Sociol.* **1959**, *5*, 132–146. [CrossRef]
75. Catton, W.R., Jr. *Overshoot*; University of Illinois Press: Urbana, IL, USA, 1982.
76. Catton, W.R., Jr. Foundations of human ecology. *Sociol. Perspect.* **1994**, *37*, 75–95. [CrossRef]
77. Dunlap, R.E.; Catton, W. What environmental sociologists have in common (whether concerned with 'built' or 'natural' environments). *Sociol. Inq.* **1983**, *53*, 113–135. [CrossRef]
78. York, R.; Mancus, P. Critical human ecology: Historical materialism and natural laws. *Sociol. Theory* **2009**, *27*, 122–149. [CrossRef]
79. Smil, V. *Energy in World History*; Westview: Boulder, CO, USA, 1994.
80. Cohen, J.E. *How Many People Can the Earth Support?*; W. W. Norton & Co.: New York, NY, USA, 1995.
81. Meadows, D.; Randers, J.; Meadows, D. *Limits to Growth: The 30-Year Update*; Chelsea Green: White River Junction, VT, USA, 2004.
82. Jorgenson, A.K.; Clark, B. Assessing the temporal stability of the population/environment relationship in comparative perspective. *Popul. Environ.* **2010**, *32*, 27–41. [CrossRef]
83. Jorgenson, A.K.; Clark, B. The relationship between national-level carbon dioxide emissions and population size. *PLoS ONE* **2013**, *8*, e57107. [CrossRef] [PubMed]
84. Mazur, A. How does population growth contribute to rising energy consumption in America? *Popul. Environ.* **1994**, *15*, 371–378. [CrossRef]
85. Rosa, E.A.; York, R.; Dietz, T. Tracking the anthropogenic drivers of ecological impacts. *Ambio* **2004**, *33*, 509–512. [CrossRef] [PubMed]
86. York, R.; Rosa, E.A.; Dietz, T. Footprints on the earth: The environmental consequences of modernity. *Am. Sociol. Rev.* **2003**, *68*, 279–300. [CrossRef]
87. York, R.; Rosa, E.A.; Dietz, T. A rift in modernity? *Int. J. Sociol. Soc. Policy* **2003**, *23*, 31–51. [CrossRef]
88. Schnaiberg, A. *The Environment*; Oxford University Press: New York, NY, USA, 1980.
89. Gould, K.A.; Pellow, D.N.; Schnaiberg, A. Interrogating the treadmill of production. *Organ. Environ.* **2004**, *17*, 296–316. [CrossRef]
90. Burkett, P. Marx's vision of sustainable human development. *Mon. Rev.* **2005**, *57*, 34–62. [CrossRef]
91. Schnaiberg, A.; Gould, K.A. *Environment and Society*; St. Martin's Press: New York, NY, USA, 1994.
92. Gould, K.A.; Pellow, D.N.; Schnaiberg, A. *The Treadmill of Production*; Paradigm Publishers: Boulder, CO, USA; London, UK, 2008.
93. Foster, J.B. *The Vulnerable Planet*; Monthly Review Press: New York, NY, USA, 1994.
94. Pellow, D. *Resisting Global Toxins*; MIT Press: Cambridge, MA, USA, 2007.
95. Foster, J.B. Marx's theory of metabolic rift. *Am. J. Sociol.* **1999**, *105*, 366–405. [CrossRef]
96. Foster, J.B. *Ecology against Capitalism*; Monthly Review Press: New York, NY, USA, 2002.
97. Foster, J.B. Marx and the rift in the universal metabolism of nature. *Mon. Rev.* **2013**, *65*, 1–19. [CrossRef]
98. Clark, B.; York, R. Carbon metabolism: Global capitalism, climate change, and the biospheric rift. *Theory Soc.* **2005**, *34*, 391–428. [CrossRef]
99. Foster, J.B.; Clark, B. The planetary emergency. *Mon. Rev.* **2012**, *64*, 1–25. [CrossRef]
100. Clausen, R.; Clark, B. The metabolic rift and marine ecology: An analysis of the ocean crisis within capitalist production. *Organ. Environ.* **2005**, *18*, 422–444. [CrossRef]
101. Longo, S.B. Mediterranean rift: Socio-ecological transformations in the Sicilian bluefin tuna fishery. *Crit. Sociol.* **2012**, *38*, 417–436. [CrossRef]

102. Longo, S.B.; Clark, B. The commodification of bluefin tuna: The historical transformation of the Mediterranean fishery. *J. Agrar. Chang.* **2012**, *12*, 204–226. [CrossRef]
103. Longo, S.B.; Clausen, R.; Clark, B. *The Tragedy of the Commodity*; Rutgers University Press: New Brunswick, NJ, USA, 2015.
104. Pauly, D.; Christiensen, V.; Dalsgaard, J.; Freese, R.; Torres, F. Fishing down marine food webs. *Science* **1998**, *279*, 860–863. [CrossRef] [PubMed]
105. Clark, B.; Foster, J.B. William Stanley Jevons and the coal question. *Organ. Environ.* **2001**, *14*, 93–98. [CrossRef]
106. Jevons, W.S. *The Coal Question*; Macmillan: London, UK, 1906.
107. Jorgenson, A.K. The transnational organization of production, the scale of degradation, and ecoefficiency. *Hum. Ecol. Rev.* **2009**, *16*, 64–74.
108. Polimeni, J.; Mayumi, K.; Giampietro, M.; Alcott, B. *The Jevons Paradox and the Myth of Resource Efficiency Improvements*; Earthscan: London, UK, 2008.
109. York, R. The paradox at the heart of modernity. *Int. J. Sociol.* **2010**, *40*, 6–22. [CrossRef]
110. York, R. Three lessons from trends in CO_2 emissions and energy use in the United States. *Soc. Nat. Resour.* **2010**, *23*, 1244–1252. [CrossRef]
111. York, R.; Rosa, E.A.; Dietz, T. The ecological footprint intensity of national economies. *J. Ind. Ecol.* **2004**, *8*, 139–154. [CrossRef]
112. Kates, R.W.; Clark, W.C.; Corell, R.; Hall, J.M.; Jaeger, C.C.; Lowe, I.; McCarthy, J.J.; Schellnhuber, H.J.; Bolin, B.; Dickson, N.M.; *et al.* Sustainability science. *Science* **2001**, *292*, 641–642. [CrossRef] [PubMed]
113. Liu, J.; Dietz, T.; Carpenter, S.R.; Alberti, M.; Folke, C.; Moran, E.; Pell, A.N.; Deadman, P.; Kratz, T.; Lubchenco, J. Complexity of coupled human and natural systems. *Science* **2007**, *317*, 1513–1516. [CrossRef] [PubMed]
114. Liu, J.; Mooney, H.; Hull, V.; Davis, S.J.; Gaskell, J.; Hertel, T.; Lubchenco, J.; Seto, K.C.; Gleick, P.; Kremen, C.; *et al.* Sustainability: Systems integration for global sustainability. *Science* **2015**, *347*, 1258832. [CrossRef] [PubMed]
115. Folke, C. Resilience: The emergence of a perspective for social-ecological systems analyses. *Glob. Environ. Chang.* **2006**, *16*, 253–267. [CrossRef]
116. Walker, B.; Salt, D. *Resilience Thinking: Sustaining Ecosystems and People in a Changing World*; Island Press: Washington, DC, USA, 2006.
117. Adger, W.N. Social and ecological resilience: Are they related? *Prog. Hum. Geogr.* **2000**, *24*, 347–364. [CrossRef]
118. Adger, W.N.; Hughes, T.P.; Folke, C.; Carpenter, S.R.; Rockström, J. Social-ecological resilience to coastal disasters. *Science* **2005**, *309*, 1036–1039. [CrossRef] [PubMed]
119. Holling, C.S. Resilience and stability of ecological systems. *Annu. Rev. Ecol. Syst.* **1973**, *4*, 1–23. [CrossRef]
120. Muradian, R. Ecological thresholds: A survey. *Ecol. Econ.* **2001**, *38*, 7–24. [CrossRef]
121. Scheffer, M.; Carpenter, S.; Foley, J.A.; Folke, C.; Walker, W. Catastrophic shifts in ecosystems. *Nature* **2001**, *413*, 591–596. [CrossRef] [PubMed]
122. Olsson, L.; Jerneck, A.; Thoren, H.; Persson, J.; OByrne, D. Why resilience is unappealing to social science: Theoretical and empirical investigations of the scientific use of resilience. *Sci. Adv.* **2015**, *1*, e1400217. [CrossRef] [PubMed]

sustainability

MDPI

Article

The Anthropocenic Turn: Theorizing Sustainability in a Postnatural Age

Manuel Arias-Maldonado

Área de Ciencia Política, Facultad Derecho UMA, University of Málaga, Campus Teatinos s/n., Málaga 29071, Spain; marias@uma.es; Tel.: +34-615-31-78-13

Academic Editor: Md Saidul Islam
Received: 30 October 2015; Accepted: 30 November 2015; Published: 24 December 2015

Abstract: So long as sustainability represents the attempt to pacify the relationship between societies and their natural environments, the concept must remain attentive to any findings about the character of such relation. In this regard, the rise of the Anthropocene cannot be ignored by environmental sociologists if a realistic understanding of sustainability is to be produced. The Anthropocene is a scientific notion, grounded on geology and Earth-system science, that plausibly suggests that human beings have colonized nature in a degree that has irreversibly altered the functioning of planetary systems. As a result, social and natural systems have become "coupled". This paper tries to elucidate the consequences that an "Anthropocenic turn" would have for sustainability studies. To such end, it will explore the related notions of hybridity and relational agency as key aspects of a renewed view of nature. Correspondingly, it argues that cultivated capital (rather than natural or manmade) must be the most important unit for measuring sustainability and devising sustainable policies in a postnatural age.

Keywords: sustainability; environmental sociology; Anthropocene; nature; conservation; hybridity; socioecological metabolism; technology

1. Introduction

Sustainability has been defined in many different ways throughout the years, but it seems fair to assume that it can be said to represent the attempt to pacify the relationship between a society and its natural environment—irrespective of the way in which the subsequent pacification is pursued. Of course, there is no single definition of sustainability, but, more importantly, there is no single way of being sustainable either: socionatural relations possess a plasticity that allow for a number of possibilities depending, for instance, on the amount of natural capital (as opposed to manmade capital) to be preserved or the reliance on technological solutions. Hence the ideological clashes around this seemingly technical concept. However, the study of sustainability must remain attentive to any findings on the character of socionatural relations. This is why environmental sociologists cannot ignore the rise of the Anthropocene if a realistic understanding of sustainability is to be produced. Such is the subject of this paper.

However, what is the Anthropocene? Broadly speaking, it is a geological concept that tries to capture the change experienced in socionatural relations on a global scale after human activity has exerted a huge influence on natural systems for millenia. The term itself suggests that the Holocene is over: we would now be living in a new geological age due to our impact on the natural environment [1]. As an overarching scientific notion, grounded on geology and Earth-system science, it thus plausibly suggests that human beings have colonized nature in a degree that has irreversibly altered the functioning of planetary systems.

Yet two different meanings can be discerned. On the one hand, the Anthropocene is a given historical period during which certain events have taken place. On the other, the Anthropocene is a result, the consequence of those socionatural events and processes. In other words, it is both a chronology (marked by the anthropogenic influence on natural systems during time) and a given state of relations between society and nature. One that, in fact, blurs their separation.

Needless to say, such a hypothesis resonates strongly as far as sustainability is concerned: as a general principle and as a particular technique that implements the latter in a given fashion. Scientists are suggesting that natural and social systems are now coupled, since the extent of the anthropogenic influence on ecological systems and natural processes is unprecedented [2,3]. This shift knows a large number of manifestations, among which climate change is surely the most popular one. But others, from urbanization to desertification, from species invasion to species extinction, must be added. Synthetic biology and growing genetic experimentation will just make this list longer. That is why, even if geologists do not give official recognition to the new geological epoch, this complex reality is not going to disappear. The Anthropocene is here to stay and environmental sociology has to take it seriously, especially when dealing with sustainability.

After all, the Anthropocene seems to confirm that society and nature are not two separate entities influencing each other, rather there exists a socionatural entanglement—that is, an irreversible, complex, and increasingly hybrid socionatural system. However, paradoxically, this does not mean that there remains no *separation* between human beings and nature. Ironically, it is because we have separated ourselves from nature in a certain way throughout history that this deep entanglement has been produced. In fact, that very separation allows us to be aware of this entanglement and offers us the chance to re-arrange socionatural relations in a new, more refined way. It is the delusion of naturalness that fades.

It should be noted, however, that not everyone is convinced that we are entering into a new era. Not every geologist is a defender of the new chronology, to begin with. On the other hand, the Anthropocene may be seen as a way of advancing a capitalistic narrative of colonization and humanization, making abstract declarations about the role of "humanity" while sidelining more particular questions about the responsibility of particular human groups over others, or just as a continuation of the grand narratives of modernity that still hinders the realization of a more radical, ecocentric political agenda. However, the grounds on which the hypothesis of the Anthropocene rest are firm enough, irrespective of how the geological question will be answered. These critical objections should rather be seen as part of the discussion on the Anthropocene, even though a possible response to them is that the Anthropocene is in itself a way to frame socionatural relations that creates a biased discussion in the first place.

The Anthropocene poses indeed a number of questions. What are the implications for environmental sociology? How does the Anthropocene change environmental sociology's self-perception? Are there intellectual resources that the discipline can turn to in order to deal with this theoretical challenge? Moreover, what does the Anthropocene say about nature and socionatural relations, and how does it affect sustainability? This paper will try to answer to these interrogations in the following sections. It will suggest that the Anthropocene debate deeply affects—or should affect—how sustainability is conceived, perceived, and searched for.

The paper is structured as follows. Section 2 will begin by considering how should environmental sociology react to the Anthropocene and what are the resources that it can employ in dealing with it. A classical concept advocated by Marx, that of socionatural metabolism, later developed into that of socioecological metabolism in order to explain the particular relationship established between a given society and its physical environment, is defended. Section 3 will turn to sustainability, exposing the key question of substitutability, *i.e.*, the degree in which natural capital may be replaced by manufactured capital. Section 4 suggests that this frame may be becoming obsolete, as the notion of natural capital loses its naturalness due to a general process of socionatural hybridization that, for this very reason, must be taken into account by sustainability studies. Finally, Section 5 ponders what exactly the impact of hybridization in our understanding of sustainability may be, suggesting that natural capital is doomed to lose prominence in favor of manufactured capital. At the same time, this will diminish the force of the service argument in defense of protecting natural capital, leaving moral arguments for the social desirability of such protection as the most decisive ones. Section 6 offers a conclusion.

2. The Anthropocenic Turn in Environmental Sociology

It is well-known that environmental sociology was born as a correction of the environmental deficit of classical sociology—a correction from within that however represents "a major departure from sociology's traditional neglect of environmental phenomena" [4]. According to this prevailing view, modern classical sociology stressed human exceptionalism and was consolidated around a humanistic worldview. For Catton and Dunlap, founding fathers of environmental sociology, this "old exemptionalist paradigm" was to be contrasted with the "new environmental paradigm" advanced by the new subdiscipline [5]. If classical sociology had been "constructed as if nature didn't matter" [6] rooted as it was in a "socio-cultural determinism" [7], environmental sociology was to put the environment in its center.

Three major theoretical risks derive from these assumptions—risks that to some extent have been confirmed in the practice of environmental sociology in the last decades, despite having also found resistance within the discipline. Firstly, the neglect of classical sociology, understood as an Anthropocenic enterprise oblivious to the presence of nature in social evolution—a neglect that does not only create an unnecessary antagonism between classical and environmental sociology but also made it difficult to find insights about the socionatural relation in the work of classical sociologists. Secondly, the rejection of socio-cultural determinism was easily turned into an ecological determinism that reduced the role of human agency within the socionatural relation and overvalued the strength of ecological constraints *vis-à-vis* human societies. In so many predictions, the latter were reduced to the condition of passive witnesses of an unavoidable ecological catastrophe, thus reducing the range of possibilities allowed in the socionatural interaction as well as the ability of human societies to shape its environment. In turn, this view exerted a major influence on sustainability, seen as a human retreat from nature rather than as a reflective continuation of a complex process of evolutionary adaptation that involve—in the case of the human species—the transformation of nature. Thirdly, in close connection to this, the role of human beings was diminished as the weakest part in the socionatural relation, a paradoxical position if we recall that the ecological crisis itself would have been provoked by human beings themselves in their meddling with natural systems throughout time. If anthropocentrism was strongly rejected, exemptionalism and exceptionalism too: there was nothing special about human beings, as they are a part of nature and remains linked to and dependent of the latter.

In hindsight, such unbalances are understandable. Environmental sociology had to compensate for a prevailing anthropocentrism—almost an anthropocentrism by default—that had marked the development of the discipline from the outset. Predictably, it has also been within the latter that those risks have been, if not avoided, signaled. As a result, environmental sociology has grown more and more sophisticated and is now ready to deal with the Anthropocene, which however represents a major challenge to some core assumptions of environmental sociology as has been practiced so far. If the assumptions behind the Anthropocene are taken seriously, it is plain to see how they do relate to the theoretical risks just mentioned—confirming that some amendments are in order within the discipline if it is to keep close to the reality of socionatural relations.

In this vein, it is revealing that the Anthropocene lends credit to those classical sociologists that did pay attention to socionatural relations and took into account the environmental factor as a key one in societal development. It does so in an unexpected way, namely, restoring the plausibility of human exceptionalism. If humans have turned into great forces in natural evolution and have transformed planetary systems while adapting to their environment, it cannot be denied that they belong to a species that, belonging to nature, separates itself from it. Human/nature dualism is not ontological, but historical [8]. Ironically, this separation involves a deeper integration of human beings into nature via the transformation of the latter, actually turned into human environment. This intimate relationship between human beings and nature, that happens also to be a deeply transformative one, was anticipated by Karl Marx [9] through his concept of "socio-ecological metabolism" (*Stoffwechsel*). If we read Marx's explanation under the new hermeneutical context provided by the Anthropocene, he appears clearly as a contemporary:

> "Labour is, first of all, a process between man and nature, a process by which man, through his own actions, mediates, regulates and controls the metabolism between himself and nature. *He confronts the materials of nature as a force of nature.* He sets in motion the natural forces which belong to his own body, his arms, legs, head and hands, in order to appropriate the materials of nature *in a form adapted to his own needs.* Through this movement he acts upon external nature and changes it, and in this way he simultaneously changes his own nature. (...) It (the labor process) is the universal condition for the metabolic interaction (*Stoffwechsel*) between man and nature, the everlasting nature-imposed condition of human existence" (my emphasis) [9].

To some extent, then, the Anthropocene might help to revive the old sociological paradigm, or rather those aspects of it that are more helpful when the massive transformation of the environment is to be explained. A successful transformation, it should be added: as the ecological catastrophes announced by classical environmentalism have failed to materialize, contemporary environmental sociology has to consider the possibility that there is no collapse at hand, but rather a complex socionatural relationship that has be to regulated through sustainable policies not so much oriented towards a human retreat from nature (which is not feasible anymore given the reciprocal imbrication of social and natural systems) than towards a recognition of the multi-layered and hybrid nature of the socionatural entanglement. Marx's metabolic rift is not a flawless notion, but it points to the right direction: that of recognizing the transformative powers of human agency and the dynamic character of a relationship that ultimately responds to the particular way of being of the human species. As Bellamy Foster, commenting Marx, points out:

> "The material exchanges and regulatory action associated with the concept of metabolism encompassed both 'nature-imposed conditions' and the capacity of human beings to affect this process" [10].

It should be noted that Marx, albeit capturing the essential features of the socionatural relation, failed to take some important aspects of it into account. On the one hand, human labor produces a number of unintended, unforeseen consequences that in turn come to affect "nature-imposed conditions", as climate change so clearly shows. Moreover, the Anthropocene itself, mostly a product of the industrial acceleration of economic growth, is one of those consequences. On the other, while recognizing ecological constrictions, Marx did not go far enough in recognizing nature's agency and the reciprocal character of the influence that humanity and nature exert on each other. However, he was quite interested in Darwin's theory and it is reasonable to think that he might have endorsed a Darwinian reading of the Anthropocene, *i.e.*, one that explains socionatural history as the product of human exceptionalism.

Other thinkers concerned with the future of environmental sociology have also stressed the importance of metabolism and the nature/culture dichotomy, as well as the need to combine structural and cultural perspectives, emphasizing both the centrality of human subjects and the materiality of nature [11]. In the words of Frederick Buttel, a reasonable sociological approach must consider "both 'structure' and 'agency', and the material and the symbolic" [12]. To such end, environmental sociology is forced nowadays to make interdisciplinary contact with other disciplines beyond the realm of social sciences, if the Anthropocene challenge is to be met. That is what Fischer-Kowalski and Weisz do when looking for insights in anthropology and biology that may help to sustain their concept of "socioeconomic metabolism", in itself an elaboration of Marx's [13]. They search for a realistic account of socionatural relations, one that

> "must take into consideration the possibility of biophysically relevant interactions between symbolic (cultural) systems and the material world that are historically variable and that are

not compensated for by natural adaptation. This implies a conception of cultural evolution beyond anthropological adaptationism" [13].

As I have suggested elsewhere [14], there are currents of evolutionary theory that provide an explanation of human beings ways on Earth that meets these requirements and should not be overlooked by environmental sociologists. They may serve as underlying foundations for an environmental sociology that puts an updated notion of socionatural metabolism at its core. On the one hand, Historical Ecology claims that changes in socionatural relations are caused by history rather than evolution, focusing on the interaction between cultures and environments, the latter beinf adapted *to* societies and not the opposite way [15]. Human agency is thus fully recognized as a transformative force [16,17]. On its part, niche-construction theory holds that organisms do adapt to their environments changing them through a strategy of niche-construction that in the case of human beings is turbocharged by cultural accumulation and transmission [18–20]. That is why adaptationism should be reconsidered as a model for explaining human-environment relations, replacing it instead with a model that puts emphasis on a transformative human agency that is both the problem (creating the ecological crisis in the first place) and the remedy (providing solutions through a number of ways) [21].

Moreover, niche-construction theory, grounded as it is in the operations of human culture in a given environmental context, leaves room for recognizing that any socionatural interaction is socially bounded and culturally constrained. Instead of possessing unique features in any given time and space, socionatural relations vary *relatively* in different social settings, thus producing disparate "socio-ecological regimes" [22]. This is an advantage over rigidly anti-essentialist views of nature, since that plurality of regimes is encompassed under the broader context of a universal drive towards aggressive adaptation—a species drive. It should be added that this interplay between structure and variation belongs quite naturally to the province of sociology.

As a result, the universal human drive towards adaptation manages exhibits a good deal of variability, since the social re-construction of nature in which it consists leads to different socio-ecological regimes depending on local circumstances as much as on cultural representations. In fact, they interact continuously. However, it is also true that national and local societies are gradually converging around a "Western" *Weltanschaaung* that unifies values and technologies alike. Globalization and digiticization are accelerating this process, by which particular patterns of socionatural interaccion are being eroded.

Nevertheless, an important side-effect of the Anthropocenic narrative is the normalization of capitalism. By stressing the species ways of being rather than blaming particular episodes of human history, the Anthropocene suggest that the aggressive adaptation to the natural environment is a universal human drive, a permanent feature of socionatural relations—not a trait of capitalism or the particular effect of the capitalistic treatment of nature. The idea that capitalism "produces" nature, thus, is misleading [23]. It is true that, in Smith's own phrasing, "first nature" is replaced by an entirely different, produced, "nature". However, capitalism is an accelerator of an otherwise unavoidable process of human colonization and transformation of the environment, not its prime *cause*. Needless to say, the way in which capitalism "produces" nature merits research as a particular—and particularly intense—episode of a wider story [24].

Another feature of socionatural relations that the Anthropocene has reinforced and thus highlighted is hybridity. Environmental sociologists have claimed for long a time now that hybrids are the new normal in this realm, so that the discipline is now well-prepared for recognizing the degree in which this feature has been intensified in the Anthropocene. As Bruno Latour argued, hybrids do not fall into either of the competing categories of social or natural but instead weave together elements of both. Humans and non-humans cannot be separated anymore, rather they are embedded in networks where their respective boundaries become increasingly diffuse [25]. The social and the natural, in other words, should be seen as "actively generated co-constructions" [26]. Society and nature influence each other at all levels, making society a natural force as much as making nature a societal force [27]. The Anthropocene shows, in a number of ways, how difficult it has become to disentangle the natural and the social—but mostly because the natural has been colonized by the social, influenced by it, transformed in different degrees. There is no pure nature anymore, but different scales of "naturalness" depending on the particular history of a living being, a natural process, an ecosystem. Although society and nature are now coupled and they form actually a socionatural entanglement, social and natural agency should not be conflated: human actors are more powerful agents of change than natural actants. An Anthropocenic turn in environmental sociology must recognize the primacy of human transformative powers, both intentional and unintentional.

An environmental sociology for the Anthropocene age should then be one that, as Fischer-Kowalski and Weisz suggested seventeen years ago, should distinguish between two key processes of socionatural interactions: socioeconomic metabolism (refered to the material exchange between social and natural systems) and colonization of natural processes (that is, the intentional and unintentional alteration of natural beings and processes through different means) [13]. It is the extent to which this colonization has taken place that merits such a powerful label as the Anthropocene. Crucially, the recognition of human transformative powers involve a more open approach towards sustainability, namely, one that is not so concerned with *technical* limits to human action or social development but rather more interested in the discussion about the most *desirable* social organization *vis-à-vis* the natural environment. The plasticity of socionatural relations, which arguably was always greater than classical environmental sociology acknowledged, makes supposedly rigid ecological limits rather contestable. For instance, when Dunlap and Catton discuss the functions performed by ecosystems, they may well claim that "exceeding the capacity of a given ecosystem to fulfill one of the three functions may disrupt not only its ability to fulfill the other two, but also its ability to continue to function at all" [28], thus leading to ecological unsustainability and socionatural collapse—yet it debatable whether an ecosystem possesses anymore a fixed, unchangeable capacity, or whether this capacity can be altered, or stretched, through social intervention.

In sum, a postnatural environmental sociology, *i.e.*, a sociology that undergoes an Anthropocenic turn, should then be grounded on the following premises:

(i) A recognition of the extent to which society has colonized nature, up to the point where the proposition that the latter has ended is not incongruous—if we understand this ending as the pervasive influence of society on nature. The most important result of such colonization is hybridization, the mixture of the natural with the social.

(ii) Nature and society thus form a socionatural entanglement, a reflection of their co-evolution and co-construction throughout history—a process, which, however, does not preclude the primacy of the human agency due to the potency of human transformative powers.

(iii) This process of colonization and transformation is in itself the product of the human species particular way of being, according to which there exists a universal drive towards adaptation that takes the form of an aggressive adaptation wherein the physical environment is transformed: a niche-construction process that in the case of humans is accelerated by cultural transmission and admits, despite its universality, regional and local variations.

(iv) Socionatural relations are thus expressed through a number of socio-ecological regimes that organize them in different settings, depending on the way in which each society relates itself with its natural environment (type of socionatural metabolism, degree and forms of colonization, system of beliefs). Globalization, economic growth and the diffusion of technology are increasingly reducing those differences in the Anthropocene age.

(v) This means that socionatural relations are not marked by the logic of limits emphasized by classical environmentalism and early environmental sociology, but defined on the contrary by a plasticity that is scientifically and technologically enforced. As a result, there is no single sustainability, but a plurality of potential sustainabilities—or sustainable socioecological regimes—depending on what value choices are made when deciding which is the most desirable relationship with natural systems, forms and beings.

As Field *et al.* have noted, much environmental sociology is done in a "grand theory" mode, reflecting a tendency to approach the environment in relatively general terms rather than seeing it as endowed with particular features that vary in different settings with consequences for populations or communities [29]. Ideally, an environmental sociology for the Anthropocene should be able to combine both perspectives, so that middle-range work supports and illuminates what grand theories have to say. This is especially the case in the field of sustainability studies, where a delicate balance has to be achieved wherein theoretical insights are not made in an empirical vacuum, while at the same time the latter do not impose an straitjacket on sociological imagination, all the more now that technological acceleration opens up new perspectives on what is technically possible in this realm with alarming speed.

If we turn our attention to sustainability, the question is: How does a postnatural understanding of nature affect it? Why is the Anthropocene a game-changer that makes old views on the subject obsolete and forces environmentalism to reframe the sustainability question, abandoning survivalism in favor of a vision of the good sustainable society? The key question concerns how substitutability—the core concept in advanced notions of sustainability—is affected by increasing hybridity.

3. The Substitutability Question

Once a powerful but ultimately vague notion associated with issues of social justice and fair economic development, sustainability has become a complex and refined concept that revolves around a number of neatly identified issues—from different forms of capital to ecosystem services. Information technologies are helping to make more precise measurements and to monitorize the outcomes of environmental policies and conservation programs. At the same time, sustainability has been relocated into a wider framework, namely, the social response to global warming—in itself another, yet the most spectacular, manifestation of the Anthropocene. Therefore, sustainability is linked to strategies of mitigation and adaptation that emphasize the consumption of energy and the production of food for an increasing population as key aspects of any sustainable global society. For classical environmentalists, sustainability is still a tool for political transformation, yet the association between sustainability and a radically democratized, postcapitalistic, frugal society is weaker than ever. In fact, climate change and the related notion of the Anthropocene expose the flaws of the classical green view of sustainability.

In this regard, social justice used to be seen as an important basis for sustainability [30], but, despite the usefulness of multi-dimensional approaches and the political attraction that exerts its association to human development [31,32], it is fair to say that socionatural patterns of interaction are now considered the most relevant indicator for identifying different modes of sustainability. After all, if every sustainable society must operate according to a particular socioecological regime [13], it makes sense to focus on the way in which the socionatural interaction is to be dealt with.

Essentially, what has to be decided in each case—but also globally if problems such as climate change are to be tackled—is how much nature must be protected. Or, in other words, the degree to which natural capital is to be substituted by human-made capital. How much natural capital can be substituted becomes thus the key question. In other words, we have to decide how to keep a sufficient pool of resources irrespective of whether we are willing to preserve nature for its own sake, to pass on a high level of welfare to future generations, or to avoid climate apocalypse. Such reasons are technically irrelevant, although they are of course politically crucial.

Introduced by economists in the early 1990s [33,34], the concept of capital natural has become a key one insustainability studies. That is hardly surprising, since it is very helpful for describing the social uses of nature, thus making it easier to bring sustainability issues into economic reasoning and political decision-making [35]. As such, capital natural is made up of all natural resources that humans employ for human ends. This usage is what makes nature a *resource*, despite the fact that sometimes nature performs functions for humans without even being touched (the satisfaction of aesthetic or moral needs, for instance). Properly speaking, natural capital is a meta-concept that identifies different functions and benefits that nature provides to humans [36].

Therefore, natural capital is not directly observable. It has to be humanly used as such to be included in that category. Likewise, particular materials or chemical processes might be excluded from such consideration for a long time, until a way of using them for human needs is discovered. Hence the contingency of nature *as* capital. On the other hand, usages can change or become multiple, as biofuels come to show. However, if such a wide concept is to be operational, it has to be refined by distinguishing between different subtypes of natural capital, as well as by sorting out their relative importance for human beings according to the usage they make of it.

Natural capital is thus a part of the capital on which human depend for surviving and thriving. Other forms of capital are *human-made* (comprising human creations and human capital itself) and *cultivated* ones (the latter encompassing domesticated animals, cultivated plants, and all kinds of manipulated beings and habitats) [37]. Both categories can also be unified under the label "manufactured capital". Ultimately, both human-made capital and cultivated capital depend on nature: either by using directly natural stuff or re-orienting living natural entities and components for human ends. Different socio-ecological regimes will use differently and in disparate degrees these varieties of capital.

Likewise, within natural capital writ large, the most relevant criterion for further distinctions is the relative importance of the contribution that different segments of the former make to human survival and welfare. Thus the distinction between *disposable* (irrelevant), *fungible* (important but not crucial), and *critical* (irreplaceable) natural capital. An important warning to be made is that *irreversibility* should not be conflated with *criticality*. They are two different qualities: the loss of a natural being or habitat may be irreversible, but not necessarily critical. Properly speaking, criticality should refer to those natural components or entities whose loss might seriously compromise human survival or means a severe curtailment of its welfare. They have to be irreversible and irreplaceable, the environmental functions they perform not being amenable to substitution by cultivated or human-made capital. Unfortunately, it is as of yet difficult to know for certain which ecosystems and functions are critical and which are not [38]. Moreover, the definition of critical natural capital relies not only on our capacity to supply factual knowledge about socio-ecological systems, but also on discussions about the values that underline our use of natural capital [39]. Thus, neither irreplaceability nor criticality is an absolute category.

Another way of looking at this problem is to distinguish between natural *stocks* and *funds* [39,40]. As their names suggest, stocks are consumed after being used, while funds can be used indefinitely. Whereas coal and oil belong to the former category and adopt the form of *flows*, the latter include living (air, water, sun) and non-living nature (plants or animals) that comes in the form of *service* provision. This should make it easier to measure irreversibility and irreplaceability. However, a relational approach incorporating *ecosystems* is advisable, since both flows and services depend on the systemic relacions that make up for the latter.

Yet an interesting addition to this taxonomy is provided by environmentalists who are worried about things other than human survival and welfare. Non-economic usages of nature should be added to the concept of natural capital, lest the reductionist view of neo-classical economics is reproduced [41]. Thus, if we leave functionality aside, what needs recognition is the intrinsic value of nature. The latter is represented by those natural units -whether beings or habitats- that represent unique ecological associations [42]. These units are like a living memory of natural evolution and as such not substitutable. Holland suggests calling them "units of significance" [42]. Their value lies in their meaning, a living meaning that disappears with them.

However, how are we going to decide? Where does the critical quality of natural capital lie? Is every natural form meaningful, or we have to choose between greater and lesser "meanings", due to the impossibility of protecting everything? A further distinction has to be introduced between goods, functions and services, as provided by natural systems. A good is resource provided by ecosystem *components*, whereas functions and services are structural conditions for human life and are performed by ecosystem *processes* [34]. In turn, four functions can be identified: *regulation* of ecological processes and systems of life support; *production* of raw materials, food, genetic resources; *habitat provision*, so that plants and animals have an habitat to live in, thus helping to preserve biodiversity; *information functions* that are related to moral or aesthetic human needs, as well as to the provision of research, cultural, or historical data. However, the latter category does not do enough to prevent the critique that the services argument cannot explain why those caring for nature want to preserve it and is thus ultimately harmful [43]. We will come back to this.

A particular sustainability will be defined by how it does decide upon the protection and management of these different categories of capital—their combination and reciprocal relations. This will produce a given level of human welfare, as well as a given degree of natural protection. However, being self-evident that some amount of natural capital is to be preserved, it is unclear which is the precise degree of protection that is required. There is no consensus on the right level of protection. If we could determine which assets are critical, in fact, there would be no conflict between different positions on substitutability whatsoever [44]. This is especially clear regarding information functions, since people may have different views about them, *i.e.*, about the amount of nature that has to be preserved for them to be provided. The same goes for units of significance.

Moreover, the absence of scientific certainty about how much natural capital should be protected on account of its criticality just adds to the confusion. To suggest that a given amount of natural capital *might* be substituted is different than saying it *must* be kept at all cost. Usually this opposite views are summarized in the distinction between *weak* and *strong* models of sustainability. The weak ones accepts high rates of substitution, whereas the strong ones restrict the substitution of natural capital, thus enlarging the scope of criticality [45]. But it is impossible to know for sure which is right. A further, fundamental reason for that uncertainty is that we cannot know in advance what parts of nature that now seem to be critical or just important will be substitutable inb the future—so that, leaving normative claims *and* historical meanings aside, the irreversible natural capital of today may just be the disposable or fungible natural capital of tomorrow.

Somehow, sustainability studies assume a stable nature as an object of analysis and a measure for calculations regarding the provision of goods and the performing of functions and services to humankind. However, what if nature has changed? What if the process of hybridization that is the side-effect of the human colonization of nature has altered the workings of natural systems? What if the human ability to transform nature has fewer limits than expected?

4. The Great Hybridization and Its Consequences

The central claim in the Anthropocene hypothesis is that human beings have been a major agent in natural evolution, and increasingly so since the Industrial Revolution began. As a result, it has become more and more difficult to distinguish between societies and environments, since they have merged into a complex socionatural entanglement. This process of human colonization and appropriation makes sense from an evolutionary perspective. It is not thus an aberration, but the logical result of the aggressive adaptation that distinguishes the species way of being.

This historical process is also a process of hybridization by which nature is gradually losing its autonomy from society. Habitats, natural processes, animals become increasingly affected by human activity, irrespective of whether that influence is visible or not. It makes sense to see the Anthropocene as culminating The Great Hybridization of society and nature. Whereas the best metaphor for the human recombination is provided by Haraway's cyborg [46], natural hybridization has been defined by Latour a process by which nature is transformed into nature-culture objects resulting from social activities and practices [47]. This in turn means that assemblages are more important than ontologies, since the category of "the natural" becomes increasingly meaningless [48]. Moreover, the process by which hybridization is produced counts more than the hybrid itself [49], because the latter cannot be simply reduced to its components, but produces an emergent novelty whose qualities are the result of an association [50,51]. Old dualisms that separate nature from culture or the mind from the body are thus not the best tools to understand this new material reality.

It is the Anthropocene that comes to show that hybridization is the new normal. Ellis and Ramankutty have introduced the notion of "anthropogenic biomes" to describe the lack of "naturalness" of basic ecological units [52]. Natural purity was already under attack from cultural historians [53], but the idea that what looks like nature *is* nature remains firmly entrenched. That is wrong:

> "Anthropogenic biomes are best characterized as heterogeneous landscape mosaics, combining a variety of different land uses and land covers. Urban areas are embedded within agricultural areas, trees are interspersed with croplands and housing, and managed vegetation is mixed with semi-natural vegetation (e.g., croplands are embedded within rangelands and forests)" [52].

This comes to demonstrate how the social and the natural are intermingled in a deep fashion all over the Earth, an embeddedness of ecosystems and social systems that actually "couples" them [2, 54]. At the same time, new patterns of biodiversity are emerging, as this logic of intentional and unintentional recombination is intensified. Examples abound, from rocks made of plastics and corals in Hawaii to species invasions and alterations in phenotypic characteristics of exploited species [53,55]. Moreover, some talk of "Homogocene" to reflect how those species that best adapt to human systems become predominant over those specialist ones that have lived in isolation [56]. Others, though, make a different point, emphasizing how this general alteration in biodiversity will actually produce new habitats and species or put into contact species formerly separated [57]. Needless to say, this is of the utmost importance for any conception of sustainability that includes the preservation of natural forms among its goals: if it is not clear what is natural, or what does it mean to preserve original habitats, to decide upon how to organize conservation becomes more confusing in the Anthropocene. Classical environmental management does not look like a reasonable guide anymore [58]. Restoration is an increasingly meaningless notion, unless historical fidelity is taken as a guiding principle rather than a rigid goal [59] and new socioecological conditions are integrated into the analysis [60]. Leaving nature to restore itself isolating particular habitats from social contamination ("wild adaptation") [61] or reintroducing lost species in new environments ("rewilding) [62] have also been proposed as alternatives for a climate-changed world.

Yet there is a different pathway, namely that of forgetting about this old grammar and departing from the assumption that nature as it used to be does not exist anymore: natural spaces are entities that lie in the past and cannot be recovered. A shift should thus be fostered towards ecological design and management, a strategy of conservation that forgets about prehuman landscapes and goes for a more diverse and promiscuous human-nature entanglement that is, at the same time, more human-friendly [63]. However, the complexity of the Anthropocene should be kept in mind when dealing with this delicate issue, so that a case-by-case approach should be privileged lest we prevent ecosystems from performing their functions and providing their services. Intact ecosystems still active in remote areas, for instance, should be preserved [64]. Them aside, though, a postnatural ontology is the signature of the Anthropocene [65]. As Baldwin has suggested, nature is now but a socio-political sphere produced and made sense of thanks to technological and ideational practices [66]. Such view is now common within the social sciences [67].

To embrace the idea of the Anthropocene means thus to advance towards an understanding of nature that takes human influence on it seriously—before considering the moral implications of that influence. What the Anthropocene hypothesis states is that there is no way back for human beings, because we are not just *embedded* in nature, but also *entangled* with it in an irreversible and complex way. Sustainability is not, cannot be, what it used to be.

5. Rethinking Sustainability in the Anthropocene

In order to reflect on sustainability in the postnaural age, the classical distinction between weak and strong sustainabilities does not seem the most appropriate under the new light provided by the Anthropocene and the hybridization process it aptly summarizes. The main reason is that the latter is troublesome for a meaningful distinction between different types of capital. Despite its apparent clarity, categories such as human-made and natural capital serve to obscure a relational, bizarre material reality where mutual contamination and recombination are actually prevalent. It also creates a misunderstanding about the possibility to devise an ethics of substitutability or even about the feasibility of a strong sustainability. That is also why the notion of *cultivated* capital should be privileged in this context as the one that captures best the paradoxes of sustainability in the Anthropocene.

Sustainability involves designing a socionatural relation that can be maintained in the long term. It concerns that relation as such, but the latter are by definition dynamic and changing. But if human beings have become the main agent of change with the passing of time, how useful it is to employ the category of natural capital? Nature is a living and reproductive entity [68]. Nature is not to be seen anymore as an entity affected by the damaging influence of an external agent (humanity), but rather as a dynamic compound of influences of which human beings are an increasingly important one. There is even talk of "technonatures", a term that tries to convey the fact that this socionatural entanglement involve the mediation of diverse hybrid materialities and non-human [69].

Yet could it not be the case that substitutability itself has become a flawed category? As such, it presupposes the idea of nature being *replaced* by something made by humans. But what we find is rather the mixing of the natural and the social, their reciprocal assimilation. An alternative concept is that of "biofacts", describing entities originated or influenced by human activity or design, be the latter outwardly visible or not [66]. An important question is thus whether we wish to protect nature's *integrity* or just their *apparent* naturalness.

But the natural is not an absolute category, but a quality that can be measured in degrees. There is a *genetic* nature and a *qualitative* nature, as Dieter Birnbacher has argued: one is historial and alludes to the existence of nature without human intervention, the other is phenomenological and comprises the natural entities that can be affected by human activity [70]. Which one is to be protected? And what that does mean exactly? Three possibilities arise: (i) protecting *environmental functions,* which might entail preserving entire ecosystems or substituting them in different degreesm depending on their criticality; (ii) preserving nature *in a genetic sense,* that is, the integrity of beings and habitats and systems; or (iii) taking care of nature *in a qualitative sense,* which involves maintaining natural forms whichever it is the true anthropogenical impact that lies behind that appearance of naturalness.

These are the main dimensions along which a postnatural sustainability should be defined. The discussion about different forms of capital, including subtypes of natural capital, must also relate to them. Apparently, we should pay attention to the environmental functions performed by nature as a whole, so that those that are essential to our welfare are sufficiently provided [33,71]. But if enough stock is to be preserved in order to guarantee the provision of those functions into the future, the question remains as to how much natural stock is that and of which kind. At this point, the multi-functionality of natural capital must be underlined. Pelenc and Ballet sum up this idea through a simple example:

> "natural capital is multifunctional *i.e.,* in certain situations it can provide several services simultaneously. For example, the flow of water in a river can provide biological services (the reproduction of fish), economic services (the fish can be caught or the flow can be used to produce hydroelectricity), and recreational services (bathing in the river). This multidimensional aspect of natural capital means that it is unlikely for manufactured capital to act as an appropriate substitute" [72].

Different forms of capital should thus be seen as complementary in producing human wellbeing [73]. For those who argue that those functions depend on the relations within and between ecosystems, replacements and substitutions are actually dangerous options. Moreover, due to our lack of knowledge about how natural systems function, we cannot know for certain what the effects of destroying natural capital will be on human well [74]. Therefore, the strong sustainability approach assumes that the substitutability between natural capital and other forms of capital should be strictly limited to the circumstances where the use of the services provided by natural capital does not lead to the irreversible destruction of this capital because its depletion cannot be compensated for by investing in other forms of capital [73].

However, we might very well develop the ability to replace those functions. What if today's irreversibility is tomorrow's replaceability? Substitutability would then be less a danger than a matter of *choice*: a choice involving a decision about the amount of nature that we wish to protect on grounds other than its contribution to human survival and welfare, *i.e.*, aesthetical, moral, philosophical, and recreational. The moral one appears to be the strongest and is surely the one most clearly associated with environmentalism. It is ultimately a fairly simple one: nature should be conceived in non-instrumental terms. Were this rule to be followed, substitution would be severely restricted on ontological terms [75]. It is history and historical processes what matter accordingo to Holland, not the attributes displayed by a particular natural entity: meaning belongs to nature in a historic sense, irrespective of its external qualities [76]. That is, nature in a historic sense, irrespective of its external qualities. A theoretical path that leads to the ecocentric mandate according to which even when substitution is possible, should not be made [36]. According to this view, there is no such thing as a substitution—only a loss.

This moral position notwithstanding, the maing goal of a sustainable society is to maintain a durable socionatural relations into the future, not conservation *as such*. It may include conservation, trying to protect nature in either (or both) a historic or qualitative sense—but properly speaking it does not have to. Nothing prevents that a given sustainable society protects irreversible natural capital without doing the same with nature's "units of significance": the meaning could be lost while the functions are kept. Paradoxically, though, substitution might help to protect nature's intrinsic value, as in preserving whales by replacing whale oil with electric light.

Enters hybridization, though. Natural capital will arguably maintain its relevance for the time being, but it is important to take into account how much natural systems and processes have mixed up with social ones. Therefore, an enlarged conception of cultivated capital seems much more central to the discussion on sustainability in a postnatural age. This includes not only domesticated animals, as well as livestock, fisheries, and plants that are submitted to human control and breeding, but also reconstituted materials of any kind that involve a conscious human recombination of natural beings. Critical natural capital is by no means to be dismissed, partly because it possesses a great political value as a reminder of nature's otherness and intrinsic value. At the same time, though, it should be acknowledged that it is not a closed category: what is critical today might stop being so in the future on account of scientific and technnological innovations. Moreover, critical nature may also keep its functionality despite it being altered or intervened upon without being thoroughly substituted. It would thus not be less critical, but certainly less natural. It can become cultivated capital as well, since it is not ludicrous to expect that natural capital as such will be *cultivated* in the Anthropocene age: synthetic biology is just in its infancy. Sustainability in the Anthropocene will not be viable without science and technology acting as key facilitators of environmental adaptation.

By itself, this does not answer how much nature should be protected *aside* from that which is considered in a given moment unsubstitutable on account of its criticality as provider of basic environmental functions. There are two different questions at stake: the amount of nature that we *have to* protect and that which we *would like* to see preserved: whereas the former should be answered on technical grounds, the latter depends on normative judgments. And how do we answer the first one will most likely influence how do we answer the second. It is bound to be a complex, forever ongoing discussion. Conclusions will always be temporary, for at least two reasons.

Firstly, apparent limits to replaceability will be moved forward science and technology provides new solutions. Irreversible natural capital may turn into fungible or disposable one, so that the amount of nature deemed as critical may *diminish* with the passing of time. This likelihood introduces a key difference in the debate on sustainability, because the design of the latter will not depend on whether a particular version of it is *feasible* or not, but rather on whether it is *desirable*. Proponents of strong versions of sustainability may conflate these two sets of arguments: technical feasibility and moral desirability. In fact, they do just that whenever the cultural functions of the environment are taken as critical or certain technical limits are taken as absolute rather than contingent.

Secondly, social prevalent views on nature may very well change too, thus changing in turn a given consensus on how much nature should be protected. As Holland remarks, we may develop new assumptions about natural capital utility [42]. However, this value change could make the protection of qualitative nature more desirable. There is no reason to think that a greater substitutability (in technical terms) will necessarily diminish the preference for naturalness (in moral or aesthetical or recreational terms). In fact, there are reasons to think that both fondness for charismatic animals and for recreational nature (mountains, natural reserves, beaches) are strong incentives to protect nature in a qualitative sense—at least in a degree that makes the satisfaction of these human needs possible. Ironically, global tourism may help to preserve those landscapes that it came to endanger in the first place.

In sum, notions such as substitution and criticality, as well as the traditional distinction between two types of sustainability depending on whether it provides a strong or a weak defence of nature should not be taken for granted in a postnatural age. In the Anthropocene, the question of nature should be discussed in a framework that recognizes the fact of hybridization and gives a more prominent role to cultivated capital in an extended sense.

6. Conclusions

This paper has explored how the Anthropocene—namely, those insights about the current state of the socionatural relation that the Anthropocene hypothesis lays on the table—affects currents understanding of sustainability from the vantage point of environmental sociology, which also involves considering the impact of the Anthropocene for environmental sociology itself. In this regard, the discipline's self-perception has revolved around the distinction between an ecological paradigm and an anthropocentric one, the former trying to correct classical sociology's inclination to dismiss natural constraints in favor of a sociocultural determinism, behind which an anthropocentric worldview is discernible. However, the Anthropocene involves a restoration of human exceptionalism, by showing how the human species way of adapting to the environment is grounded on the ability to transform the latter through a process of niche construction that cultural transmission (between and through generations) provides with unparalleled potency. Nature's transformation and domination is thus not a choice or an evolutionary accident, but an expression of the species way of being. As a result, after the modern acceleration, further intensified after World War II, human colonization of nature has reached unprecedented degrees, ushering a process of socionatural hybridization that—leaving appearances aside—involves a ubiquitous social influence on natural forms, beings, and habitats. The corresponding plasticity of socionatural relations, made possible by scientific and technological intervention, should be taken into consideration when reflecting upon sustainability.

As for sustainability itself, this paper has argued that it cannot be conceived anymore as if hybridization and advanced technologies did not exist. Its reliance on the notion of natural capital should be disputed, since it is not clear for how long this category maintain will its vigor in the face of an increasing substitutability. It is rather manufactured capital, and above all cultivated capital (that which is based upon the intended hybridization of social and natural forms and beings), that should be put at the center of a reinvigorated, Anthropocene-friendly understanding of sustainability. However, this does not mean that those parts of nature that remain less touched by human influence or express the latter in less visible—more "natural"—ways (such as pets, landscapes, forests, and beaches) is to be abandoned to the hands of a destructive technology. On the contrary, there are good reasons to expect a further displacement of social values in the direction of preservation. Most likely, though, moral arguments will have less relevance than emotional, aesthetic and recreational ones. Furthermore, this preservation will not be accepted if it hinders current and future levels of welfare—global middle classes are on the rise. Thus it seems advisable that environmental sociology, as well as environmentalism writ large, exerts an Anthropocenic turn. Only by doing so will they remain relevant in the redefinition of socionatural relations and hence sustainability that is already taking place.

Conflicts of Interest: The author declares no conflict of interest.

References

1. Steffen, W.; Grinevald, J.; Crutzen, P.; McNeill, J. The Anthropocene: Conceptual and Historical Perspectives. *Philos. Trans. R. Soc.* **2011**, *369*, 842–867.
2. Liu, J.; Dietz, T.; Carpenter, S.R.; Alberti, M.; Folke, C.; Moran, E.; Pell, A.N.; Deadman, P.; Kratz, T.; Lubchenco, J.; *et al.* Complexity of Coupled Human and Natural Systems. *Science* **2007**, *317*, 1513–1516. [PubMed]
3. Ellis, E.; Goldewijk, K.; Siebert, S.; Lightman, D.; Ramankutty, N. Anthropogenic transformation of the biomes, 1700 to 2000. *Glob. Ecol. Biogeogr.* **2010**, *19*, 589–606.
4. Dunlap, R.E.; Marshall, C. Environmental sociology. In *The Handbook of 21st Century Sociology*; Bryant, D., Peck, D., Eds.; Sage: Thousand Oaks, CA, USA, 2006; pp. 329–340.
5. Catton, W.R.; Dunlap, R.E. Environmental Sociology: A new paradigm. *Am. Sociol.* **1978**, *13*, 41–49.
6. Murphy, R. *Sociology and Nature*; Westview: Boulder, CO, USA, 1996; p. 10.
7. Dunlap, R.E.; Martin, K.M. Bringing Environment into the Study of Agriculture. *Rural Sociol.* **1983**, *48*, 201–218.
8. Arias-Maldonado, M. *Real Green. Sustainability after the End of Nature*; Ashgate: London, UK, 2012.
9. Marx, K. *Capital, Volume 1*; Vintage: New York, NY, USA, 1976; pp. 283–290.
10. Foster, J.B. Marx's Theory of Metabolic Rift: Classical Foundations for Environmental Sociology. *Am. J. Sociol.* **1999**, *105*, 366–405.
11. Woodgate, G.; Redclift, M.R. *New Developments in Environmental Sociology*; Redflict, M.R., Woodgate, G., Eds.; Edward Elgar: Cheltenham, UK, 2005; pp. xv–xxvi.
12. Buttel, F.H. Classical theory and contemporary environmental sociology: Some reflections on the antecedents and prospects for reflexive modernization theories in the study of environment and society. In *Environment and Global Modernity*; Spaargaren, G., Mol, A.P., Buttel, F.H., Eds.; Sage: London, UK, 2000; pp. 17–39.
13. Fischer-Kowalski, M.; Weisz, H. Society as hybrid between symbolic and material realms: Toward a theoretical framework of society-nature interaction. In *Advances in Human Ecology*; Freese, L., Ed.; JAI Press: Stanford, CA, USA, 1999; Volume 8, pp. 215–251.
14. Arias-Maldonado, M. *Environment and Society. Socionatural Relations in the Anthropocene*; Springer: Heidelberg, Germany, 2015.
15. Balée, W. Historical ecology: Premises and postulates. In *Advances in Historical Ecology*; Balée, W., Ed.; Columbia University Press: New York, NY, USA, 1998; pp. 13–29.
16. Balée, W. The Research Program of Historical Ecology. *Annu. Rev. Anthropol.* **2006**, *35*, 1–24. [CrossRef]
17. Laland, K.; Brown, G. Niche-construction, Human Behavior, and the Adaptive-Lag Hypothesis. *Evol. Anthropol.* **2006**, *15*, 95–104. [CrossRef]
18. Laland, K.; O'Brien, M. Cultural Niche-construction. An Introduction. *Biol. Theory* **2012**, *6*, 191–202. [CrossRef]
19. Smith, B. Niche-construction and the behavioral context of plant and animal domestication. *Evol. Anthropol.* **2007**, *16*, 188–199. [CrossRef]
20. Kendal, J.; Tehrani, J.; Odling-Smee, F. Human niche-construction in interdisciplinary focus. *Philos. Trans. R. Soc. B* **2011**, *366*, 785–792. [CrossRef] [PubMed]
21. Isendahl, C. The Anthropocene forces us to reconsider adaptationist models of human-environment interactions. *Environ. Sci. Technol. Lett.* **2010**, *44*, 6007. [CrossRef] [PubMed]
22. Fischer-Kowalski, M.; Haberl, H. Conceptualizing, observing and comparing socio-ecological transitions. In *Socio-Ecological Transitions and Global Change—Trajectories of Social Metabolism and Land Use*; Fischer-Kowalski, M., Haberl, H., Eds.; Edward Elgar: Cheltenham, UK, 2007; pp. 1–30.
23. Smith, N. *Uneven Development: Nature, Capital and the Production of Space*; Blackwell: Oxford, UK, 1984.
24. Castree, N.; Braun, B. The construction of nature and the nature of construction. Analytical and political tools for building survivable futures. In *Remaking Reality: Nature at the Millenium*; Braun, B., Castree, N., Eds.; Routledge: London, UK; New York, NY, USA, 1998; pp. 3–42.

25. Latour, B. *We Have Never been Modern*; Harvester Wheatsheaf: London, UK, 1992.
26. Irwin, A. Society, nature, knowledge: Co-constructing the social and the natural. In *Sociology and the Environment: A Critical Introduction to Society, Nature and Knowledge*; Polity: Cambridge, UK, 2001; pp. 161–187.
27. Russell, E. *Evolutionary History: Uniting History and Biology to Understand Life on Earth*; Cambridge University Press: Cambridge, UK, 2011.
28. Dunlap, R.E.; Catton, W.R. Which Function(s) of the Environment Do We Study? A Comparison of Environmental and Natural Resource Sociology. *Soc. Nat. Resour.* **2002**, *15*, 239–249. [CrossRef]
29. Field, D.R.; Luloff, A.E.; Krannich, R.S. Revisiting the Origins and Distinctions between Natural Resource Sociology and Environmental Sociology. *Soc. Nat. Resour.* **2013**, *26*, 211–225. [CrossRef]
30. Dobson, A. *Justice and the Environment. Conceptions of Environmental Sustainability and Theories of Distributive Justice*; Oxford University Press: Oxford, UK, 1998.
31. Baker, S. *Sustainable Development*; Routledge: London, UK, 2006.
32. Seghezzo, L. The five dimensions of sustainability. *Environ. Politics* **2009**, *18*, 539–556. [CrossRef]
33. Costanza, R.; Daly, H.E. Natural capital and sustainable development. *J. Conserv. Biol.* **1992**, *6*, 37–46. [CrossRef]
34. De Groot, R.S. *Functions of Nature*; Wolters-Noordhoff: Amsterdam, The Netherlands, 1992.
35. Ekins, P.; Simon, S.; Deutsch, L.; Folke, C.; de Groot, R. A framework for the practical aplication of the concepts of critical natural capital and strong sustainability. *Ecol. Econ.* **2003**, *44*, 165–185. [CrossRef]
36. Ott, K.; Döring, R. *Theorie und Praxis Starker Nachhaltigkeit*; Metropolis-Verlag: Marburg, Germany, 2004; p. 176.
37. Holland, A. Sustainability: Should we start from here? In *Fairness and Futurity. Essays on Environmental Sustainability and Social Justice*; Dobson, A., Ed.; Oxford University Press: Oxford, UK, 1999; pp. 46–68.
38. Dedeurwaerdere, T. *Sustainability Science for Strong Sustainability*; Edward Elgar: Northampton, UK, 2014.
39. Döring, R. Natural Capital—What's the difference? In *Sustainability, Natural Capital and Nature Conservation*; Metropolis-Verlag: Marburg, Germany, 2009; pp. 123–142.
40. Faber, M.; Proops, J.; Manstetten, R. *Evolution, Time, Production and the Environment*; Springer: New York, NY, USA, 1997.
41. Chiesura, A.; de Groot, R. Critical natural capital: A socio-cultural perspective. *Ecol. Econ.* **2003**, *44*, 219–231. [CrossRef]
42. Holland, A. Natural Capital. In *Philosophy and the Natural Environment*; Atfield, R., Belsey, A., Eds.; Cambridge University Press: Cambridge, UK, 1994; pp. 169–182.
43. Deliège, G.; Neuteleers, S. Should biodiversity be useful? Scope and limits of of ecosystem services as an argument for biodiversity conservation. *Environ. Values* **2015**, *24*, 165–182. [CrossRef]
44. Atkinson, G.; Dietz, S.; Neumayer, E. *Handbook of Sustainable Development*; Edward Elgar: Cheltenham, UK; Northampton, UK, 2007; p. 4.
45. Neumayer, E. *Weak versus Strong Sustainability: Exploring the Limits of Two Opposing Paradigms*; Edward Elgar: London, UK, 2010.
46. Haraway, D. *Simians, Cyborgs, and Women: The Reinvention of Nature*; Routledge: New York, NY, USA, 1991.
47. Latour, B. *Politics of Nature. How to Bring the Sciences into Democracy*; Harvard University Press: Cambridge, UK, 2004.
48. Latour, B. *Reassembling the Social: An Introduction to Actor-Network Theory*; Oxford University Press: Oxford, UK, 2005.
49. Swyngedow, E. Modernity and hybridity: Nature, *Regeneracionismo*, and the production of the Spanish waterscape, 1890–1930. *Ann. Assoc. Am. Geogr.* **1999**, *89*, 443–465. [CrossRef]
50. Hinchliffe, S. *Geographies of Nature. Societies, Environments, Ecologies*; Sage: London, UK, 2007; p. 51.
51. Bakker, K.; Bridge, G. Material worlds? Resource geographies and the "matter of nature". *Prog. Hum. Geogr.* **2006**, *30*, 5–27. [CrossRef]
52. Ellis, E.; Ramankutty, N. Putting people in the map: Anthropogenic biomes of the world. *Front. Ecol. Environ.* **2008**, *6*, 439–447. [CrossRef]
53. Corcoran, P.; Moore, C.; Jazvac, K. An anthropogenic marker horizon in the future rock record. *Geol. Soc. Am. Today* **2014**, *24*, 4–8. [CrossRef]
54. Cronon, W. *Uncommon Ground. Rethinking the Human Place in Nature*; W.W. Norton & Company: New York, NY, USA, 1996.

55. Darimont, C.T.; Carlson, S.M.; Kinnison, M.T.; Paquet, P.C.; Reimchen, T.E.; Wilmers, C.C. Human predators outpace other agents of trait change in the wild. *Proc. Natl. Acad. Sci. USA* **2009**, *106*, 952–954. [CrossRef] [PubMed]

56. Rosenzweig, M. The four questions: What does the introduction of exotic species do to diversity? *Evol. Ecol. Res.* **2001**, *3*, 361–367.

57. Thomas, C. The Anthropocene could raise biological diversity. *Nature* **2013**. [CrossRef] [PubMed]

58. Schlosberg, D. For the animals that didn't have a dad to put them in the boat: Environmental Management in the Anthropocene. In Proceedings of the ECPR General Conference Sciences Po, Bourdeaux, France, 4–7 September 2013.

59. Higgs, E. History, novelty, and virtue in ecological restoration. In *Ethical Adaptation to Climate Change: Human Virtues of the Future*; Thompson, A., Bendik-Keymer, J., Eds.; MIT Press: Cambridge, MA, USA, 2012; pp. 81–102.

60. Sandler, R. Global warming and virtues of ecological restoration. In *Ethical Adaptation to Climate Change: Human Virtues of the Future*; Thompson, A., Bendik-Keymer, J., Eds.; MIT Press: Cambridge, MA, USA, 2012; pp. 63–80.

61. Meiklejohn, K.; Ament, R.; Tabor, G. Habitat Corridors & Landscape Connectivity: Clarifying the Terminology. Center for Large Landscape Conservation. 2009. Available online: http://www.twp.org/sites/default/files/terminology%20CLLC.pdf (accessed on 3 November 2010).

62. Monbiot, G. A Manifesto for Rewilding the World. 2013. Available online: http://pm22100.net/01_PDF_THEMES/98_Monbiot/130527_A_Manifesto_for_Rewilding_the_World.pdf (accessed on 1 December 2015).

63. Kareiva, P.; Robert, L.; Marvier, M. Conservation in the anthropocene: Beyond solitude and fragility. In *Love Your Monsters. Postenvironmentalism and the Anthropocene*; Shellenberg, M., Nordhaus, T., Eds.; The Breakthrough Institute: San Francisco, CA, USA, 2011; pp. 24–32.

64. Caro, T.; Darwin, J.; Forrester, T.; Ledoux-Bloom, C.; Wells, C. Conservation in the Anthropocene. *Conserv. Biol.* **2011**, *26*, 185–188. [CrossRef] [PubMed]

65. Barry, J.; Mol, A.; Zito, A. Climate change ethics, rights, and policies: An introduction. *Environ. Politics* **2013**, *22*, 361–376. [CrossRef]

66. Karafyllis, N. *Biofakte-Versuch Über den Menschen Zwischen Artefakf und Lebewesen*; Karafyllis, N., Ed.; Mentis: Paderborn, Germany, 2003; pp. 11–26.

67. Baldwin, A. The nature of the boreal forest: Governmentality and forest-nature. *Space Cult.* **2003**, *6*, 415–428. [CrossRef]

68. Biesecker, A.; Hofmeister, S. Starke nachhaltigkeit fordert eine Ökonomie der (Re)produktivität. In *Die Greifswalder Theorie starker Nachhaltigkeit*; Egan-Krieger, T., Schultz, J., Pratap, P., Voget, L., Eds.; Metropolis-Verlag: Marburg, Germany, 2009; pp. 169–192.

69. White, D.F.; Wilbert, C. *Technonatures. Environments, Technologies, Spaces, and Places in the Twenty-First Century*; Wilfrid Laurier University Press: Waterloo, ON, Canada, 2009; p. 6.

70. Birnbacher, D. *Natürlichkeit*; Walter de Gruyter: Berlin, Germany, 2006.

71. Cleveland, C.J.; Hall, C.A.S.; Kaufmann, R. Energy and the US Economy: A biophysical perspective. *Science* **1984**, *255*, 890–897. [CrossRef] [PubMed]

72. Pelenc, J.; Ballet, J. Strong sustainability, critical natural capital and the capabilities approach. *Ecol. Econ.* **2015**, *112*, 36–44. [CrossRef]

73. Dietz, S.; Neumayer, E. Weak and strong sustainability in the SEEA: Concepts and measurements. *Ecol. Econ.* **2007**, *61*, 617–626. [CrossRef]

74. Brand, F. Critical natural capital revisited: Ecological resilience and sustainable development. *Ecol. Econ.* **2009**, *68*, 605–612. [CrossRef]

75. Neumayer, E. Human development and sustainability. *J. Hum. Dev. Capab.* **2012**, *13*, 561–579. [CrossRef]

76. O'Neill, J. Sustainability, welfare and value over time. In *Governing Sustainability*; Adler, W.N., Jordan, A., Eds.; Cambridge University Press: Cambdirge, UK, 2009; pp. 283–304.

sustainability

MDPI

Article

Managing Nature–Business as Usual: Resource Extraction Companies and Their Representations of Natural Landscapes

Mark Brown

Department of Communication and Culture, BI Norwegian Business School, Nydalsveien 37, 0484 Oslo, Norway; mark.brown@bi.no; Tel.: +47-4641-0698; Fax: +47-2326-4781

Academic Editor: Md Saidul Islam
Received: 20 July 2015; Accepted: 24 November 2015; Published: 30 November 2015

Abstract: This article contributes to knowledge of how one category of business organization, very large, British-based, natural resource extraction corporations, has begun to manage its operations for sustainability. The object of study is a large volume of texts that make representations of the managing-for-sustainability practices of these multinational corporations (MNCs). The macro-level textual analysis identifies patterns in the wording of the representations of practice. Hajer's understanding of discourse, in which ideas are contextualized within social processes of practice, provides the theoretical approach for discourse analysis that gives an insight into how they understand and practice sustainability. Through this large-scale discourse analysis, illustrated in the article with specific textual examples, one can see that these natural resource MNCs are developing a vocabulary and a "grammar" which enables them to manage natural spaces in the same way that they are able to manage their own far-flung business operations. They make simplified representations of the much more complex natural landscapes in which their operations are sited and these models of nature can then be incorporated into the corporations' operational management processes. Their journey towards sustainability delivers, in practice, the management of nature as business continues as usual.

Keywords: sustainable development; corporate sustainability; operating management; managing nature; discourse

1. Introduction

Although the term *sustainable development* had then been in wide circulation for almost two decades, Dryzek [1] insisted in 2005 that it still referred "not to any accomplishment, still less to a precise set of structures and measures to achieve collectively desirable outcomes" but that it remained "a discourse". Another decade on, and confirming his view that meanings of the term are still being explored, this special issue of *Sustainability* will provide "an environmental sociology approach to understanding and achieving the widely used notion of sustainability" [2]. The particular focus of this article is on how business corporations with a commitment to sustainable development are beginning to understand what sustainability means for them and what their particular practice of the term is achieving.

For some time already, scholars have expressed their misgivings that the corporations' implementation of what they choose to call *sustainability* is not that at all. In 2004, Gray and Milne [3] point out that triple-bottom-line (TBL) accounting is in danger of being confused with sustainability reporting, a key argument for which is the organization-level *vs.* system-level perspective that separates the two. In 2010, Gray's [4] literature survey is able to observe that "most business reporting on sustainability and much business reporting on activity around sustainability has little, if anything, to do with sustainability" (p. 48). Nine years later, in 2013, Milne and Gray [5] question whether TBL is helpful at all in moving us towards sustainability. I shall discuss these misgivings in the methodological rationale. My own view is that such terms as *sustainability* start their lives as labels for new ideas that will move the arguments forwards. Then meanings (plural) develop as a consequence of practices (plural)—both intellectual and behavioral. We can learn about how the term is being understood by its practitioners, as we study their practice; and there may be value in that knowledge. Understanding how business corporations might move towards more sustainable modes of operation is one of many research projects within environmental sociology. However, according to a leading researcher, scholars have not yet developed comprehensive theories for sustainability management; we do not have a clear view of where to get to or how to get there [6].

Shifting focus from "there" to "here", knowledge about how far business has gotten in managing for sustainability is also in short supply. In a recent article reviewing research in the field, Zollo, Cennamo, and Neumann [7] observe that a great deal of research effort to date has been allocated to the two "broad sets of questions: why should companies move beyond serving merely economic purposes and what makes a company more sustainable" (p. 242). As a supplement to work on the theoretical, long-term models, they argue, we should also study the stepwise process of organizational evolution towards sustainability *i.e.*, where "sustainable" business practice is now and where it may be going. This article makes a contribution to knowledge of how one category of business organization, very large, British-based, natural resource extraction corporations, has begun to manage its operations for sustainability.

2. Hypothesis Development

The corporations in the object of study include British Petroleum, Rio Tinto, Shell, and Anglo-American, (see Section 4. Method for details of selection criteria). These companies have made public commitments to pursuing a sustainable future and they make use of nature both in terms of resource extraction and as a sink for unwanted byproducts of production processes. Their intention to move towards "sustainable" modes of operation means that they will also be at the forefront of any measures undertaken to organize for a sustainable relationship with the natural environment. Studying how they are implementing practices whose aim is a sustainable co-existence with nature and how they represent these practices' interaction with nature, will provide an insight into how they understand the term *sustainability* and where this understanding may be taking us.

The primary object of study was an electronic text database of social and environmental reports and press releases describing how 25 different business corporations, dominated by the extractive MNCs previously mentioned, are managing for sustainability. In order to identify patterns in the selection and arrangement of words, very large volumes of text, ideally running upwards of millions of words, are necessary. Only then will individual words occur often enough in the database for usage patterns to become apparent. As one among many tools of discourse analysis, the corpus linguistic approach can provide findings at a macro-level. However, with such volumes of text to be analyzed, a computer search method has to be employed. The researcher must approach the corpus having worked out some hypothesis regarding what may (or may not) be found in the object of study. This procedure is described below.

Such words as *targets*, *reporting*, and *controls*, have long been key words in the discourse of modern business, and it is no surprise that these corporations take a similarly managerial approach in the implementation of their environmental commitments. Taking British Petroleum's annual sustainability report [8] as an example of the genre and word searching through the downloaded PDF document, one finds the word *management* in regular use. BP cares "about the safe *management* of the environment" (emphasis added) (p. 3). Drilling down from this overview perspective, the report includes representations of more specific aspects of managing the environment:

"We take steps to assess and *manage* potential impacts on biodiversity, such as compiling a wildlife or biodiversity *management* plan or consulting with relevant experts and agencies to assess suitable actions" (emphasis added) (p. 45).

From the perspective of this article, it is also interesting to note how management, understood as a social, organizational process, combines the various objectives of the corporation in one system:

"BP's operating *management* system (OMS) (.) brings together BP requirements on health, safety, security, the environment, social responsibility and operational reliability, as well as related issues, such as maintenance, contractor relations and organizational learning, into a common *management* system" (emphasis added) (p. 25).

If we interpret this statement through the lens of environmental sociology, it would seem that BP's perception of the natural landscape is mediated through its operational management system (OMS). If the OMS includes a similarly managerial approach to sustainability as it does to the corporation's traditional, oil and gas activities, then we should expect it to contain targets for BP's sustainable relationship with the natural landscape as well as records of actual performance and accounts of the discrepancy—positive or negative—between the two. Making the safe assumption that the other so-called "green" corporations are also using the same managerial approach to their sustainability ambitions, it ought to be possible to find evidence in their textual representations that helps us to understand how nature is incorporated into their operating "management-for-sustainability" processes. In the remainder of this article, the terms *green business* and *sustainable business* are used simply as a convenient label for the corporations in this study. No normative claim regarding the achievement of sustainable goals is intended.

Hypotheses

In searching through the database of green business texts, therefore, the first hypothesis is that we should find words representing (i) the natural landscape and (ii) management processes. In addition to finding this evidence, a review of the words may provide clues to how the corporations perceive natural space through the lens of their operating management system. More significantly, the second hypothesis is that we should find textual evidence in which the corporations make representations of nature as the object of the sorts of processes of monitoring, reporting and control that one associates with a typical operational management system e.g., words from category (i) ought to appear as objects of verbs in category (ii).

The computer-based process of searching for particular words can, of course, only identify the presence of particular textual signs. The move from a word—understood purely as a textual signifier—to meaning is a necessarily interpretive process in which ambiguity can exist and must be resolved by human intervention. The linguistic evidence presented in the findings section contains no greater interpretive intervention than such avoiding of ambiguities.

In the final section, however, and responding to the call of this special issue, the "linguistic discourse" presented in the findings is interpreted within the context of corporate "managing-for-sustainability" practice. The article suggests how this social process influences corporate perceptions and understandings of the natural landscape.

3. Methodological Rationale: A Theory of Words, Meaning, and Practice

The theoretical underpinning for the interpretation of the findings brings together a theory of words, meaning and, most importantly, practice, which corresponds closely with Hajer's [9] understanding of discourse:

"Discourse is here defined as a specific ensemble of ideas, concepts, and categorizations that are produced, reproduced and transformed *in a particular set of practices* and through which meaning is given to physical and social realities" (emphasis added) (p. 44).

His understanding of discourse contextualizes its ideas within certain actions and he argues that actors assign meanings to reality through social processes of practice (see also van Leeuwen [10]). Using Hajer's meaning, a particular discourse derives some of its uniqueness from the particular practices of the actors. Assuming that the extractive MNCs in this sample are approaching the challenge of sustainable development by implementing broadly similar practices, the ideas, concepts, and categorizations circulating within their discourse of sustainability will be broadly homogeneous. Some environmental scholars have already argued along these lines *i.e.*, that terminology acquires meaning for a given group of practitioners through the operationalization of ideas [11,12]. Wenger's *Communities of Practice* [13] provides the central theory of meaning formation in this conceptualization of discourse and in this article. Sociolinguistics had previously linked language use to variables such as race, gender, age, and social class (see Labov [14], Macaulay and Trevelyan [15], and Wolfram [16]). The introduction of the concept of a community of practice, however, provides greater explanatory power in accounting for linguistic variation for example, between two ethnically British, middle-class, university-educated women, one of whom works, say, for British Petroleum while the other works for Greenpeace.

This latter point illustrates one implication of Hajer's "practice-dependent" understanding of discourse and offers one possible explanation for the often-observed phenomenon of conceptual fuzziness. Scholars in many different research environments have already pointed out the variation in meanings of particular terms or attempted normative definitions of key terminology [17–22]. Such fuzziness is not necessarily undesirable. The term *sustainable development* has been an extremely powerful driver of change, partly perhaps because its fuzziness has enabled it to appeal to a broad constituency of opinion. This article takes no normative stance on what definition should be assigned to the term. Neither, however, does it ignore the plasticity of some so-called sustainability discourse. Milne, Kearins, and Walton [23] (see also [24]) is an example of scholarship that deconstructs the discourse of self-styled sustainable business, by drawing attention to the intellectual inconsistencies that lie within the rhetorically-persuasive representations. In my reading of green business discourse, it is the more general statements—often delivered by the CEOs and/or the communications departments and often focused on goals, values, principles, or the future—that are most prone to this sort of fanciful sustainability discourse. Such discourse needs critical examination with a view to exposing its inconsistencies.

Increasingly, therefore, I have turned my attention to the nuts and bolts of the environmental management reports as a more reliable representation of what the corporation is doing. I concede that the corporations are making only partial representations of a much more complicated reality and that their decision what to represent and what not to represent is, most probably, weighted by instrumental interests. However, I think the language representations of practice are sincere albeit partial, attempts by the corporation to describe its reality.

Returning to the relationship between words and meaning, the second theoretical assumption of this article is that the meaning a community of practice associates with a word is reflected in the way in which the community uses the word. This position belongs to theory of language in use. The chronology in the study, as well as the associated development of theories of language in use, can be traced through Firth [25], Austin [26], Searle [27], and Halliday [28]. In parallel with the development of computing power in the later decades of the 20th century, corpus linguistics—the study of authentic texts or speech acts made by language communities—demonstrated, using very large quantities of text, that "meaning can be associated with a distinct formal patterning" [29] ((p. 6); see also Stubbs [30]). Thus, we have a three-way correlation in which the patterns of wording found in the texts of green business correlate with their organizational practices which, in turn, provide us with insights into the meaning which this group of social actors assigns to a particular word. One can identify both how green businesses are starting to practice sustainability and how they understand the term.

4. Method

A more comprehensive account of the method can be found in previously published work [31,32] as, in the interests of brevity, just a brief overview is provided here. The selection criterion for what qualified as a British sustainable corporation was public membership of the World Business Council for Sustainable Development [33] and/or the UN Global Compact [34]. The first stage of the methodology was to build a database of texts that were representative of this community of practice and this was done by copying material down from the websites of the 25 businesses identified and saving it as txt files. In searching for distinctive patterns of wording in a body of texts, it is necessary to have a point of comparison and two references were set up. The first was publically available; the British National Corpus (BNC) [35]. BNC had several advantages as a benchmark. First, it was created by a group of highly-respected project partners, that included the British Library Research and Development Department, Oxford University Computing Services, Lancaster University, Oxford University Press, and Longman Group Ltd. Second, since one of its design goals was to construct a language corpus typical of British English, it provided a very good match for British sustainable businesses; national differences in the usage of the English language could be eliminated as a possible variable. A third advantage with the BNC was its ready availability. Finally, its very large size, 90 million words, provided confidence that it was representative of typical English, against which other databases could be compared.

The second point of comparison was a database composed of texts written by British social and environmental NGOs; 37 in total. In designing language databases that are to be compared, there are two mutually exclusive design objectives which must be reconciled as best one can. On the one hand, it is important that each body of texts is representative of the organizations that have provided the material. On the other hand, one wants to be able to compare the texts of the two databases, with a view to saying something interesting about them. If the communities of practice that have produced the text have very different representations of experience, then one runs the risk of merely demonstrating that different people talk about different things. The compromise solution was to define, in advance of the text selection activity, what ideational content I was looking for in order for a text to be downloaded into its respective database (see Brown [31] for a more detailed treatment of this process).

The PC program I used is called Wordsmith and is marketed by Oxford University Press [36]. The author of the software has published work that describes the linguistic phenomena which Wordsmith can identify [37]. There are also many previous studies that have used a similar keyword approach to that presented in this article [38–41]. In its first processing of the texts of the two communities—sustainable businesses and the environmental NGOs—it made a list of words that appeared in each of the two sample databases, ranking them in order of frequency. A moment's reflection is all that is necessary to realize that the most frequently used words in any wordlist are ones that we all use: "the, and, a, of, but" *etc.* and that these simple frequency-based word lists were of no value. However, Wordsmith then compared the two frequency-based word lists in turn with the corresponding reference list for the benchmark of "typical English" provided by the BNC, which also ranks words such as "the, and, a, of, but" *etc.* as the most frequent. This first comparison procedure generated a list of statistical keywords, ranking highest those words that appeared much more frequently in the green business and environmental NGO word lists than when they were used in the "typical English" of the BNC.

The first running of this process produced two keyword lists in which the representation of sustainability management was recognizable. However, Wordsmith is simply a very fast counting machine and, lacking any form of human intelligence, it records the appearance of absolutely everything in the texts. Consequently, this first processing included keywords that were statistically key to these environmental reports but not semantically significant to the process of identifying the representations of managing-for-sustainability practice. A cleaning up procedure was necessary in order to remove different categories of these uninformative keywords. Examples included proper nouns (Shell, Rio Tinto, GreenPeace, Africa, US, Doha), units of measurement (tonnes, GWH, litres), products and materials (gasoline, platinum, bauxite), acronyms (ACCP, WBCSD), and terms referring to the internal organization of reports (appendix, pdf, section). The top 20 key words used by the green businesses in their representations of management-for-sustainability practice are presented in Table 1 to give an impression of the results from this mechanistic process. The Wordsmith software generates the keyness value shown in the table. It is an indicator of how much more frequent the usage of a word is, compared with its usage in the benchmark database. The value 1.0 would indicate that it is no more frequent, so one can see that keyness values in the tens of thousands indicate that these particular terms are used massively more often by sustainable business than is the case for "typical English".

Table 1. The top twenty one-word keywords used by sustainable business.

	Sustainable Business—the Top 20 One-Word Keywords				
N	Keyword	Keyness	N	Keyword	Keyness
1	ENVIRONMENTAL	50,282.01	11	ENVIRONMENT	17,173.10
2	BUSINESS	33,236.84	12	BIODIVERSITY	17,137.79
3	ENERGY	32,561.70	13	COMPANIES	16,551.74
4	SUSTAINABLE	28,694.50	14	DEVELOPMENT	15,605.07
5	EMISSIONS	27,957.12	15	GLOBAL	15,575.64
6	EMPLOYEES	21,345.17	16	REPORT	15,331.68
7	SAFETY	21,059.48	17	STAKEHOLDERS	15,162.41
8	MANAGEMENT	20,525.46	18	GROUP	14,986.09
9	WASTE	19,852.47	19	CORPORATE	14,716.54
10	PERFORMANCE	19,044.42	20	OPERATIONS	14,448.74

Table 1 presents just the top 20 keywords of green business in a list that extends to several hundred whose keyness is statistically very significant. For example, the 500th most key word for sustainable business: wetlands, had a keyness of over 1000 in comparison with the "typical English" of the BNC. Using the same process, I also generated keyword lists for the top 100 two-word and top 50 three-word keywords. The top 20 of these keywords are presented in Table 2.

Table 2. The top twenty two- and three-word keywords used by sustainable business.

	Sustainable Business—the Top Twenty Two-Word and Three-Word Keywords				
N	Keyword	Keyness	N	Keyword	Keyness
1	SUSTAINABLE DEVELOPMENT	18,026.69	11	GROUP COMPANIES	4885.16
2	HEALTH AND SAFETY	9373.65	12	HIV AIDS	4822.87
3	CLIMATE CHANGE	8229.32	13	CORPORATE SOCIAL	4781.19
4	ENVIRONMENTAL PERFORMANCE	7138.26	14	ENVIRONMENTAL AND SOCIAL	4558.15
5	CORPORATE RESPONSIBILITY	7077.07	15	BEST PRACTICE	4237.95
6	ENVIRONMENTAL MANAGEMENT	6598.33	16	MANAGEMENT SYSTEMS	4156.79
7	BUSINESS PRINCIPLES	6256.01	17	CORPORATE SOCIAL RESPONSIBILITY	4134.68
8	GREENHOUSE GAS	5405.60	18	NATURAL GAS	4090.75
9	ENERGY EFFICIENCY	5316,.6	19	RESPONSIBILITY REPORT	3804.73
10	SOCIAL RESPONSIBILITY	5283.72	20	HUMAN RIGHTS	3798.92

At this point in the method, the search technique changed from Wordsmith's efficient but mechanistic process to a humanly slow, hopefully intelligent and certainly interpretive approach. These lists of the top 500 one-word, top 100 two-word and top 50 three-word keywords of green business and the environmental NGOs became the objects of study for identifying groups of words that shared certain semantic similarities, some of which I report in the findings.

The first hypothesis that I proposed involved representations of (i) the natural landscape and (for green business) (ii) representations of managing for sustainability in such locations. The first semantic search, therefore, was to look for words that made representations of some aspect of the natural landscape. Some of the words were easily identifiable—forest, for example—whereas others with a more ambiguous meaning were checked for their intended meaning. In order to do this, Wordsmith generated a list of 20 randomly picked occurrences of the word. From a careful reading of the context in which it was used, I interpreted the meaning. Using this technique, for example, the word *growth* was not included in the semantic field of representations of natural space because it was clear that business used the word exclusively as a representation for economic expansion. Pursuing evidence to test part (ii) of the first hypothesis, the second search was to look for words that made representations of a process of management that might be applied to the natural landscape. As an aid to identifying the sorts of words for which I would be searching, I prepared a schematic describing the different stages of a process of operational management-for-sustainability. The schematic evolved out of my reading various environmental reports that were a part of the database the circular process of monitoring, reporting, analyzing, deciding, planning, and implementing is presented in Figure 1. It functioned as a sort of "semantic template" as I searched through the two- and three-word keywords of sustainable business looking for representations of a process of management.

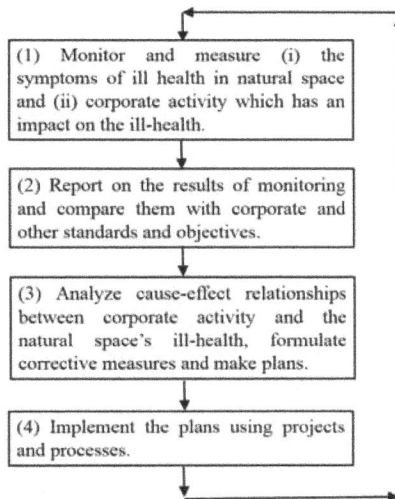

Figure 1. A process of operational management for sustainability.

Summarizing, this procedure identified words in the green business database representing (i) the natural landscape and (ii) processes of operational management. The logical place to look for evidence that sustainable businesses might have started the process of managing the natural landscape, was where the words in category (i) functioned, grammatically as objects of the processes in (ii). Wordsmith has a function that enables one to study this phenomenon. It picks out 20 random occurrences of the same operator-selected word from the entire database. For each of the 20 occurrences it extracts the string of text in which the word appears with a cut off of 200 characters of text on each side of the selected word. This is normally sufficient contextualizing text for the operator to check for any possible ambiguities and confirm the writer's intended meaning. This process is discussed further in the findings in which just such a problem is illustrated.

5. Findings

These findings provide a response to the hypotheses and empirical questions which sought linguistic evidence showing how green businesses relate to natural space as they attempt to manage for sustainability. Here I present three distinct vocabularies of keywords from the text databases and then demonstrate the textual interaction of two of them within the corporations' texts.

5.1. Finding One: Two Different Representations of Natural Space

From a comparison of the top 500 one-word keywords of the two communities of practice, I found that the environmental NGOs use two different types of words to represent nature, whereas the corporations use just one. For want of a better term, the NGOs use a "natural" (*sic*) vocabulary to represent the natural landscape. These are words that make a direct reference to some physical—fleshy or fibrous—part of nature. Examples in their top 500 keywords (see Table 3) include, *crops*, *forest*, *soil*, *whale*, and *villagers*.

Table 3. "Natural" representations of the natural landscape among the top 500 one-word keywords of the environmental NGOs, ranked in descending "keyness".

CROPS	LIVELIHOODS	WATER	FOODS	MARINE
FARMERS	POOREST	PEOPLE'S	FEED	VILLAGERS
FOOD	CROPS	SOIL	WHALE	REINDEER
FOREST	PEOPLES	MAHOGONY	RAINFOREST	PLANTS
POOR	RURAL	WHALES	RAIN	WOMEN'S
FORESTS	SEED	COUNTRYSIDE	BEET	POULTRY

Reading Wordsmith's randomly generated occurrences of such words in the texts in which they appear, one gains the immediate impression of environmental NGOs that position themselves in specific places in the natural landscape with the intention of bearing witness on their condition and the threats to which they are exposed.

Such words cannot be found in the top 500 keywords of green business. The texts of their environmental reports do make representations of natural spaces. However, they do this with what I would dub a "vocabulary for being managed". Examples of these terms are *areas*, *biodiversity*, *community*, *habitats*, and *site* (see Table 4).

Table 4. A "vocabulary for being managed"—representations of nature among the top 500 one-word keywords of both the environmental NGOs and green business (ranked alphabetically).

AREAS	COMMUNITY	HABITATS	RESOURCES
BIODIVERSITY	ENVIRONMENT	HEALTH	SITE
COMMUNITIES	HABITAT	RESOURCE	SITES

These words are the preferred textual representations of the natural landscape for sustainable business. The great advantage of them, from a green business point of view, is that they lend themselves to being defined and measured numerically. This enables information about the natural landscape to be incorporated into the operational management system. Although the preferred terms of sustainable business are abstractions of nature rather than the more immediate representations that the NGOs use, it is important to underline that a reading of the corporations' usage reveals that they are nonetheless concerned for the well-being of the natural landscape. I discuss in more detail the way in which businesses make use of this feature of their vocabulary in the interpretation and conclusions section.

5.2. Finding Two: Sustainable Business Representations of Operational Management-For-Sustainability

The second category of evidence in the first hypothesis was representations in the green business texts of business processes, whose goal, presumably, is the sustainable usage of nature. Accordingly, the second search I conducted among the keywords of sustainable business was for representations of some aspect of operational management. Examples from the top one-word keywords of green business that might represent operational management are *reporting*, *plan*, *impact*, and *indicators*. However, careful reading of the usage of these one-word keywords in their environmental reports, revealed that they normally occurred as part of two- or three-word units of meaning, e.g., *environmental reporting*, *social impact*, *habitat action plan*, and *GRI indicators* that I found in the top 150 two- and three-word keywords. I therefore transferred the search process to this list and found that 63—just under one half—might be placed in this "operational management-for-sustainability" category, shown in Table 5. In addition to the four keywords just mentioned, other examples included *biodiversity action plans*, *environmental management system*, *frequency rate*, and *key performance indicators*. I have deliberately included the last two examples because the terms *frequency rate* and *key performance indicators*, and many other of the examples, are representations that one could just as easily find in business-as-usual operational management systems.

Table 5. The two- and three-word keywords of sustainable business whose representation is a part of the process of operating management.

ACTION PLAN	GOOD CORPORATE CONDUCT
ACTION PLANS	GRI INDICATORS
ANNUAL REPORT	HABITAT ACTION PLAN
BEST PRACTICE	HIGH STANDARDS
BIODIVERSITY ACTION	INJURY FREQUENCY
BIODIVERSITY ACTION PLAN	INJURY FREQUENCY RATE
BIODIVERSITY ACTION PLANS	INTERNATIONAL MARKETING STANDARDS
BUSINESS CONDUCT	KEY PERFORMANCE
BUSINESS PRINCIPLES	KEY PERFORMANCE INDICATORS
CODE OF BUSINESS	LAWS AND REGULATIONS
CODE OF CONDUCT	MANAGEMENT SYSTEM
CONTINUOUS IMPROVEMENT	MANAGEMENT SYSTEMS
CORPORATE GOVERNANCE	OBJECTIVES AND TARGETS
CORPORATE RESPONSIBILITY	PERFORMANCE DATA
CORPORATE RESPONSIBILITY REPORT	PERFORMANCE INDICATORS
CORPORATE SOCIAL RESPONSIBILITY	PRODUCT STEWARDSHIP
DOW JONES SUSTAINABILITY	REPORTING INITIATIVE
ENVIRONMENT REPORT	RESPONSIBILITY REPORT
ENVIRONMENTAL IMPACT	RISK ASSESSMENT
ENVIRONMENTAL IMPACTS	RISK MANAGEMENT
ENVIRONMENTAL MANAGEMENT	SAFETY MANAGEMENT
ENVIRONMENTAL MANAGEMENT SYSTEM	SAFETY PERFORMANCE
ENVIRONMENTAL MANAGEMENT SYSTEMS	SOCIAL IMPACT
ENVIRONMENTAL PERFORMANCE	SOCIAL INVESTMENT
ENVIRONMENTAL PERFORMANCE REPORT	SOCIAL PERFORMANCE
ENVIRONMENTAL REPORT	SOCIAL REPORT
ENVIRONMENTAL REPORT APPENDICES	SOCIAL REPORTING
ENVIRONMENTAL SUSTAINABILITY REPORT	SOCIAL RESPONSIBILITY
FREQUENCY RATE	SOCIALLY RESPONSIBLE
GLOBAL COMPACT	SUSTAINABILITY REPORT
GLOBAL REPORTING	SUSTAINABILITY REPORTING
GLOBAL REPORTING INITIATIVE	

In order to validate hypothesis two, it was necessary to look for evidence that these operational management processes are being applied to the natural landscape rather than representing the business processes with which the green corporations manage their traditional operations. Wordsmith can generate many reports that each show different usages of words by a particular community of practice. The most important function with regard to the findings in this article is its ability to generate a randomly selected sample of lines of text in which a selected keyword appears. This allows the researcher to read authentic examples of the usage of the keyword by the community of practice. In the theoretical section, I made the case that the way in which a community of practice uses a word reflects the meaning that the community associates with it and further, that this meaning is a consequence of practice and then reflection and social negotiation about that practice. I present the findings from this search and an interpretation that connects word usage with meaning and meaning with practice in the next section.

5.3. Finding 3: Operational Management of the Natural Landscape

The findings presented from the green business database are of (i) a "vocabulary for being managed" and (ii) a ubiquitous language of operational management. The logical place to look for evidence that sustainable businesses have started the process of managing the natural landscape was where the vocabulary for being managed takes center stage. If these terms function as the objects of the green business language of operational management-for-sustainability, then the findings support the second hypothesis. As described previously, Wordsmith provides a reporting function that can generate such text examples. Appendix contains four tables, each of which contains 20 randomly-selected text extracts from the green business database. The four words from the "vocabulary for being managed" that were selected for study are *habitats*, *sites*, *areas*, and *communities*. Once the operator has selected the central word and the amount of text to be retrieved either side of it, Wordsmith generates the reports itself. The only modifications I have made to these findings is to put the central term in boldface to draw attention to it, and to shade in grey the text that represents some aspect of the green business operational management-for-sustainability.

Reading of the tables for *habitats*, *sites*, and *communities* reveals that their intended representation is always of some natural, geographic space. However, the table for *areas* reveals this word's ambiguity; green business uses this term in two ways. In some examples, the referent is natural space; "Opencast mining areas and discard dump sites are key areas of focus" (Table A3, line 15). In other lines the word's referent is an abstract aspect of corporate attention; "Where elements and indicators in the GRI are relevant to Rio Tinto, we aim to report against them. In other areas we have reported against indicators that are more relevant to driving performance improvements" (Table A3, line 9). From my reading of Table A3, I concluded that nine of the 20 usages of *areas* are references to the natural landscape. This means that Appendix contains 69 examples of such a representation.

These findings demonstrate that textual representations of an operational management-for-sustainability are usually present to contextualize the abstract vocabulary of natural space. There are representations of monitoring and measuring the natural landscape; "Develop an information management system which records and manages improved grassland data alongside other priority habitats" (Table A1, line 10). There are representations of reporting and comparing the information with corporate or other standards; "An assessment of alignment against these three stages and against the individual process steps of AA1000 is made. In addition VeriSEAAR© can highlight areas where improvement has been achieved" (Table A3, line 11). Planning for making improvements to natural space is very common; "Have Biodiversity Action Plans (BAPS) in place at all sites where Shell operates in areas of high biodiversity value" (Table A3, line 1). Finally, there are representations of the implementation of plans using projects and processes; "We carry out Social Impact Assessments (SIA) to identify potential impacts on communities and to develop strategies to manage these" (Table A4, line 17). From a close study of sustainable business' textual representations of practice, one can see that it is introducing business processes whose goal is the sustainable management of nature.

6. Discursive Interpretation

In this section, the linguistic findings are interpreted within corporate managing-for- sustainability practice in an attempt to flesh out its discourse. First, however, I need to make explicit the conceptualization of business corporations which is assumed in this interpretation. I conceive of them as being the business processes which are managed by the corporation's officers. For example, I conceive of a sustainable MNC as a process engineer's technical drawings, dictating the material and energy flows from which useful products are synthesized from raw materials; or as the financial controller's spreadsheets modeling the flows of assets into, around and out of the P and L and balance sheet. In this particular understanding of the organization, free human agency and human moral consciousness have no role; the employees of the corporation are reduced to instruments who manage the execution of business processes. I shall return to this assumption in the closing section.

6.1. The Vantage Point from Which One Views the Natural Landscape Affects Its Representation

In their role of guardians of nature, the environmental NGOs are in the privileged position of being able to position themselves—rhetorically—within the natural landscape. From this local vantage point, they see the detail of the natural spaces and this accounts for representations such as *crops*, *forest*, *soil*, *whale*, and *villagers* that are reported in the findings. In the sustainability reports of green business, however, the natural spaces in which their operations are located are viewed from corporate headquarters, often on the other side of the globe. This affects the language of representation in two ways. First, there are so many natural spaces, corporate headquarters must aggregate them by making abstractions of their physical reality. Second, since the physical materiality of a particular location cannot be known by the corporation's senior management at first hand, a form of representation of the natural space must be chosen that communicates meaningful information to these senior officers. At present, the meaningfulness has to be quantifiable. This is a "linguistic" discourse that is understood within the corporation and connects with the next point.

6.2. Centralized Management Systems Require Numbers to Be Able to Manage

In order for a global enterprise to manage its far-flung operations effectively from HQ, it has its own internal models of the different installations. For example, the mining company's primary concentrator, which operates in the sand dunes running along South Africa's Atlantic coast, is controlled from a head office in central London. Day-to-day operational control is under the local plant manager and her engineers, but she reports the key numbers back to London on a regular basis in a spreadsheet. These numbers provide senior management with a sufficiently detailed representation of the complex physical materiality of the primary concentrator. At HQ they are compared in the spreadsheet with the expected year-to-date numbers and any variations are analyzed before instructions are sent back to South Africa. In short, the corporation's productive installations are operationally managed from headquarters using numbers.

Confirming one of the conclusions of Bansal and Knox-Hayes ([42] p. 77), if the MNC now commits itself to becoming a sustainable MNC, it is likely to utilize its existing management processes in its attempts to manage for sustainability. In order to be able to manage them, the natural spaces—just like the MNC's productive installations—need to be represented in a set of key numbers that can be recorded in a spreadsheet so that comparisons can be made between actual and expected year-to-date, variations analyzed, and new instructions issued to the local manager. This interpretation accounts for the second type of representation of nature that was presented in the findings; a vocabulary for being managed. Words such as *habitats*, *sites*, *areas*, and *communities* have the advantage—for an MNC—of being quantifiable.

Returning to the mining company's challenge, the unique physical materiality of the natural space around the primary concentrator is represented, for example, as a habitat. Within the MNC's corporate model, the habitat is "understood" by a selection of key measurements; its area, the air quality measured daily at specific GPS coordinates, a series of biodiversity indicators *etc*. These are measured, recorded in the spreadsheet and, along with the production numbers for the concentrator, communicated back to London where they can be analyzed, decisions made, instructions issued, and action taken. One way, then, in which green corporations relate to natural spaces in their managing-for-sustainability project, is by constructing representations of nature that can be incorporated within a spreadsheet.

With its perception based on an admittedly limited range of indicators, green business is able to describe the health of a natural space. One might balk at the notion suggested in the previous section, that MNCs are engaging in a process of managing nature. However, such a response might be mitigated by changing the representation from the term *management* to *nurturing* or *stewardship*. Continued corporate practice and its study will help us understand which of the practices are desirable and clarify what meanings the green corporations are investing in *sustainability* through their actions.

6.3. Incorporating the Local within the Global?

The abstract "vocabulary for being managed" is not, necessarily, a sign that these global green businesses are wholly ignorant of the local natural landscape around their productive installations. Certainly, when examining the macro-level characteristics of their texts, I was unable to find natural language representations. However, they do occasionally focus their gaze on a particular, local, natural space. For example, in line 15 of the report for the usage of *habitats* (Table A1, Appendix), the MNC describes its "employment of two rangers to help protect and enhance the wildlife habitats at our Musselburgh and Valleyfield ash lagoons". Pursuing this line of inquiry, on the webpages of ScottishPower [43], there are several natural representations of specific plants and animals, which belong to a particular habitat; not enough to register at the macro-level among the sustainable corporations' keywords, but some nevertheless. This practice is an example of Crane *et al.*'s idea [44] of corporate ecological citizenship which "requires corporate managers to be "embedded" in local environments to foster sustainable behaviors" (p.95). The two rangers who are employed by ScottishPower to protect and enhance the wildlife habitats at the Musselburgh and Valleyfield ash lagoons might be considered as "embedded" in the ash lagoons. This idea raises the possibility that within an MNC, pockets of local sustainability practice might be established by the corporation.

Pursuing this idea further, one intuitive impression I have gained from reading the corporate environmental reports is of occasional sustainability projects that are spared from the "vocabulary-for-managing" representation. These projects appear to be characterized by their being selected and then presumably funded, from the very highest levels of the corporation. Another avenue for further research would be to test for the existence of such a category of project. The working hypothesis is that corporate managing-for-sustainability practice might be divided into different categories. One category, following the main findings of this article, could be called operational sustainability practice. These would be all of the practices that are the consequences of the green corporation seeking to maintain a sustainable balance between its particular business operations and the natural landscapes around them. Their representation in reports would be characterized by the managerial vocabulary presented in this article. A second category, which might be called intrinsic sustainability practice, would be interventions in natural space that were not necessarily related to specific business processes but which the MNC, nevertheless, wished to implement. The representations of these projects might not be reduced to a language that Bell and Morse [17] characterize as having "quantification at its heart" (p. 42). The study of such projects in managing-for-sustainability business practice might point the way to finding practices that can know the local natural landscape in a more holistic way than the managerial techniques that are in evidence in these findings.

In closing this interpretive section, I revisit my assumed conceptualization of an MNC as different business processes from which human agency and moral consciousness are absent. In this article, I have referred to *nature*, *natural space*, and the *natural landscape* while deliberately avoiding using the term *place*. Without wishing to appear to be summarizing a very large debate, one generally-agreed distinction between *space* and *place* is the role of human sense-making in the latter. Place "has a history and meaning. Place incarnates the experiences and aspirations of a people. Place is not only a fact to be explained in the broader frame of space, but it is also a reality to be clarified and understood from the perspective of the people who have given it meaning" ([45], p. 387). Following Tuan, and assuming the conceptualization of an MNC that I advanced at the start of this section, when the sustainable corporation implements its operational managing-for-sustainability practices, the best we can get is a mechanistic sensitivity to natural space. Some researchers have argued that one promising avenue towards sustainability is for global MNCs to make a transition towards a place-based sensitivity to the natural landscape [46]. The process of knowing place is a practice that I think belongs in the Milne and Gray project for growing *ecological literacy* [5]. My tentative response to this challenge is that the MNC needs to find a way in which its business processes merge with its human stakeholders' moral consciousness so that the corporation can truly be said to give meaning to natural spaces in the construction of meaningful places.

7. Limitations and Future Research

It is, perhaps, not surprising that some form of direct management of nature will be a part of the sustainability response by these natural resource extraction companies. It is harder to imagine that an international bank's environmental reports would have so much focus on managing natural space. Further work needs to be done to find out how generalizable these findings are.

A second limitation is that the companies that were selected to be representatives of sustainable business were selected, because they had made a commitment to working towards a sustainable future. Making a commitment to a goal, however, is not the same as successful, authentic practice. A great deal of research using different methodologies for measuring sustainability practice in different types of organizations must be done in order to test the managing-nature hypothesis.

The third limitation concerns a distinction between operational, as opposed to strategic, management processes. There is no space here to do more than point out the general organizational challenge of making connections between the two, both in theory and practice. Non-alignment between the corporation's strategic objectives and its Performance Management System (PMS) has been a recognized phenomenon in the literature for over 40 years [47]. Business-as-usual strategic goals are usually fairly easy to articulate, so the focus is normally on bringing the PMS into alignment with long term strategy. In contrast, managing-for-sustainability strategic models are, as mentioned in the introduction, only vaguely understood [6]. Recognizing that nature itself is in a constant state of flux, the likelihood is that long-term models of the sustainable firm will co-evolve with changing natural landscapes [48]. It is at this stage, therefore, very difficult to ascertain with certainty if the sort of findings identified in this article—where we are—help or hinder a corporation's transition to more strategically-sustainable forms of operation and organization.

Acknowledgments: The author would like to thank the academic editor, the editorial team and the two anonymous reviewers for their constructive criticism in the writing of this article.

Conflicts of Interest: The author declares no conflict of interest.

Appendix A. Representations of Operational Management of the Natural Landscape in the Texts of Sustainable Business

On the following pages are four tables each containing 20 lines of text. For each table the same word selected from the vocabulary of being managed is centrally placed. The occurrences have been randomly extracted by Wordsmith from the texts of green business and the lines of text are sufficiently long to be able to make an informed interpretation of the context in which the particular term is used by the corporation concerned. The terms selected are *habitats*, *sites*, *areas*, and *communities*.

Table A1. *Habitats*.

Line	Extract from Texts of the MNCs with Keyword *Habitats* Placed Centrally
1	ss raising, action plan monitoring and audit. XXXXX Hare The XXXXX hare is found throughout Britain. It is mainly nocturnal and is generally found in open grassland habitats with nearby woodland and hedgerows which provide resting places or "forms" during the day. The hare's diet consists mainly of herbs and grasses in summer and
2	Spotted Flycatcher Muscicapa striata Action Plan 1. Introduction 1.1 The spotted flycatcher is a summer migrant to the UK. The species prefers habitats of open woodland, hedgerows with mature trees, parkland and large gardens. The species feeds almost entirely on insects. 1.2 The spotted flycatcher generally a
3	icroscopic plants and animals to larger species such as common frogs and dragonflies. In addition, standing open water habitats are commonly fringed by important wetland habitats such as marshy and swamp, which are covered in a separate swamp and marsh Habitat Action Plan. This broad standing open water Habitat Action Plan has been pr
4	nhance selected scrub habitats and create new scrub areas in appropriate locations with low current nature conservation particularly where these border existing woodland habitats. WT ? ? ? ? ? 4.2 Action Potential Partners Year Meets Objective No. 2003 2004 2005 2006 2007 2013 5.2.4 Develop generic management prescri
5	Swamp And Marsh Habitat Action Plan 1. Habitat Description 1.1 The term "swamp and marsh" covers a range of wetland habitats where water is at or near the ground surface for most of the year. Swamp and marsh vegetation tends to be associated with ponds, floodplains and lakesides. 1.2
6	ion plan is reviewed and updated. WT ? ? 4.3 5.4 Awareness 5.4.1 Produce an internal communication so that best practice for the conservation of improved grassland habitats can be used by staff where they occur on site. WT ? ? ? 4.4 5.4.2 Make areas of improved grassland available for recreation and amenity where this does not
7	ecological survey programme. WT ? ? ? 4.1 5.1.2 Develop an information management system which records and manages improved grassland data alongside other priority habitats. WT ? 4.1 5.1.3 Implement data management system, recording improved grassland data and keep up-to-date as appropriate. WT ? ? ? ? ? 4.1 5.1.4 Research
8	animals and provides corridors allowing dispersal and movement between other habitats. Such habitats are becoming increasingly valuable for wildlife, as other grassland habitats are lost. 2.2 Current threats to the improved grassland habitat include: • Recreational pressure. • Adding nutrients through leaving grass cuttings 'in-s
9	on the broadleaved woodland habitat action plan. LA, WT ? ? ? ? ? ? 4.4 5.4.3 Encourage appropriate public access for study and enjoyment of broadleaf woodland habitats. WT ? ? ? ? 4.4 5.4.4 Publish results of broadleaved woodland habitat action plan. - ? ? ? ? ? 4.4 6. Partners EN—English Nature LA—
10	cological survey process. WT ? ? ? 4.1 5.1.2 Develop an information management system, which records and manages broadleaved woodland data alongside other priority habitats. WT ? 4.1 5.1.3 Implement data management system, recording broadleaved woodland data and keep up-to-date as appropriate. WT ? ? ? ? ? 4.1 5.1.4 Suppor

Table A1. *Cont.*

Line	Extract from Texts of the MNCs with Keyword *Habitats* Placed Centrally
11	Partners Year Meets Objective No. 2003 2004 2005 2006 2007 2013 5.1 Data Collection and Information Management 5.1.1 Determine the extent of broadleaved woodland habitats on all Biffa landholdings through the ecological survey process. WT ? ? ? 4.1 5.1.2 Develop an information management system, which records and manages bro
12	ollection Site Management Monitoring Biodiversity Noxious Invasive Pest and Management and Audit Awareness Weeds Species SPECIES and HABITAT ACTION. PLANS SPECIES HABITATS Skylark Woodland Nature Conservation Amenity Linnet Broadleaf Woodland Lowland Meadows Improved Grassland XXXXX Hare Scrub Swamp and Mar
13	contribution Montgomeryshire Lake Vyrnwy—3 year population monitoring project undertaken with recommendations for future management of the moorland and woodland edge habitats Peak National Park UDV—UK's first major re-introduction project launched in October 2003 Black Poplar LBAP Target STW action STW contribution Derbyshire
14	ientific Interest (SSSIs) in our region. A number of these have been identified as being of European importance, and under EC Directives on the conservation of natural habitats, flora and fauna, and birds, are nominated as Special Areas of Conservation (SACs) and Special Protection Areas (SPAs). Special Areas of Conservation are strictl
15	ent Plans; and ● Implementation/asset management of the Management Plans. Generation supports the employment of two rangers to help protect and enhance the wildlife habitats at our Musselburgh and Valleyfield ash lagoons. These rangers actively manage the reserves and help the public to enjoy the facilities provided. A valuable bi
16	tegy is to minimise damage to biodiversity when we develop new projects or as part of our maintenance operations and to positively enhance our landholding for species, habitats and heritage through measures such as our Rural Care Programme. ScottishPower aims to set a good example to other energy users by actively seeking to minimise
17	graze on Bridger Coal Company's successful reclamation site, also located in Wyoming. PacifiCorp regards mining activities as an opportunity proactively to enhance habitats. Taking part in biodiversity Our Environment Policy states that " … we will strive to continue to be regarded as a good and trusted neighbour … ". We engage ou
18	tishPower Environmental Performance Report 2003/04 Performance Review Our strategy for the management of biodiversity issues is to minimise impact, positively enhance habitats, ensure that planned restoration enhances habitat and species where environmental impact is unavoidable and contribute to biodiversity processes such as Local
19	into sites in ten countries started the BirdLife International/Rio Tinto partnership. The organisations share the aim of enhancing the conservation of birds and their habitats as a means of contributing to sustainable development. The fourth annual Rio Tinto mine site birdwatching events organised by BirdLife and Rio Tinto during 200
20	y five of the world's people rely on plants for primary health care, and plants help regulate our climate and bind our soils. They provide food, fibres, timber, fuel and habitats for the wildlife, birds and insects that keep our fragile ecosystem in balance. Our Investing in Nature programme, through Botanic Gardens Conservation Internat

Note: Processes of Managing Shaded in Grey.

Table A2. *Sites*.

Line	Extract from Texts of the MNCs with Keyword *Sites* Placed Centrally
1	ified to the International Standards Organisation's ISO 14001 environmental management systems standard by the end of 2003. At the end of 2001 over a quarter of our sites were certified. We have a team of experts who provide specialist environmental help and guidance to our manufacturing plants throughout the world on issues rang
2	l efficiency of our manufacturing operations, and to incorporating environmental factors in the design and re-design of our products—eco-innovation. We now have 103 sites certified to the international environmental management standard ISO 14001. Our goal is to have all our lead sites certified by 2003. For more information see th
3	ISO 14001 by the end of 2003. During 2002, a further 23 sites were certified, but 20 certified sites were closed or divested. At the end of 2002 we had 114 certified sites. We are continuing with the certification of individual sites but we are likely to fall short of 100% certification of our lead sites by end 2003. This is large
4	lection and reporting of environmental performance data via a global electronic system. Highlights for 2004: • 100% of sites reported environmental data • 98.6% of sites reported on all key environmental parameters, apart from COD • 93.4% of sites reported COD data • following feedback from the business, the pro-forma used for
5	ncy in manufacturing Environmental impact data In addition to reducing our impact per tonne of production, in 2004 the total environmental impact of our manufacturing sites decreased for most of our key performance indicators. Unilever manufacturing performance 2000–2004: trends in absolute load to the environment 00 01 02 03 04 23
6	in 2001 and made additional commitments with regard to protected areas in 2003, including a commitment not to explore or drill for oil and gas in natural World Heritage Sites (see Shell and protected areas) and in 2005 developed a Biodiversity strategy through to 2010 (see our Plans for 2005). The Shell Group Biodiversity Standard We
7	Year Meets Objective No. 2003 2004 2005 2006 2007 2013 5.2.4 Ensure that common frog management prescriptions are included in Biodiversity Management Modules for sites where the species occurs. WT ? ? ? 4.2 5.3.1 Monitoring and Audit 5.3.1 Develop a monitoring protocol for common frogs on Biffa sites. WT, LA, HCT ? 4.3 5.3.
8	s Year Meets Objective No. 2003 2004 2005 2006 2007 2013 5.2.4 Develop generic management prescriptions to improve scrub areas for biodiversity across all Biffa sites. WT ? 4.2 5.2.5 Ensure that scrub management prescriptions are included in Biodiversity Management Modules for sites where the habitat occurs. WT ? ? ? 4.2
9	Meets Objective No. 2003 2004 2005 2006 2007 2013 5.2.5 Develop generic management prescriptions to improve lowland meadow areas for biodiversity across all Biffa sites whilst retaining any amenity features. WT ? 4.2 5.2.6 Ensure that lowland meadow management prescriptions are included in Biodiversity Management Modules for si
10	ected sites. EN, WT, UNI ? ? ? ? 4.2 5.2.5 Develop generic management prescriptions to improve eutrophic standing water areas for biodiversity across all Biffa sites. WT ? 4.2 5.2.6 Ensure that eutrophic standing water management prescriptions are included in Biodiversity Management Modules for sites where it occurs. WT ?

Table A2. *Cont.*

Line	Extract from Texts of the MNCs with Keyword *Sites* Placed Centrally
11	treatment sites, almost doubling existing capacity. Case study: Minworth Sewage Treatment Works package, GaSSim, for calculating methane gas emissions from landfill sites. Mitigation To mitigate our environmental impacts we are concentrating on our most significant emissions relating to energy use and transport. Energy managemen
12	s up. This reflects improvements in reporting quality. One incident required regulatory notification but the Company has had no prosecutions or fines. * EMS—number of sites certified to ISO 14001 has remained at 9 with all sites aiming to be certified by 2006. * External benchmarking- the business increased its score in the Business i
13	taken for project staff in order to increase environmental awareness and use of PEA checklist. Implement contaminated land risk ranking system, targeting high risk sites for investigation. Undertake environmental initiatives which support the UK Biodiversity Plan. Monitor agreed environmental mitigations for construction activ
14	cies, habitats and heritage. This involves the following concrete activities: • Ensure our actions do not cause significant adverse effects on the biodiversity of the sites within which we operate. • Where features of strategic biodiversity importance occur on our larger land holdings, to protect this biodiversity and contribute to
15	ween the two wetlands. Mondi is a dedicated supporter of the Natural Heritage Programme and is proud to be a part of this worthy programme. The Mondi Natural Heritage Sites are very valuable for the conservation of biodiversity but they also play a vital role in education and recreation.
16	it is our intention to implement consistent guidelines for the preparation of these reports. Local site reports detailing our environmental performance at 60 operating sites are also published on our website (page 64). These are currently prepared in accordance with the ISO 14001 terminology. In 2005, we intend to review the format of t
17	Caspian and Mediterranean, one of the issues studied was how the region's cultural heritage should be managed. Surveys of the proposed route identified approximately 500 sites of potential heritage interest. More detailed investigations identified appropriate measures for each site, including surface investigations, trial pits, full excav
18	ween the two wetlands. Mondi is a dedicated supporter of the Natural Heritage Programme and is proud to be a part of this worthy programme. The Mondi Natural Heritage Sites are very valuable for the conservation of biodiversity but they also play a vital role in education and recreation. ion.
19	MONDI NATURAL HERITAGE SITES ? Mondi Forests, ? PO Box 37 Johannesburg 2000, ? Tel +27 11 647 0400 ? Fax +27 11 647 0568 ? a member of Anglo American plc group Mondi Ltd. Bookings or enqui
20	for community benefi t. BIODIVERSITY Biodiversity action plans in place Anglo American has set targets for the development of biodiversity action plans (BAPs) at all sites

Note: Processes of Managing Shaded in Grey.

Table A3. Areas.

Line	Extract from Texts of the MNCs with Keyword *Areas* Placed Centrally
1	ojects that are employed to protect biodiversity and sensitive environments. 2. Have Biodiversity Action Plans (BAPs) in place at all sites where Shell operates in areas of high biodiversity value. Related target: * By end 2005 a clear understanding of what a High Biodiversity Value Areas (HBVA) means for Shell. * Per 1 January
2	to renovate flood plain habitats north of the landfill. Meanwhile, sheep graze on the restored sections of the landfill close to the site's boundaries. Non-operational areas are progressively restored to grassland. Rob Sanders Registration Number UK-S-0000019 Restored area on Redhill landfill site t w e n t y I SLE OF W I G H T Th
3	wider aims and will seek out opportunities to create local nature reserves and publicly accessible green space, joining together with others where practical. In urban areas we will assess the potential for planting and landscaping schemes that will enhance biodiversity, create green corridors, stepping stones and havens for wildlife.
4	nt techniques for standing open waters at selected sites. EN, WT, UNI ? ? ? ? 4.2 5.2.6 Develop generic management prescriptions to improve standing open water areas for biodiversity across all Biffa sites. WT ? 4.2 5.2.7 Ensure that standing open water management prescriptions are included in Biodiversity Management Modules
5	2 Action Potential Partners Year Meets Objective No. 2003 2004 2005 2006 2007 2013 5.2.5 Develop generic management prescriptions to improve running water areas for biodiversity across all Biffa sites. WT ? 4.2 5.2.6 Ensure that running water management prescriptions are included in Biodiversity Management Modules for s
6	t hours each year. An assessment of Sagit's ice cream factory in Caivano near Naples also identified areas where energy could be saved, and a system for targeting these areas and monitoring progress has been set up. This resulted in a reduction in total energy use at Caivano of 8.7% up to the end of 2003, compared with 2001. The amount o
7	rs, using the new Reputation Tracker survey (page 11). Respondents were asked to assess Shell's overall "environmental responsibility" and our performance in specific areas (e.g., minimising impacts from our operations, offering cleaner fuels and developing renewable energy). Environmental responsibility was found to be one of the top
8	h our European emissions reporting requirements and adds consistency across our reporting. We are pleased that ScottishPower continues to make progress in a range of areas. This is reinforced by the variety of awards and recognition we have accepted from our stakeholders and peers, many of which are mentioned in this report. We als
9	activities incorporate the principles outlined by the GRI. Where elements and indicators in the GRI are relevant to Rio Tinto, we aim to report against them. In other areas we have reported against indicators and elements that are more relevant to driving performance improvements within our business. In addition to this Group level
10	dership Survey in the website for more details. We monitor the questions employees put to senior managers through the QandA pages on myGSK to ensure we pick up potential areas of concern. We also track readership of news stories on myGSK to help improve the relevance and interest of the content. www.gsk.com—GlaxoSmithKline Corporate

Table A3. *Cont.*

Line	Extract from Texts of the MNCs with Keyword *Areas* Placed Centrally
11	e Social Report. An assessment of alignment against these three stages and against the individual process steps of AA1000 is made. In addition VeriSEAAR© can highlight areas where improvement has been achieved, or areas for further improvement in the future. There is more information at the Bureau Veritas website www.bureauveritas.co.u
12	e areas. (e.g., IUCN protected area categories 1–4, world heritage sites, biosphere reserves). While a small number of companies operate facilities in or near protected areas, no impacts have been determined. www.bat.com/socialreport APPENDIX A A4 www.bat.com/socialreport APPENDIX A EN26 Changes to natural habitats resulting from a
13	place in environmentally sensitive areas. Some are officially protected, but many are not. BP believes it is for governments to decide whether sensitive or protected areas should be open to development and, if open, what measures should be taken to protect them. We will operate in sensitive areas only if we are convinced we can proper
14	ase; helping them to become more competitive at home and in global markets. In addition, we also invest in community investment programmes: focusing our support in three areas: enterprise development, education and energy access. More directly, our business benefits our employees and shareholders. In 2004 we paid wages and salaries, soc
15	l environmental management plans. This involves the wise use of natural resources and, where possible, the prevention of adverse environmental impacts. Opencast mining areas and discard dump sites are key areas of focus. ACSA's approach to implementing this policy includes: _ environmental baseline studies; _ environmental impact ass
16	fatalities are most likely to occur have been identified as falls of ground, moving machinery and transportation. Coal dust and possible methane explosions also remain areas of high focus. ACSA's Lost-Time Injury Frequency Rate (LTIFR), which reflects the number of shifts lost due to injuries for every 200,000 hours worked, includes i
17	er, one of the three operation areas N A M A K W A S A N D S F O O T P R I N T 2 0 0 2 18 Africa in October 2002; ● The Company as a whole and all three operational areas attained NOSA Platinum 5 Star gradings on the NOSA Integrated SHE System in October 2002; ● 47 workplace sections within the Company have worked 2909 days wit
18	om specialist environmental consultants and Anglo Coal Environmental Services, provide us with the practical means to retain or enhance the biodiversity of sensitive areas managed by our operations. Actions for biodiversity management that meet the requirements of the 'White Paper on Conservation and Sustainable Use of South Africa
19	cts which are critical to AWG's sustainability performance. At present, the systems for data collection are not sufficiently robust to allow complete reporting for all areas. AWG should focus on embedding its targets into all business units and on developing complementary management systems to enable reporting against these targets. T
20	o produce a strengthened health and safety policy, with particular emphasis on improvements to standards and responsibilities. Further updating of Group-wide standards in areas of potential risk identified at the conference is being carried out by seven working groups. This work will be completed by the end of 2003 and will ensure that b

Note: Processes of Managing Shaded in Grey.

Table A4. *Communities*.

Line	Extract from Texts of the MNCs with Keyword *Communities* Placed Centrally
1	civil society groups and government authorities and begin building partnerships that strengthen the long term sustainability of community projects". Social Benefiting communities Nigeria update Community agitation for greater and rapid development of the Niger Delta region remains high. The Nigerian government has taken steps toward
2	ustainable energy future. It has two core objectives: reducing environmental impact of fossil fuel use, and increasing access to sustainable energy, particularly in poor communities of developing countries. It's an innovative effort, which allies social investment with a business approach and develops projects that build on Shell's unique
3	upport the government's target of achieving a 20% reduction in CO_2 emissions by 2010. The Trust is delivering investment in new renewable projects of benefit to local communities enabled by customers who pay a small premium through our Green Tariff. Further information about our Green Energy Fund or our other activities within the c
4	west Territories of Canada. No one was relocated for this development and detailed agreements have been established during the five year consultation with neighbouring communities and approval process. • Sustainable urban land development: Kennecott Utah Copper (KUC), holds land surrounding the mine in excess of its needs and there i
5	included in the performance section, page 22. Commitment: "Develop measurement tools that reflect the effectiveness of our contribution of resources to neighbouring communities and of our relationship with the communities". Rio Tinto has signed a memorandum of understanding with Warwick University Corporate Citizenship Unit, UK,
6	inues, with completion targeted for early 2005. Establishment of an effective process to address the compensation claims improved relationships between KEM and local communities and has enabled the implementation of a number of important community development and capacity building initiatives which seek to secure sustainable solut
7	SandE reports Does your company have any statistics on implementation of community involvement policies? * 2004 SD Review: results data is provided for contributions to communities, contribution to the economy and input/output measurements * See also all local SandE reports Do you have any information on your land agreements? * 2004 S
8	n, particularly for children from disadvantaged backgrounds, is the central aim of our education initiatives. Community Encouraging our employees to engage with the communities in which we do business is a vital part of CSR. FAQs Read the frequently asked questions about Corporate Social Responsibility. CSR report Download the la
9	sources in an efficient manner, we will move further down the track of sustainable and responsible manufacturing. This is good for shareholders, employees, communities and the environment—and hence good for our business. This will continue to be a focus in the future". 13 the assessment of the health, safety (excludin
10	ganization and the World Bank to local schools and community-based organisations. Where possible, we ensure that our programmes are sustainable and can be repeated in communities with similar needs. Our programmes comprise major initiatives in public health, support for education, product donations, and support for employee involve

Table A4. *Cont.*

Line	Extract from Texts of the MNCs with Keyword *Communities* Placed Centrally
11	cal libraries and providing them with book vouchers Powergen Environment Fund The Powergen Environment Fund provides £50,000 a year to environmental projects in local communities. In 2001/2002, 16 projects received support, including hedge planting, nature trail improvements and composting schemes. The fund was launched in 1999 and i
12	ommunity, sensitive to all the impacts of our business. SC: The pressures you've mentioned have derived from regulators, government, stakeholders, shareholders, local communities and environmental impacts—all influencing the strategic decisions that you make. Do you think your stakeholders recognise just how many factors you are havi
13	s we are not unique. But China's social needs remain vast. No individual company is likely to make a huge difference by itself. But by working with government, partners, communities and other stakeholders, and by sustaining the commitment over time, foreign investors can have an impact within their areas of expertise. And it is for that r
14	ESPONSIBILITY REPORT 2004 2 Chief Executive's statement continued BG Group seeks to identify and manage potential impacts. We believe that, where relationships with communities are built on openness and trust, projects can proceed with mutual benefit. BG Group respects the right of our host countries to decide their own developme
15	ternity pay, paternity leave and adoption leave, including for same-sex couples. Corporate Social Responsibility 30 INTRODUCTION SERVING AUDIENCES IN BUSINESS COMMUNITIES THE ENVIRONMENT FEEDBACK The corporation won the 2004 Zayed International Prize for Environment, with The World series Earth Report—
16	included in the performance section, page 22. Commitment: "Develop measurement tools that reflect the effectiveness of our contribution of resources to neighbouring communities and of our relationship with the communities". Rio Tinto has signed a memorandum of understanding with Warwick University Corporate Citizenship Unit, UK, f
17	t, human rights and social investment. UNDERSTANDING OUR IMPACTS AND RESPONSIBILITIES We carry out Social Impact Assessments (SIA) to identify potential impacts on communities and to develop strategies to manage these. We also engage with local, national and international non-governmental organisations (NGOs). Understanding the p
18	ment and capacity building in Northern KwaZulu-Natal. The organisation's ongoing training initiatives, allied to relationship building and a deep understanding of the communities in which it operates, forms the basis of the development of sound small businesses in these rural areas. These include small-scale farming projects and rur
19	reports to the Board of directors. 7. Stakeholder engagement: Promote and maintain open and constructive dialogue and good working relationships with employees, local communities, regulatory agencies, business organisations and other affected and interested parties, to increase knowledge and enhance mutual understanding in matters o
20	tening Stakeholder engagement We engage widely with a range of stakeholders to ensure our CSR programme addresses their key concerns. This includes investors, NGOs, communities, suppliers, customers and employees. Last year we made a commitment to complete an independent review of our Group-level stakeholder engagement

Note: Processes of Managing Shaded in Grey.

References

1. Dryzek, J.S. *The Politics of the Earth: Environmental Discourses*, 2nd ed.; Oxford University Press: Oxford, UK, 2005; p. 145.
2. Sustainability through the Lens of Environmental Sociology. Available online: http://www.mdpi.com/journal/sustainability/special_issues/EnvironmentalSociology (accessed on 17 July 2015).
3. Gray, R.H.; Milne, M.J. Towards reporting on the triple bottom line: Mirages, methods and myths. In *The Triple Bottom Line: Does It All Add Up?*; Henriques, A., Richardson, J., Eds.; Earthscan: London, UK, 2004.
4. Gray, R. Is accounting for sustainability actually accounting for sustainability... and how would we know? An exploration of narratives of organizations and the planet. *Account. Organ. Soc.* **2010**, *35*, 47–62. [CrossRef]
5. Milne, M.J.; Gray, R.H. W(h)ither ecology? The triple bottom line, the global reporting initiative, and corporate sustainability reporting. *J. Bus. Ethics* **2013**, *118*, 13–29. [CrossRef]
6. Starik, M.; Kanashiro, P. Toward a Theory of Sustainability management: Uncovering and Integrating the Nearly Obvious. *Organ. Environ.* **2013**, *26*, 7–30. [CrossRef]
7. Zollo, M.; Cennamo, C.; Neumann, K. Beyond What and Why: Understanding Organizational Evolution towards Sustainable Enterprise Models. *Organ. Environ.* **2013**, *26*, 241–259. [CrossRef]
8. Building a Stronger, Safer BP. Available online: http://www.bp.com/content/dam/bp/pdf/sustainability/group-reports/Sustainability_Report_2014.pdf (accessed on 17 July 2015).
9. Hajer, M.A. *The Politics of Environmental Discourse: Ecological Modernization and the Policy Process*, 1st ed.; Clarendon Press: Oxford, UK, 1995.
10. Van Leeuwen, T. *Discourse and Practice: New Tools for Critical Discourse Analysis*, 1st ed.; Oxford University Press: Oxford, UK, 2008.
11. Franceschi, D.; Kahn, J.R. Beyond Strong Sustainability. *Int. J. Sustain. Dev. World Ecol.* **2003**, *10*, 211–220. [CrossRef]
12. Özkaynak, B.; Devine, P.; Rigby, D. Operationalising Strong Sustainability: Definitions, Methodologies and Outcomes. *Environ. Values* **2004**, *13*, 279–303. [CrossRef]
13. Wenger, E. *Communities of Practice: Learning, Meaning, and Identity*, 1st ed.; Cambridge University Press: Cambridge, UK, 1998.
14. Labov, W. *The Social Stratification of English in New York City*, 2nd ed.; Cambridge University Press: Cambridge, UK, 2006.
15. Macaulay, R.K.S.; Trevelyan, G.D. *Language, Social Class and Education: A Glasgow Study*, 1st ed.; University of Edinburgh Press: Edinburgh, UK, 1977.
16. Wolfram, W. *A Sociolinguistic Description of Detroit Negro Speech*, 1st ed.; Center for Applied Linguistics: Washington, DC, USA, 1969.
17. Bell, S.; Morse, S. *Sustainability Indicators: Measuring the Immeasurable*, 2nd ed.; Earthscan: London, UK, 2008.
18. Brand, F.S.; Jax, K. Focusing the meaning(s) of resilience: Resilience as a descriptive concept and a boundary object. *Ecol. Soc.* **2007**, *12*, 23–39.
19. Fergus, A.H.T.; Rowney, J.I.A. Sustainable Development: Lost Meaning and Opportunity? *J. Bus. Ethics* **2005**, *60*, 17–27. [CrossRef]
20. Hacking, T.; Guthrie, P. A Framework for Clarifying the Meaning of Triple Bottom-Line, Integrated, and Sustainability Assessment. *Environ. Impact Assess. Rev.* **2008**, *28*, 73–89. [CrossRef]
21. Hopwood, B.; Mellor, M.; O'Brien, G. Sustainable development: Mapping different approaches. *Sustain. Dev.* **2005**, *13*, 38–52. [CrossRef]
22. Waas, T.; Hugé, J.; Verbruggen, A.; Wright, T. Sustainable Development: A Bird's Eye View. *Sustainability* **2011**, *3*, 1637–1661. [CrossRef]
23. Milne, M.J.; Kearins, K.; Walton, S. Creating adventures in Wonderland? The journey metaphor and environmental sustainability. *Organization* **2006**, *13*, 801–839. [CrossRef]
24. Milne, M.J.; Tregidga, H.M.; Walton, S. Words not actions! The ideological role of sustainable development reporting. *Account. Audit. Account. J.* **2009**, *22*, 1211–1257. [CrossRef]
25. Firth, J.R. *Papers in Linguistics 1934–1951*, 1st ed.; Oxford University Press: London, UK, 1957.
26. Austin, J.L. *How to Do Things with Words*, 1st ed.; Oxford University Press: London, UK, 1962.
27. Searle, J.R. *Expression and Meaning: Studies in the Theory of Speech Acts*, 1st ed.; Cambridge University Press: Cambridge, UK, 1979.

Sustainability **2015**, *7*, 15900–15922

28. Halliday, M.A.K. *An Introduction to Functional Grammar*, 2nd ed.; Edwin Arnold: London, UK, 1994.

29. Sinclair, J. *Corpus, Concordance, Collocation*, 1st ed.; Oxford University Press: Oxford, UK, 1991.

30. Stubbs, M. *Text and Corpus Analysis*, 1st ed.; Blackwell Publishers Ltd.: Oxford, UK, 1996.

31. Brown, M. *Managing Nature-Business as Usual: Patterns of Wording and Patterns of Meaning in Corporate Environmental Discourse*; Acta Humaniora: Oslo, Norway, 2008.

32. Brown, M. A Methodology for Mapping Meanings in Text-based Sustainability Communication. *Sustainability* **2013**. [CrossRef]

33. WBCSD. Business Solutions for a Sustainable Word. Available online: http://www.wbcsd.org/home.aspx (accessed on 17 July 2015).

34. UN Global Compact. Available online: https://www.unglobalcompact.org/ (accessed on 17 July 2015).

35. British National Corpus. Available online: http://www.natcorp.ox.ac.uk/ (accessed on 17 July 2015).

36. Wordsmith Tools. Available online: https://elt.oup.com/catalogue/items/global/multimedia_digital/9780194505161?cc=no&selLanguage=en&mode=hub (accessed on 17 July 2015).

37. Scott, M.; Tribble, C. *Textual Patterns: Key Words and Corpus Analysis in Language and Education*, 1st ed.; John Benjamins Publishing Company: Amsterdam, The Netherlands, 2006.

38. Archer, D., Ed.; *What's in a Word-List? Investigating Word Frequency and Keyword Extraction*, 1st ed.; Ashgate Publishing: Farnham, UK, 2009.

39. Kemppanen, H. Keywords and Ideology in Translated History Texts: A Corpus-based Analysis. *Across Lang. Cult.* **2004**, *1*, 89–106. [CrossRef]

40. Rayson, P. From key words to semantic domains. *Int. J. Corpus Linguist.* **2008**, *13*, 519–549. [CrossRef]

41. Toolan, M. Values are Descriptions; or, from Literature to Linguistics and back again by way of Keywords. *Belg. J. Engl. Lang. Lit.* **2004**, *1*, 11–30.

42. Bansal, P.; Knox-Hayes, J. The Time and Space of Materiality in Organizations and the Natural Environment. *Organ. Environ.* **2013**, *26*, 61–82. [CrossRef]

43. ScottishPower. Available online: http://www.scottishpower.com/pages/reputation_and_sustainability.asp (accessed on 17 July 2015).

44. Crane, A.; Matten, D.; Moon, J. Ecological citizenship and the corporation: Politicizing the new corporate environmentalism. *Organ. Environ.* **2008**, *21*, 371–389. [CrossRef]

45. Tuan, Y. Space and Place: Humanistic Perspective. In *Philosophy in Geography*; Gale, S., Olsson, G., Eds.; D. Reidel: Dordrecht, The Netherlands, 1979.

46. Shrivastava, P.; Kennelly, J.J. Sustainability and Place-Based Enterprise. *Organ. Environ.* **2013**, *26*, 83–101. [CrossRef]

47. Skinner, W. Manufacturing—Missing Link in the Corporate Strategy. *Harv. Bus. Rev.* **1969**, *47*, 136–145.

48. Stead, J.G.; Stead, W.E. The Coevolution of Sustainable Strategic Management in the Global Marketplace. *Organ. Environ.* **2013**, *26*, 162–183. [CrossRef]

sustainability

MDPI

Article

Nature–Culture Relations: Early Globalization, Climate Changes, and System Crisis

Sing C. Chew [1],* and Daniel Sarabia [2]

[1] Department of Urban and Environmental Sociology, Helmholtz Centre for Environmental Research—UFZ, Leipzig 04105 Germany
[2] Department of Sociology, Roanoke College, 221 College Lane, Salem, VA 24153, USA; sarabia@roanoke.edu
*   Correspondence: sing.chew@humboldt.edu; Tel.: +49-341-235-1746; Fax: +49-341-235-1836

Academic Editor: Md Saidul Islam
Received: 19 October 2015; Accepted: 11 January 2016; Published: 14 January 2016

Abstract: Globalization has been on everyone's lips in light of the contemporary conditions. It has been viewed mostly as a stage reached as a result of long-term societal changes over the course of world history. For us, globalization has been an ongoing process for at least the last 5000 years. Little attention has been paid to the socioeconomic and natural processes that led to the current transformation. With the exception of historical sociologists, there is less interest in examining the long-term past as it is often assumed that the past has nothing to teach us, and it is the future that we have to turn our intellectual gaze. This paper will argue the opposite. We believe a long-term tracing of the socioeconomic and political processes of the making of the modern world will allow us to have a more incisive understanding of the current trajectory of world development and transformations. To plead our case, we outline the emergence of the *first* Eurasian World Economy linking seven regions (Europe, the Arabian Peninsula, East Africa, the Persian Gulf, Central Asia, South Asia, Ceylon, Southeast Asia, and China) of the world, with the exception of the Americas, starting as early as 200 BC, and the sequence of structural crises and transformations (trading networks and commodities) that has circumscribed the structures and trends of the current global system. Such consideration in our view is limited if we do not also include the relations between social systems and Nature, and the rhythms of the climate. For the latter, an awareness of the natural rhythms of the climate as well as human induced changes or climate forcing have triggered system-wide level collapses during certain early historical periods.

Keywords: social system crisis; social change; globalization

---

1. Introduction

Materialistically, world history has always been a history of the relationship between Culture and Nature. For at least the last five thousand years, this relationship has been an enduring one, whereby human communities from the least transformed in terms of complexity to the most advanced postmodern social systems have required Nature to meet their basic needs of survival and reproduction. If such is the case, then for the discipline of sociology this, therefore, should be *the overall dimension* that sociology should expend its overall efforts in understanding and explaining the processes and structures that reproduces this relationship. A relationship described by Marx [1] in his early writings as "the humanization of nature" and "the naturalization of man". Unfortunately, the history of sociological scholarship does not reflect this interest nor accept it as *the* dimension that should be its focus. With the global concern now focusing on global warming with environmental changes affecting social, political and economic processes, notwithstanding the "fashionability" of studying environmental changes these days, sociologists have over the last decade or so come around to addressing this dimensional relationship with much fervor. Rather than using a broad brush to paint the absence and neglect of sociological scholarship in examining Nature and Culture relations, we wish to note of the early scholarship of Catton and Dunlap [2], O'Connor [3,4] and Schnaiberg [5] that did try to explore this Nature–culture dimension in depth, and especially Catton and Dunlap's call for an end to the human exemptionalist paradigm that has been underlining sociological scholarship. The outcome of this early intervention did not result in sociological scholarship focusing on biophysical factors as causal and dependent/independent variables and dealing with Nature as the phenomenon all to be examined *holistically and in a dialectical fashion*; instead what resulted was to treat such calls as *a* dimension of study (environmental sociology) in sociology like the various dimensions that sociologists were already focused on (sociology of the military, economic sociology, political sociology, *etc.*).

Our endeavor in this exercise is not to revisit every direction that sociological scholarship has taken in the area of the environment since the call of Catton and Dunlap [2], but rather to review only briefly the sociological scholarship focusing on political economy and system crisis that is the focus of our paper dealing with Nature–culture relations. The latter scholarship we are referring to are those theoretical analyses that utilize political economy models (for example, see Foster [6,7] and Moore [8] that have their anchors dropped in Karl Marx's ocean in their attempts to account for the social, political and economic processes in the reproduction of "capitalist mode of production" systems of the last 500 years. In doing this, we suggest that these Marxian political economy studies have not gone far enough in their frameworks ontologically and theoretically, for they have refrained from putting Nature front and center in their analytical calculations. Instead, these analyses treat Nature as a benign substrate as the base by which capital and its various modalities of modes of production utilize and desecrates in telling their versions of Marxist inspired history. We believe, they continue to treat Nature still within the human exemptionalist paradigm, and in certain cases Nature is viewed as a social construction, or in our view anthropocentrically, instead of seeing Nature and other natural processes as having independent and causal dimensions that have impacted on the reproduction of human societies over world history, thus, generating system crisis that these political economy analyses have endeavored to explain.

Our response to the aforementioned paragraph, without relying on theoretical or ontological rationalizations, will be through a historical materialistic exercise, propose our *ecocentric* framework through our articulation of world history whereby we combine "theoretically informed history and historically informed theory" to explain system crisis over *la longue durée*. Our *la longue durée* is not the 500-year time span starting from supposedly capitalist Western Europe that some like Foster [7] and Moore [8] present, but one that stretches over 5000 years of world history. Historically, in this context, we like to focus specifically on globalization as a phenomenon that has also gained wide currency and popularity these days, and specify the evolution of this global process in relation to the occurrence of natural phenomenon such as climate changes resulting in social system crisis over world history. We assert that Nature can be both an independent and dependent variable in determining the evolution of human societies. We suggest that Nature can be the "agent" in long-term social change (see for example, [9–12]). In this context, perhaps it is not "capitalist" socioeconomic relation that is the determinant in the last instance that shapes the trends and tendencies of world history over *la longue durée* like the Marxian political economists have asserted [13] (p. 9).

2. The Attempt to Green Marx

Among the Neo-Marxists, James O'Connor's [3,4] contribution was the first major reinterpretation of Marx's political economy with his insertion of Nature into the overall equation of understanding the contradiction of the process of capital accumulation in late industrial social formations. With the first contradiction of capital between labor and the owners of production ultimately leading to system crisis, O'Connor's insisted that there was a second contradiction of capital as well that has to be added and considered. This being the expansionism of capital tends to cause environmental problems that will ultimately lead to crisis in the reproduction of capital. The significance of O'Connor's writings is his infusion and highlighting of Nature as the substrate for the reproduction of capital accumulation and that there is the tendency to cause social system crises with the exploitative use of Nature as a consequence of capital's inherent expansionism and incessant drive for increasing accumulation. Unlike other Neo-Marxists who at that time (1970s–1980s) were still preoccupied towards addressing system crisis via the process of accumulation and the social relations of production along with the role of the Capitalist state in this process, the insertion of Nature by O'Connor's stress on the second contradiction of capital shifted attention to the conditions of the times then. For that his intervention has to be applauded.

We can see that O'Connor's work influenced numerous Neo-Marxists leading to the development of a number of Neo-Marxist frameworks today. Instead of commenting on these developed frameworks *in toto* which will take up too much space, we propose to examine one that has garnered Neo-Marxist support in terms of publication and citations in sociology that we feel is pertinent to our article's proposals and objective. Here we are referring to the Metabolic Rift framework of Foster's [6,7] and extended by Moore [8,14].

In this model, Foster [6,7,15] proceeded by mining the seams of Marx's mother lode, especially *Das Kapital*, even to the level of footnotes to support Foster's attribution that Marx had always paid attention to Nature in his writings. In response to numerous critics of Marx who have stated that Marx did not in any systematic, exhaustive way address issues such as "the exploitation of Nature, Nature's role in the creation of value, the existence of distinct natural limits, Nature's changing character and the impact of this on human society, the role of technology in environmental degradation, and the inability of mere economic abundance to solve environmental problems" [7] (p. 168); Foster [7] argued that the ecological blinders are not present in Marx's thought, and that each of the aforementioned criticisms was addressed in his theory of metabolic rift. Delivering in his analyses coverage of environmental issues of salience during his time period, Marx addressed the problem of agriculture under capitalism and in particular made observations regarding the soil, while also turning his attention to deforestation, pollution in urban centers, and overpopulation [13] (p. 168). Through this particular lens, Marx came to understand sustainability and the tensions brought about by a rural urban divide, a nexus of relations some analysts conclude Marx captured in his model of social ecological metabolism [8].

Foster shows that in Marx's *Das Kapital* there is a critique of capitalist arrangements linked to the depletion of soil [13]. Marx, according to Foster [13], presented an explanation of "how large scale-industry and large-scale agriculture combined to impoverish the soil and the worker" (pp. 174–175). Foster [13] contends that both Marx's retort of Darwinian evolutionary theory and his formulation of social and ecological metabolism significantly explains the dialectic of capital accumulation and the exploitation of Nature and its various consequences while at the same time considering the relations of production.

In more recent publications, Foster [13] in keeping with the times updated his diagnosis and prognosis of late capitalist systems by alerting us to the supra level that we should be concerned with of the dangers that have now evolved to the level of the world system and the *planetary rift* that we are experiencing, and will face in the future. Foster [13] (pp. 213, 229) however true to his Marxian roots, continues to maintain that we should focus on the relations of production that is undermining ecological sustainability, and that "long-term prospects demand truly revolutionary change, especially a rupture with the accumulation/growth imperative of capitalism". Clearly one can see that Foster's metabolic rift is still placing its emphasis on the paramount status of Culture, with Nature being the substrate that provides value formation for capitalist social relations that are extended to other social structures such as urban and rural areas spanning geo-spatial boundaries on a planetary scale. By Culture's actions (capitalist social relations) Nature is impacted and degraded. This model just follows O'Connor's second contradiction of Capitalism type of argument nothing more nothing less theoretically. Anthropocentric in orientation, it is simply following the basic Marxian argument of the theory of value and crisis dressed up with specific references to natural elements and environmental sustainability.

Moore [8,16–24] satisfied with the Neo-Marxian ecological foundations that Foster had carefully constructed and legitimated via the publication of Foster's seminal piece in the *American Journal of Sociology* in 1999, proceeded to utilize this theoretical model to fashion a defense of Wallerstein's world system perspective accused of being neglectful of Nature [25–28]. Keeping within the dogmas of leftist understanding of the emergence Capitalism in Western Europe over a certain time period from either the 16th century to the late 18th century, Moore [16] undertook an exercise to "naturalize" the work of Wallerstein's analysis. By then, Wallerstein [29,30] himself had started to include Nature as a factor in his continuing analysis of the evolution of the modern world-system.

In trying to extend a Marxian analysis of Nature–culture relations, Moore's [8,17] recent attempt is to criticize the metabolic rift theory of Foster's by suggesting that Foster's model is dualistic, based on a Cartesian binary that puts "biophysical (nature) crises in one box and accumulation (culture/socioeconomic) crises in another" [8] (p. 2). For Moore, this leads to the conclusion that biophysical problems are outcomes of capitalist accumulation "but not constitutive of capitalism as a historical system" as he puts it [8] (p. 2). What this means is that Capitalism "develops through nature-society relations" and does not act on Nature. By fiat, Moore [8] (p. 2) declares capitalism is an "ecological regime"! Our reply to this is so is everything else! By such a theoretical move, Moore has flattened and collapsed society and nature into one with no distinguishing features, and not as young Marx [1] and Habermas [31] had distinguished of the different objectifications: that of the natural world and that of social/individual worlds, and in praxis, nature is socialized and the human is naturalized. This attribution of Moore's that Capitalism as World Ecology advances Capitalism—which is a theoretical concept—into a concretized entity having nature-like properties and rhythms. Therefore, Nature to Moore is a *social construction* devoid of its natural properties, and instead Capitalism is the living, breathing organism oozing the sweat of labor. Again, instead of giving *equal* weight to the dimension of Nature and the dimension of Culture, and acknowledging the distinct characteristics of Nature as subject/object, Moore's model remains within the framework of awarding paramount status to Culture, in this case for Moore, Capitalism; and hence anthropocentric. To further this notion of Capitalism as an ecological regime, the theory of value of Marx was imported by Moore to furnish the scaffolding for his model of Capitalism as World Ecology. Distinguishing distinctions in Marx's theory of value under conditions of overproduction and underproduction, Moore proceeds to sketch out the different historical periods that can be explained according to the different types of the logics of capital accumulation in order not to flatten the historical epochs nor generalize the logic of capital accumulation. Such a theoretical stance is then used to turn the different logics of capital accumulation to account for the history of capitalism or historical capitalism.

We have several reservations concerning Moore's model: Firstly, inspite of Moore's persistent reference to *the dialectic*, we do not see any dialectical process in Moore's model. With his constant resort to *the dialectic* to explain the logic of the model he is proposing, one wonders *what is the interplay between the "particular" and "universal" or of the "moment" and totality for him?* By in the "moment", we mean a phase or aspect of a cumulative dialectical process; and not just a period in time. Moore for wanting to be dialectical chose in referring to the logic of capital accumulation process from *Das Kapital* as his methodological reference point instead of Marx's *Grundrisse*. *We are confused as to what is the particular in relation to the totality for Moore.* For him, we think, Capitalism is his totality and everything else gets subsumed. It is rather difficult for us to accept ontologically that Capital is considered the totality and everything else are particulars. But if one wants to employ the dialectical process, the universal or totality and the particular proceed through history in a constant interplay resulting in a transcendence (*Aufhebung*) with no end (no complete identity). For Moore the culture/nature relation is Historical Capitalism, therefore the totality and particular are one, or as he [8] (p. 34) puts it: "Capitalism does not have an ecological regime; it is an ecological regime." Collapsing the "universal" and the "particular" is by no means dialectical nor transcending, it annihilates the course of history (*Auflösung*); hence no freedom and no history! We are puzzled that Moore would subscribe to this philosophy of history for he is undertaking a Marxist dialectical analysis or at least that is his intention.

Secondly, Moore [8] (p. 17) suggests that his project moves "from the analysis of what *makes* capital to what capital makes, from the logic of capital to the history of capitalism". This means a rigorous inspection of historical events of accumulation on a world ecological basis. Despite Moore's enthusiasm [8,16–24] for the work of Marx and the Neo-Marxists and their analyses of Capitalism, Moore seems to be unaware of the vigorous debate on the nature of "capitalism", its time emergence and its duration that have taken place in world-systems analyses. Moore's fervent attempt to theorize Capitalism as historical capitalism with its long cycles never grapples with the numerous works that have been produced over the last two decades under the umbrella term of world system history whereby there are concentrated discussions on what is "capitalism", the duration of the world system, and the various dynamics underlining such a system from hegemonic rivalry, core-periphery relations, the global accumulation process and long waves (see for example, [9,10,32–41]). Following Wallerstein [42] for Moore, the emergence of capitalism and the capitalist mode of production, its duration and nature all began in the 15th century in Western Europe. World history for Moore starts then, and there was no history before then. Moore's historical capitalism began then and evolved, anything else before this watershed moment were ignored or considered unnecessary to our understanding of culture/nature relations over world history. Moore's historical analysis therefore is limited and this reservation on our part can also be applied to Foster's work [7]. The adherence to the 15th century for the rise of the "capitalist" mode of production in Western Europe, Moore's model suffers from the same accusation of being Eurocentric that the late Frank [36] had leveled at the world-system history of Wallerstein [42].

3. Historical Globalization and System Crisis

For the Left, in trying to explain globalization there are basically two narratives. The first argues that globalization is the *stage* reached with the process of accumulation starting at the nation-state level followed with the export of capital in the form of imperialism, and over time the accumulation of capital covers the globe. The other states that globalization is a process that has occurred through world history, and that it is not a stage reached but one that started at different time periods. For Wallerstein [42], it was the 15th century, whereas for Frank [36] and others [9,10,37,40,41] the process started at least 5000 years ago.

If it is not a stage reached but an evolving process, we can make the assumption of the evolution of a structure that we can categorize as a world economy/system. We use the term world economy instead of world-economy as the latter has been utilized by world-systems specialists for a historical structure that has a certain set of socioeconomic and political attributes and trends "capitalistic" in nature that do not necessarily cover a wide geographic space. To world-system specialists, this historical structure of a world-economy is a world in itself, hence the hyphenation between world and economy [43]. In our case, a world economy is not distinguished necessarily by a mode of production but that it covers a *global geographic* space with multiple cores/regions linked at a minimum by a trading system. *It is an evolving global economy "of the world"*. Depending on the temporal sequence, an economy of the world encompassing different chiefdoms, kingdoms, civilizations, empires, and states in a global division of labor, technology, and knowledge circumscribed by different cultural patterns.

Regardless of time or geographic space, this historical globalization process is punctuated with system crisis that impacts on the regions of the world. The spread of such crisis conditions and what it covers is determined by the systemic connections of this world economic structure that continues to evolve and transform to overcome these systemic crisis phases. These systemic crisis phases termed as Dark Ages are not only distinguished by downturns in socioeconomic conditions and political rivalry but are also characterized by climate changes and natural environmental degradations [9–11]. These crisis phases—occurring in the past and more to come in the future—were by no means continuously exhibiting increasing ecological degradation. These periods also led to lesser ecological degradative practices because of the socioeconomic decline [10]. Climate changes should also be considered as a major factor in the precipitation of system crisis [11].

If we examine world history in terms of trade connections, we can trace the contours of a "regional" world economy encompassing the Eurasian region of Mesopotamia, the Arabian Peninsula, Levant, Anatolia, Iran, the Indus Valley, and Egypt by 3000 BC [9,10]. Beaujard [44–46] has identified three possible regional world systems from 1000 BC onwards. For him, there was the Western world system, the Eastern world system, and the Indian world system during the Iron Age with growing interactions between these systems from 350 BC onwards. Regardless of whether it is a single world system that started in the Fertile Crescent and over time encompassing other regions of the world as postulated by Frank and Gills [33] or Beaujard's [46] three regional world systems coalescing into one world system, what is clear is that by *the turn of the first century of the current era* we find a world system encompassing Europe, East Africa, and Asia (South, Southeast and East) [9,10,46]. In world history, we can conceive of it as the *first Eurasian* world economy as the only major region that has not been connected at this point in world history is the Americas.

Conceptually, the factors and processes that trigger system crises or Dark Ages over the last five thousand years of recorded human history have not been fully understood. Their identifications have relied on analyses of the political economy of the world system in the 1970s and 1980s, and on the whole, based on Marxian crisis theory. The outcome of this is that several interrelated and intrinsic factors have been distinguished providing conditions for the generation of barriers to system reproduction: overproduction/under consumption, crisis of state authority and competition, social exploitation and polarization leading to antisystemic social movements, and crisis of sustainability. Clearly, the first three factors relate to the dynamics occurring at the social (world) system level with the fourth focusing on the interaction between the social (world) system and the natural system. With the first three factors being anthropogenic in origin, system crisis and transformations are considered to have socioeconomic and political roots.

Notwithstanding such a declaration of economy as being determinant in "the last instance" in terms of system reproduction, others for example, such as O'Connor [3,4], Chew [25,26], and Roberts and Grimes [27] have declared the need to treat Nature as part of the equation towards understanding system dynamics. As we have identified in the previous section, O'Connor [3] has stated that besides the widely accepted capital/labor relation contradiction, there is also a second basic contradiction of capitalism: the capitalist system's insatiable consumption of natural resources leading to crisis conditions. If the latter is the case, on several occasions, one of us (see for example Chew [9,25]) has suggested that based on a materialist conception of history we should reintroduce Nature back into social analysis, and perhaps it is "ecology in command" in *the last instance* that induces system crisis conditions. The rationale for this is based simply on the material fact that the social system requires the natural system to reproduce itself. Along this vein, some historians have also recently raised the issue of Nature's historical agency in conditioning the transformation of human societies [12,47]. Besides Nature's agency, we would like to assert that in the long run, perhaps climate is an important driver of social system structural crisis. This we will attempt to show in this paper.

Besides the above considerations, it is also clear the duration of these world system crisis phases have not been worked out clearly. Different duration for each crisis phase has been proposed according to the different views of world system development [9,33,36,38,41,42,48]. Time-wise, the duration stretches from 50 years to a thousand year in length. Such differences are based on a number of factors and conditions such as the span of historical epochs, empirical verification of economic stagnation, sociopolitical trends, and natural system indicators such as the rate of deforestation.

What the Marxists and other world-system theorists have suggested to date for the past system crises have been recurring and contingent factors of an anthropogenic nature that condition system crises. What about those of ecological and natural origins? Ecologically stressed conditions (deforestation, soil erosion, desiccation, *etc.*), climate changes, and tectonic shifts/volcanicity have been proposed as factors that can precipitate system crises [9,25,26]. How can we be more precise in demarcating these ecological factors? We believe our precision can be enhanced by materialistically identifying system crises with recurring Dark Ages that have occurred in human world history. For it is during these Dark Ages that system transformations occur. Not only do we witness socioeconomic and political declines, but also, ecological degradation, climatological shifts, and natural disturbances. Climatological changes and natural disturbances (tectonic shifts) recalibrate our understanding of the evolution of the world system by adding natural system factors to the already declared anthropogenic ones for our understanding of system transformation.

Given the above parameters, we can abstract historically the several processes and factors that depict a Dark Age period in order to have a clearer understanding of the various factors that precipitate a system crisis and transformation. Such an abstraction starts by delineating the connections between the natural system and social system in the reproduction of the world (social) system.

Barriers to the reproduction of the world system are formed when humans induced changes to the ecology and climate. The degradative aspects of human activity are conditioned by social organizational factors (urbanization, accumulation, wars, technological innovations, and population) that impact on system reproduction. Natural disturbances such as earthquakes and volcanic eruptions also condition the reproduction and evolution of the world system, and thus work independently. Climate as well can affect precipitate and affect the crisis independently and also dependently. For the latter, human actions can cause climate changes, for example, global warming, and thus causing the crisis. We need, therefore, to consider the degree of weight these factors have in precipitating a system crisis.

Through the course of human history, system crises have appeared in the "concrete" in the form of Dark Ages. Over world history, these historical phases are rare. Between 3000 BC and AD 1000, there have been indications of only two such identified phases (2200 BC, 1700 BC and 1200 BC–700 BC (considered as one phase in terms of the crisis of the Bronze Age) and AD 400–AD 800/900) occurring in the world system from Northwestern India, West Asia, the Mediterranean, and Europe. Several scholars such as Desborough [49], Snodgrass [50–52], and Braudel [53–58] have discussed the conditions of life during past Dark Ages highlighting the economic, political and social disorder with population losses, deurbanization, *etc.* Furthermore, historical records and archaeological evidence indicate a flattening of the social hierarchy, and devolution away from a complex form of sociopolitical organization and lifestyles that existed prior to the onset of Dark Ages. The trends and patterns of Dark Ages therefore show developmental reversals: fall in population levels, decline or loss in certain material skills, deurbanization and migration, decay in cultural aspects of life, fall in living standards and thus wealth and trading contacts.

If Dark Ages are prolonged ecological crisis periods, crisis provides opportunities. In other words, crisis conditions provide the opportunities for the resolution of contradictions that have developed to such a state that inhibits the reproduction of the world system. It leads to pathways and processes that would mean system reorganization and transition. If reorganization does not occur, system collapse usually follows. This has been seen historically (see for example [9,25,26,59–61]). If this is the case, ecological limits become also the limits of the socioeconomic processes of the world system, and the interplay between ecological limits and the dynamics of the social system define the historical tendencies of the human enterprise [9,11,60].

To this extent, Dark Ages or system crises also offer opportunities for Nature. Dark Ages should be appreciated as periods for the restoration of the ecological balance that has been disrupted by centuries of intensive human exploitation of the natural system. The downscaling of socioeconomic processes during Dark Ages provides the opportunity for Nature to recover.

4. The Bronze Age Crisis

The Ancient World System

The ancient world of the Near East and Northwestern Indian subcontinent during the third millennium was characterized by a system of overlapping core regions (for example, Egypt, Mesopotamia, and Harappan Civilization of Northwestern India) (see for example [9,33,62]). Within such a political economic matrix, each core interacted with its immediate hinterland and with each other leading to certain core regions attempting to manipulate its adjacent hinterland, and at times trying to control it [9]. Given such political incursions and trading initiatives, systemic connections were established, and during moments of systemic crisis, crisis-like conditions reverberated throughout the system providing opportunities and constraints depending on the circumstances.

In the Far East, there were no systemic connections with the Near East at this point in time. It was to come later around 200 BC. System-wise, we have two subsystems in place at the start of the Bronze Age crisis. In the Near East we have a subsystem encompassing Southern Mesopotamia, Northwestern India, the Eastern Mediterranean including Egypt, and Central Europe. In the Far East, there was a subsystem with geographic coverage enclosing China and other parts of East Asia.

The accumulation of surplus, urbanization, and population growth are the prime drivers of the processes of the social (world) system, which in turn, define the social (world) system's interactions with the natural system [9]. The interacting relationships among urbanization, population and production/trade mean that resources from the natural system are utilized for the reproduction of the social systems. Thus social collapse and/or crisis of the natural world can be attributed to the excessiveness of the dynamics—accumulation of surplus, urbanization and population growth—of the social world. This is just one side of the equation. We contend that the other side of the equation that encompasses Nature–culture relations (leading to natural resource scarcities and landscape degradation) and *climate* are also key factors affecting socioeconomic and political collapse.

By the late third millennium, sailors from the Aegean were able to sail to the Syro-Palestinian coast thus linking the Aegean and Central Europe by sea with the Near East. Such types of connections foster the beginnings of a "global" division of labor from Northwestern India to the Eastern Mediterranean, and of long-distance trade articulated within a single interacting whole: the Bronze Age World System. We thus have the beginnings of a globalization process, and the emergence of the world system that started five thousand years ago.

Viewed from the perspective of Nature, such world historical processes (urbanization and accumulation) induced a continuous and degradative transformation of the landscape. Trees were removed for agriculture, and to meet the energy and material needs of urbanizing communities. The valleys were excavated for canals to provide irrigation for crops, and for the transportation of people and goods. Other lands were dug up for their natural resources and building materials. Such wide-scale human activities such as deforestation led to soil erosion in the mountains and hills, and the continuous impact of human activities further heightened the process. Rivers were dammed. In all, socioeconomic activities along with wars were transforming the landscape with scars revealing the scale of such acts.

In *World Ecological Degradation*, the level and scale of resource use by the core centers from near and afar in the third millennium BC was traced [9]. This history started on an intense trajectory from the fourth millennium onwards, and by the third millennium BC, after one millennium of drawdown of the natural capital, the natural system and social system was exhibiting signs of crisis type conditions. They emerged stretching over very long duration. Accompanying these long phases of ecological crisis were climate shifts and eruptions of natural processes that impacted on the social, political, and economic landscapes. Economic downturns followed with social-political unrest. The combination of all these conditions induces a *systemic* crisis of the world system. One such *systemic* crisis or Dark Age began around 2200 BC impacting initially Northwestern India, the Gulf, Mesopotamia, Egypt, and West Asia and had repercussions for the urbanized core areas such as Mesopotamia, Indus, and Egypt [9,63–66]. Following this phase of the crisis ending around 1700 BC, new power centers emerged in the Near East, Northern Mesopotamia, and the Eastern Mediterranean. This systemic crisis reemerged around 1200 BC at the social system level and continued until 700 BC, impacting the main areas of West Asia, Egypt, Eastern Mediterranean, and central Europe (from 800 BC onwards).

5. The Early Phase of the Near East System Crisis (2200 BC–1700 BC)

Natural System Changes

If as we have argued in the previous pages that world system crisis is also an ecological crisis accompanied with climate changes and natural disturbances (tectonic shifts, *etc.*), we should be able to find trends that reflect the ecological injuries that Nature suffered as a consequence of social (world) system dynamics of accumulation, urbanization, and population increases. The level of deforestation is a good proxy to indicate the state of the natural environment and we do find indicators of severe deforestation during the Bronze Age crisis period starting from 2200 BC onwards.

Over world history from at least 3000 BC onwards, the available forests have been intensively exploited to meet the needs of an evolving world system, starting from the core centers such as Egypt, Mesopotamia, and Harappa [9,67,68]. In the Mesopotamian case, high quality timber were sought for either through military expeditions or trade in the Zagros and Taurus mountains, the Caspian Sea area, the Eastern Mediterranean, and in Punjab [69,70]. In the Harappan case, NorthEastern Punjab (on the Siwaliks and the foothills), and the Western Ghats were the immediate areas of deforestation. Teak came from the Gir forests or from the Panch Mahals, Surat and the Dangs [71]. Timber was also sought as far as the Himalayas. The Egyptians sought their wood in neighboring areas of Lebanon and parts of the Syrian coast. For Northwestern Europe, from as early as the third millennium BC, there was extreme deforestation caused by extensive land use and animal husbandry [60] (pp. 281–292).

From an empirical analysis of the trend lines of arboreal pollen covering four geographic regions of the world: Western Europe, Central and Eastern Europe including Russia, Northern Europe and the Mediterranean, we note of severe deforestation phases [10].

The first phase of deforestation started from 3854 BC–2400 BC, and there were three/four subsequent phases of deforestation followed by reforestation that occurred towards the latter period of the course of a Dark Age. Deforestation periods are the most pertinent time points for our discussion of the ecological degradation of the early Bronze Age crisis. With one deforestation period starting around approximately 2400 BC, this dating also corresponds with Barbara Bell's [65] identification of the first Dark Age of the Ancient World. In Western Europe, arboreal pollen from areas in Belgium, Germany, and France exhibit the deforestation period starting around 2200 BC–2000 BC. In Central and Eastern Europe, trend lines of arboreal pollen show deforestation levels in areas of Hungary and Ukraine. In Northern Europe, the trend line of arboreal pollen in an area in Finland also supports this deforestation pattern. Finally, in the Mediterranean, we find areas of Greece, Spain, and Turkey exhibiting such trends. The latter area of Turkey is most pertinent for present discussion for it is where the Southern Mesopotamians sought their natural resources.

Agriculture and other anthropogenic induced changes naturally lead to forest fragmentation and deforestation, and the rise in the pollen record of indicator plants and ground weeds such as *Plantago lanceolata* [72] (p. 224); [68] (pp. 12–25). Time phases of the *rise* in the number of *Plantago* pollen when there was a *decline* in the number of arboreal pollen, supports the thesis of anthropogenic induced deforestation over five thousand years of world history.

Climate-wise, there is evidence of temperature changes (higher temperatures) and increasing drought-like conditions persisting in the Eastern Mediterranean, Egypt, West Asia, Mesopotamia, Northwestern India, Central Asia, Africa and parts of the New World starting from 2200 BC onwards during the onset of the Dark Age of the third millennium [9,73–78].

According to Fagan [77] who has argued on the impact of climate change on civilizations, this start of a warming trend again was a global event. Affected areas covered Egypt, Northern Africa, Greece, Indus, the Fertile Crescent, Crete, Russia, West Asia, and Palestine [9,65,74,77,78]. Between 2710 BC and 2345 BC, Anatolia and the Northern Crescent had arid conditions, however the Nile floodings continue to be high [79]. However, by 2205 BC, the starting time point initializing the start of the Bronze Age crisis, the Nile floods had weakened. From 2205 BC–650 BC, a period that covered the Bronze Age crisis, there was widespread aridity in Anatolia and the Northern part of the Fertile Crescent including Northern Africa [79].

For social systems with agricultural practices that are reliant on irrigation waters or from annual floods, this loss of moisture would place tremendous stress on the agricultural systems and hence, the economy and social-political stability [73]. Such was the case for the core centers of Mesopotamia, Egypt, and the Harappan civilization. Each responded differently to such stressed conditions depending on what they were facing.

The climate changes were also accompanied with the occurrences of tectonic shifts that added further strain to the social system. Tectonic shifts by themselves would not immediately impact on the reproduction of the social system unless they are in the immediate proximity of human communities or they reshape the contours of the landscape by shifting river courses. The latter is what happened in the second millennium BC By diverting watercourses, the diversions transformed some rivers into dry waterbeds that further exacerbated the already existing aridity, thus impacting on social system.

For this time period, in Northwestern India, Agrawal and Sood [80] noted of tectonic shifts that diverted the course of the Satluz and the easterly rivers away from the Ghaggar, which over time transformed into a lake-like depression during this period. The Ghaggar or Sarasvati which feeds into the Indus River was alive until the late Harappan Period (1800 BC) but was dead by the time of the Painted Grey Ware period (1000 BC). Possehl [81,82] has also confirmed this drying up of the Sarasvati and its implications for the Harappan urban complexes located on its riverbank.

Beyond the above core centers of Mesopotamia, Egypt, and the Harappan Civilization of Northwestern India, temperature changes also impacted on other ecological landscapes. In Western Asia, the introduction of Zebu cattle, which can withstand aridity, occurred during the two arid periods (2200 BC and 1200 BC) of the Bronze Age [83,84]. In central Eurasia, preliminary data also confirmed marked changes in vegetation, beginning around 2200 BC and lasting until 1700 BC [85,86]. Pollen cores indicate a sharp decrease in arboreal pollen and an increase in steppe pollen. From 2200 BC to 2000 BC, there was a severe drop in forest cover and an increase in steppification, leading to an expansion in steppe landscape from 1800 BC to 1700 BC. The pollen profiles for the region discussed in the previous section also confirmed the deforestation process. Arid conditions also affected arable land, which caused severe pressure on animal husbandry of the steppe population. The lush feather grass steppe growing on the landscape near Kalmykia for example, from 2500 BC–2200 BC gave way to dry scrubby vegetation—wormwood steppe—and even desertification by 2200 BC–1700 BC. This changed ecological landscape led to outmigration of the sedentary population from river valleys with time, and exploitation of the steppes for animal feed.

6. Socioeconomic and Political Transformations 2200 BC–1700 BC

Socioeconomic and political trends during Dark Ages are reversals of what occur during periods of expansionary growth.

6.1. Deurbanization and Migration

Tracking the reversals in socioeconomic trends during Dark Ages or over the very long-term requires considerable effort, especially when the quantitative data are sparse. Some recent attempts such as that of Modelski [37] and Thompson [40] on urbanization and economic expansion have provided us with some broad contours on these processes. In terms of urbanization, by 3500 BC for the "heartland of cities" such as Southern Mesopotamia, urban growth had progressed to such an extent that it had three cities with population at or over 10,000 [37].

By 2500 BC, during the period of Early Dynastic III, the rise of Sumer exhibited the largest urban conglomeration at 60,000 persons. Uruk by this time had been reduced to a population of 40,000 in comparison to 80,000 in 2800 BC [37] (p. 28). However, the total urban population of Mesopotamia at 2500 BC had reached 290,000 [40]. Outside of the "heartland of cities", we find Ebla located in Northern Syria with a population of about 40,000 and Mari in Northern Mesopotamia with a similar population size. Elsewhere, we have Memphis in Egypt at 30,000 persons, Mohenjo-daro and Harappa in Northwestern India at 20,000 and 15,000, respectively.

By 2200 BC, the Akkadian period and the start of the Dark Age Phase 2, the total urban population of Mesopotamia had been reduced to 210,000. This shift is also reflected in the proportion of declining urban settlement sizes. During the Early Dynastic Period II/III (2800 BC–2300 BC) the percentage of urban settlements with more than 40 hectares was about 78.4%, by the Akkadian Period (2200 BC) it had been reduced to 63.5%. Further deurbanization continued that in the Ur III and Isin-Larsa periods (2100 BC–1900 BC), and the percentage dropped further to 55.1%. This slippage continued to the Old Babylonian period, reducing further to 50.2% (1600 BC). Conversely, non-urban settlement sizes (10 hectares or less) increased. During the early Dynastic II/III period, it was about 10%, almost doubling by the Akkadian period. With the arrival of the Ur III and Isin-Larsa periods, the percentage had risen to 25%, and almost tripled to about 29.6% by the Old Babylonian period in comparison to the Early Dynastic period. This deurbanization process and migration to rural communities are also supported by the population decreases in Mesopotamian cities. From 210,000 during the Akkadian period (2200 BC), the population in Mesopotamian cities was reduced to 190,000 by the Isin-Larsa period (1900 BC). This was a loss of 10%. The population level was reduced further to 70,000 by the Old Babylonian Period (1600 BC). Overall therefore, between the start of Dark Age Phase 2 (Akkadian Period) and its end around 1700 BC (Isin-Larsa and Old Babylonian periods) we see a loss of over 66% of the urban population in Mesopotamia.

Deurbanization and population losses were also repeated in Northwestern India. According to Possehl [81,82], by the late third millennium BC there was evidence of abandonment of important buildings in the highly urbanized setting such as Mohenjo-daro where we find the Great Bath and the Granary devoid of human use. Concurrently, the Sindh region and the Baluchi Highlands also witnessed depletion and deterioration. By the early second millennium BC, Baluchistan was uninhabited. Cholistan, in Northwestern India, experienced a drop in size in terms of settled areas from an average of 6.5 hectares in 3800 BC–3200 BC to 5.1 hectares by 1900 BC–1700 BC, and finally to almost 50% less (2.6 hectares) by 1000 BC [81]. In the Sarasvati region, the shifting and drying up of the river system saw the abandonment of settlements in the inland delta of Fort Derawar. The latter area was the breadbasket of the Mature Harappan civilization.

Elsewhere for the time period of 2200 BC, similar signs of deteriorating conditions were also encounters in Anatolia, with abandonment of urban centers such as Troy II to Troy III–IV [48,87] (p. 139–152). Consequently, depopulation also resulted. Sedentary population settlements on the Anatolian plateau were also abandoned. To the west of Anatolia, Palestine also suffered such crisis conditions [88]. Walled towns were replaced by unwalled villages. There were signs of cave occupation and migratory movements. In some areas, settlements completely disappeared, and remaining settlement sites were reduced by more than half of what existed before 2200 BC [89] (pp. 1–38). Across the Mediterranean from Palestine, the Aegean experienced distress, though to a lesser extent. Between 2300 BC and 1900 BC there was a loss of sedentary population. Such losses were experienced both on mainland Greece and even Crete [90,91].

For central Eurasia similar stress conditions also prevailed. The changed ecological landscape led to out outmigration of the sedentary population from river valleys over time, and exploitation of the steppes for animal feed. Denucleation occurred with the establishment of smaller communities near oases. This spread occurred in Central Asia at Korezm (south of the Aral Sea) and Margiana (Murghab Delta) in Turkmenistan, Bactria, and Western China. This process prompted by ecological degradation and environmental changes, also occurred in Syria and Jordan. Migration out of urban centers located on the coast to the interior, and the establishment of smaller village type-settlements resulted [92] (pp. 267–273).

6.2. Political and Social Changes

If we examine the Dark Age of 2200 BC–1700 BC, political instability is one feature that highlights the political economic events. Climate changes as identified above led to famines that in turn generated political upheavals and the dissipation of central authority in Egypt. Drought conditions and lowered Nile flooding impacted on the farmers' ability to pay taxes because of lower harvest yields. This resulted in local administrators and governors, who collected taxes, having to delay their transfers to the Royal House. In turn, the King's revenues plummeted, and thus impaired his ability to pay for an army or to deal effectively with drought and famine. As a result, the stability of the political regime was affected. The sum effects of this in terms of political stability, as Bell [65] has concluded, were short reigns.

Besides political instability in Egypt during the third millennium Dark Age of 2200 BC, other reversals also occurred such as artistic degeneration and the downsizing of monumental buildings as a result of diminishing resources. The size and elaborateness of the pharaonic tombs were reduced; by this time, the tombs of kings were one-chambered affairs with less ambitious layouts [93] (pp. 316–319). Boundaries of provinces were also closed to prevent mass migration out of famine stricken areas. All these initiatives proved fruitless at times as riots broke out along with the ransacking of granaries.

In other parts of the system such as Southern Mesopotamia and Northwestern India, structural political, economic and social reversals were also occurring. These transformations were extremely impactful in view of the trading relations of the region among Southern Mesopotamia, the Gulf, and Northwestern India, and led to the demise of social systems in place with repercussions system-wide. By the third millennium BC, Southern Mesopotamia, the Gulf region and Northwestern India were linked in a trading network of commodity exchanges. Therefore, a crisis in one part of the system would also mean a translation of this stress to other parts of the system. Therefore, an ecological stress in Southern Mesopotamia would mean a lowering of agricultural output or production, and hence a drop in imports and demand. Reductions of demand in Southern Mesopotamia would impact on other regions such as Dilmun in the Gulf and the Harappan civilization through a diminished demand for their material and goods. What this type of dynamics further suggests is that supply and demand might not necessarily be a consequence of the state of the economy or based on consumer tastes and needs. But rather, supply demand dynamics are inextricably linked to the connections between the natural system and social system. Thus, anthropocentric explanations provided for systems demise have ecological roots.

Southern Mesopotamia by 2200 BC was experiencing salinization problems leading to lowered agricultural productivity and this became acute by 1700 BC. It never recovered from the disastrous decline in agricultural yields that accompanied the salinity issue. Deurbanization was the order of the day as we have indicated previously. Urban life and culture continued on a declining scale with the population concentrating only in major towns [94]. The Harappan civilization not only had to face ecological stress, climate changes with temperature increases and arid conditions like Southern Mesopotamia, it also had to undergo tectonic shifts. As the urbanized communities of the Harappan civilization were linked to the overarching Gulf trade and beyond, its infrastructure and surrounding hinterlands had therefore developed and specialized in the manufacture of products and natural resources for export. Thus, when its exports to the Gulf and beyond disappeared, it could no longer reproduce the accumulation process that had sustained its urban growth. This led to migration to the rural areas of the North and south.

7. The Final Phase of the Near East Bronze Age Crisis (1200 BC–700 BC)

The demise of Southern Mesopotamia and Northwestern India coupled with the socioeconomic and political upheavals in Egypt and their associated hinterlands from 2200 BC to 1700 BC initiated a significant system crisis of the Bronze Age. With the socioeconomic collapse of Southern Mesopotamia and Northwestern India, the demise of these economies meant also the breakdown of the Gulf trade.

After 1700 BC, at the social system level, despite the fact that ecological stress (at the natural system level) continued as reflected in the arboreal profiles listed in [10] (p. 49–53), economic recovery resumed. With recovery, other parts of the Bronze Age system such as the Eastern Mediterranean littoral (centered around Crete and mainland Greece) along with central Europe and Anatolia increasingly began to take advantage of the vacuum generated by the collapse of the Southern portion (the Gulf Trade) of the Bronze Age system. Egypt, Syria-Levant (such as Ugarit, Mari, Byblos, Ras Shamra), Crete, Cyprus and mainland Greece expanded their trading volumes utilizing the peripheral areas such as Central and Eastern Europe, and Nubia for their resource needs [9,95]. With the loss of trading dominance of Southern Mesopotamia, Mesopotamian trade shifted Northwards, thus making Anatolia an important Eastern node of this Bronze Age trading network [9]. In sum, the Eastern Mediterranean littoral became the prime axis where economic activity of the Bronze Age system concentrated during this period.

The social system adaptation and resolution of the crisis of the Bronze Age that started at 2200 BC and ending around 1700 BC was only *temporary*, for over *la longue durée*, because of the continued ecological stress and degradation including climate changes, social (world) system crisis would only appear again in 1200 BC The system crisis of 1200 BC repeated what occurred in 2200 BC except that when it finally dissipated we have system transformation with the arrival of the Iron Age.

The collapse of the Southern portion of the Bronze Age world system led to the reconfiguration of the trading networks. Shifting away from the Gulf region, the trading networks range from Crete, the Cyclades, and the Greek mainland on one side of the Aegean Sea with Troy, Cyprus, and Anatolia located across from it. Included in this configuration were the communities of Syria and Palestine, and the kingdom of Egypt. This network of socioeconomic exchanges of the Eastern Mediterranean region was also linked to communities of Western, central, and Eastern Europe, and Central Asia [61]. It was a globalized system of trade and sociopolitical exchanges.

Within this trading network, intermediary centers such as Crete increasingly played a part in the Eastern Mediterranean. The Minoan command-palace economy was involved not only in the export of surplus agricultural produce such as grains and oil, but also in the export of textiles, metal works, pottery, wood work, *etc.* [9,61]. Initially, such a diversified economic structure provided it with a competitive advantage over other regions of the Bronze Age system such as mainland Greece, the Cyclades, and Europe to the North.

Later in the millennium, the rise of Mycenaean Greece in this era increasingly eclipsed the role Crete played in the Easternmost region of the Bronze Age world economy [9]. On this trading backbone, Mycenaean Greece began to establish its economic dominance within the Aegean. Similar to the Cretan economy, Mycenaean Greece exported wine, olive oil, grains and manufactured products to Eastern and Northern Europe, and the Eastern part of the Mediterranean, and in turn, received needed natural resources such as copper, tin, and horses. To the East of this globalized Bronze Age trading system was the kingdom of Hatti with metallic resources such gold and silver whereby these precious metals were exchanged for textiles, lapis lazuli, olive oil, grain, horses, tin, *etc.* Trade contacts were established with Babylon, Mittani, Assyria, Syro-Palestine, Egypt, and Crete.

The globalizing trajectory was extended starting as early as 2000 BC onwards when these cores in the Near East as Kristiansen and Larsson [61] (p. 99) put it "turned their interest towards the barbarian peripheries in Central and Western Europe" for their natural resources and livestock such as horses. In the Caucasus, the mines supplied the copper, and there was the development of a Circum-Pontic metallurgical province that included Anatolia [61,96–98]. With such development, the central and Western European metallurgical centers were "increasingly drawn into trade relations with the palace cultures and city states of the Eastern Mediterranean and Anatolia, which reached a new flourishing after 2000 BC when the Minoan palaces were built" [61] (p. 104).

It should not be assumed that these trading networks were stable structures over time. It was a globalized system of interconnected regions and polities—a world system. Their vitality and concentration changed over time, and were conditioned by the pulsations of ecological and climate changes, notwithstanding political and economic ones. Thus, when the Dark Age returned in 1200 BC, the collapse was system-wide due to the level of connectivity.

7.1. The End of the Bronze Age in The Near East

If 2200 BC was the start of system crisis at both the natural and social systems levels, and with the natural system in continuous crisis throughout the late Bronze Age, 1200 BC signaled the beginning of social system transformation leading to the end of the Bronze Age. Starting from about 1200 BC, socioeconomic and political collapses during this period ranged from Mycenae through Egypt, the Levant, and Northern Mesopotamia to Anatolia. With the exception of parts of the periphery, the core centers of the Bronze Age system at this point in time were in crisis.

The collapse of the Bronze Age world has been explained rooted on a variety of factors. On the whole, they have been rationalized and based on anthropocentric ones such as barbarian invasions, unceasing consumption and cultural decadence, power rivalries and state competition, vagaries of development, overcentralization of authority, military and weapon innovations, and famines and diseases (see for example, [52,63,99,100]). Without a doubt, these factors at the social system level are ones to consider. However, what is lacking is a consideration of the *linkage* between the social system and the natural system, and how a disruption of this connection would ultimately induce crisis conditions, for the former (the social system) depends on the latter (natural system) for its continued reproduction. It is to underscore again the viewpoint that in the last instance it is perhaps Nature that has the final say!

For Crete, the intensive exploitation of resources for economic transactions impacted on the landscape. Deforestation generated soil erosion and flash flooding; the latter impacted on the manufacturing processes of Crete. Wood scarcity forced changes in production locations or resulted in the closure of facilities. It has even been suggested that such land deterioration contributed to the demise of Minoan Crete [101] (p. 68). These ecologically devastating trends were also repeated throughout the Bronze Age system. Mainland Greece—which provided the wood supplies to Crete when Crete's supply ran out—and other areas in Europe and Central Asia showed such scars as well. Intensification of land use and animal husbandry led to severe alteration of the landscape. Population increases along with the adoption of the ox-drawn plow further exacerbated the intensity of land utilization. Pollen record from Osmanaga Lagoon in southwest Greece in Messinia shows extreme forest removal by 2000 BC [102] (p. 5). Between 1600 BC and 1400 BC, the pine forests in Messinia were totally wiped out due to agriculture and overgrazing. Soil erosion was endemic and was controlled by terracing and the building of terrace walls. As a consequence agricultural production was affected. In the Argolid, production of cereals and olive oil generated deforestation of oak trees on the hillsides. It resulted in large amounts of earth and water draining from the slopes onto the plain of Argos and filling up stream beds leading to extensive flooding [103].

Besides terrace walls to deal with soil erosion, other technological solutions were also tried, such as the building of dams to divert water courses and dikes to facilitate drainage [104]. However, by the late second millennium BC such efforts began to fail. Erosion became uncontrollable during the Dark Age crisis of 1200 BC when socio-political life was at a standstill and population density had dropped precipitously [105].

With scarcity of wood for fuel, metallurgical and pottery works were affected which resulted in further population decline. Population migration followed the closure of these manufacturing centers, and the abandonment of Phylakopi coincided with the deforestation of Melos, where the town was located. Towns and settlements disappeared. In southwest Peloponnese, the number dropped from 150 to 14. Other regions experienced similar declines; Laconias, Argolid, Corinthia, Attica, Boeotia, Phocis, and Locris all registered losses [9,67].

In Southern Europe, there were also degradative impacts on soil formation from Urnfield settlements [59]. Kristiansen and Larsson [61] have also documented widespread ecological degradation in the Caucasus as a result of mining for metals to supply the Eastern Mediterranean and the Near East. It was also repeated for the mining area of Kargaly in the Urals that supplied metals to the whole steppe region where deforestation was the consequence according to Kristiansen and Larsson [61]. Time-series of arboreal pollen profiles of Central and Eastern Europe including Russia parallels the deforestation trajectories.

Collapse came for the Eastern Mediterranean when circumstances started to change. Ecological stress coupled with climate changes and natural disturbances impacted on Crete, Greece and the Near East. For Crete, such arid conditions impacted on agricultural production. This development was serious, as it was an important part of Cretan exports needed to offset its import of wood and other natural resources. Furthermore, geological conditions also provide grounds for the arguments made first by Marinatos [106] and followed by Chadwick [107] and Warren [108] on their impacts on Crete. The volcanic eruption on Thera following the earthquakes killed vegetation and destroyed the Minoan naval fleet. The loss of this fleet undermined Crete's power to exercise its dominant position in this region of the world system. In addition, these natural system conditions should also be considered with the political changes impacting on Crete. From 1500 BC onwards, the increasing competitive roles played by the Hittites and Kassites through their expansion and dominance of Anatolia and Mesopotamia corralled Crete's dominance. Blended into this political mixture, the ascendancy of Mycenaean Greece eclipsed the economic position that Crete enjoyed. Furthermore experiencing the loss of their sources for natural resources located on the Greek mainland that by this point in time were increasingly under Mycenaean control; Crete's reproductive capacity was stretched. Faced with these desperate conditions in the spheres of the social system and the natural system, Minoan civilization slid downhill.

What occurred in Crete was repeated in Mycenaean Greece except it was much later starting around 1200 BC By this time the natural environment was severely stretched. Rhys Carpenter's [109] thesis of climate change leading to the demise of Mycenaean Greece needs to be considered. Basically, Carpenter's proposal is that with the shift in the tracks of the cyclonic storms, which normally bring rain to Mycenaean Greece, arid conditions resulted during the 13th to the 12th centuries. As a consequence, the socioeconomic structure was impacted. Chadwick [107] and Drews [110] have challenged this thesis with Lamb [111], Braudel [58], Bryson *et al.* [112], and Bryson and Padoch [113] supporting Carpenter's position.

Along with these desperate conditions due to climate changes and natural disturbances, invading forces of Dorians and Sea Peoples made the circumstances even more dire. Such invading forces most likely have also been displaced from their habitation due to changing climate conditions and natural disasters. Climate changes and disruptions in trade routes also played a part in the overall reproductive capacity of the Hittites and the Egyptians in the other parts of this system. Lowered Nile flows affected Egyptian agriculture leading to famines. The Hittite Empire's grain shortage led to growing imports from other parts of the world system through Ugarit, and from the Syro-Palestine area. In all, the crisis was system-wide affecting the Aegean, Central and Western Europe, Egypt, Anatolia, Palestine, and Babylonia.

7.2. Socioeconomic and Political Transformations

Greece encountered a decline in socioeconomic life from 1200 BC till 700 BC, such as decline or loss of certain material skills, decay in cultural aspects of life, a fall in living standard and thus wealth, deurbanization, population losses, and loss of trading contacts within and without Greece (see for example [9,49–51,63,114–118]). The archaeological evidence unearthed suggests socioeconomic patterns that are distinctively different from the style and level of socio-cultural life prevailing prior to the onset of the Dark Age.

Population decreases occurred between 1250 BC and 1100 BC Morris [114–116] has estimated losses of about 75% followed with emigration from the core areas of the Mycenaean civilization, this trend continued for central Greece as well by 1100 BC According to Snodgrass [50] between the 12th and the 11th centuries, there was a reduction of over three quarters of the population.

Pottery and other objects recovered from excavated sites along with the architecture and design of dwellings reflect ecological stress and scarcity of natural resources. Architectural standards were lowered and there were very few signs of good stone-built construction. Small stone construction was prevalent, and we also increasingly see signs of mud-brick construction. Mud-brick structures predominated in the building structures between the 11th and 10th centuries. The emergence of a class of handmade burnished pottery, "Barbarian Ware", had few obvious links to Mycenaean styles. The appearance of this style has been attributed as an economic response to the collapse of centralized production with the demise of the palace economies, and a regression to simpler technology [119]. Furthermore, pottery styles of the period in Greece became austere, unlike the decadent style of the previous era.

Starting with the Submycenaean style of pottery (*ca.* 1125 BC–1100 BC), the austerity of the design can be seen. As Desborough [49] has put it, the standards deteriorated sharply not only to the making of the pottery but also to the painting and decoration. The design was of the simplest kind and "was a virtual bankruptcy . . . and often carelessly applied" [49] (p. 41). The variety of styles in terms of vase shapes of this type of pottery was also reduced. Rutter [120] has also suggested that luxury vases and other pottery items were quickly abandoned as necessary frills when hard times hit. There was less variety of material goods, the artifactual correlate of a less complex social order. The emergence of the Protogeometric style (*ca.* 900 BC) continued to reflect the austerity of the period [50]. Snodgrass [51] has also alerted us to the appearance of hand-made pottery during the Dark Ages. The reversion to hand-made pottery when the pottery wheel had been adopted previously suggests to us the decay of manufacturing production or even perhaps the loss of manufacturing skills. It could also mean that with social decay and collapse, there was a revival in the utilization of indigenous material in view of the disruption in trade routes.

In terms of decorations and finishing, the bulk of the pot or vase was usually left plain in the natural color of the clay and the decorations covered a third of the surface area at most [49]. The lack of intense firing also suggests to us dwindling energy supplies. The compass and the multiple brush were used for decorating the pottery. As recovery proceeds and the balance of Nature is restored, we find the plain, rectilinear or curvilinear patterns in pottery designs giving way to images depicting animals and humans. In the later Protogeometric style period, we already saw the introduction of silhouette figures of a horse or a human on the design. If we consider the decay of cultural life and the loss of the art of writing, and view pottery design as a way the potter as artist could depict sociocultural life then, the motifs that we find in these pottery designs would summarize life in Dark Age Greece. By the late Geometric style period, we find scenes of organized groups of men in uniforms, the portrayal of warfare and chariots depicting social life when the Dark Age was receding, and the return of biodiversity with animals and sea creatures being depicted.

Beyond pottery styles, other objects recovered indicate a scarcity of natural resources, especially metals, or that the supply sources had dried up. The use of obsidian, stone, and bones for blades and weapons underscores such scarcity, and also suggests that trading routes and centers for sourcing the metals might have disappeared or disrupted. Other primitive materials reappear as apparent substitutes such as bone spacer-beads for amber in jewelry, and stones were used to replace lead in sling bullets. Objects buried with the deceased increasingly were made out of iron such as iron pins and fibulae and even weapons, which all in the past were bronze, and bronze wares only returned towards the end of the Dark Age period [51]. Where bronze was used, it was found on the bulb of pins thus revealing the scarcity of bronze [49,50].

Ecological scarcity required a downscaling of material and cultural lifestyles. Such changes are reflected in burial practices that exhibited a reorganization of life along modest lines. The design of clothing and shoes was of the plainest kind [50]. A one-piece woolen garment without requiring cutting or sewing gained popularity among the female population in Submycenaean Athens and became the predominant dress design in the Protogeometric Period. Pins for dresses were scarcely used. The downscaling process is exhibited further in the formation of decentralized communities and associated population losses. The collapse of the palace driven economies with centralized monarchies were replaced by smaller political organizations dominated by an aristocrat and his family.

Whether this life-style is one that was actively sought as a consequence of ecological scarcity or occurred as an outcome of the depressive conditions of the Dark Age is difficult to gauge. It is clear however, that there was a shift from the Mycenaean way of reproducing life for they longer provided practical models. The loss of sophistication is clearly seen and as Morris [116] (p. 207) has stated, "in their funerals people seem more concerned with showing what they were *not* than with what they were". What we are sure of is that as recovery proceeded—we begin to witness this by the mid-half of the 10th century BC—trading networks were re-established and communities revived. Such an upswing was characterized by exuberance, materialistic consumption, and accumulation. As the social system recovered, we see the rise of the Bronze industry, increasing quantity of pottery buried in the tombs, the quantity of gold deposited in the burials, and signs of social cultural recovery. During the Dark Age, materialistic consumption declined, and most of the trading networks disappeared or were restricted only to the area of the Aegean Sea.

What the Dark Age of this period represented for the Mediterranean region is one where extreme degradation of the ecological landscape precipitated socioeconomic and organizational changes to meet the scarcity of resources so as to reproduce some semblance of cultural and economic life of prior times. As a consequence, systemic reorganization occurred at various levels, from the way commodities were produced to clothing fashions and designs. Hierarchical social structures disappeared during the Dark Age, as evident by burial practices, and were restored when recovery proceeded [117]. To Whitley [117] (p. 20) burial practices "may be seen as an expression both of social relations and ideology ... " During the Dark Ages, there was a shift from multiple tombs burial to single burials which reflected the change from an emphasis on heredity signifying a stratified order with ruling classes to one which reflect no expectations of descendants and little regard for extravagance [51]. The single tombs lack monumental significance and architectural quality. From the graves excavated of the Protogeometric Period (*ca.* 900 BC) there are no indications of disparities in wealth and social distinction, as exemplified in the Athenian graves. Distinction was based on age and sex rather than other social dimensions [117] (p. 115). This was to change by the Early Geometric Period (*ca.* 860 BC–840 BC) where there is an amplification of status of the person buried. Social and sexual identities of the person interred became more evident. Thus, we find the return of a hierarchical pattern and a departure from the more egalitarian structure of the Protogeometric Period. Such hierarchization continued in the Middle to Late Geometric Periods (*ca.* 770 BC–700 BC). By this period however, there was also a breakdown of the aristocratic order with the arrival of early state formation, though social hierarchical differentiation remained in place.

With the Dark Age, not only was there a loss of population, but deurbanization was also underway. The latter process continued giving rise to small communities with lower population levels [90,91]. Seen from an ecological point of view, this downscaling provided the necessary timing for Nature to restore its balance, and for socioeconomic life to start afresh when recovery returned. The collapse of the palace economies enabled the ecological landscape to restore itself that in the past were intensively exploited by the palace driven economics. In the Argolid and Messenia, according to Deger-Jalkotzy [121] (pp. 123–124), the land recovered and the tree population increased. Furthermore, with the loss of centralized control from the various palaces, not only deurbanization occurred but also decentralization. Each region thus had the opportunity to search for new mechanisms and ways to administer and reproduce socioeconomic life in general. New trends emerge following the collapse of the palaces as a consequence of the unexpected liberty that resulted from the collapse, and each region/community began to make contacts with others outside Greece towards the end of the Dark Age.

From these small communities, in the case of Greece, the preconditions for the rise of the Greek polis (cluster of villages) were put into play, and what followed was a flourishing of political and economic life as soon as the social system recovered [51]. Muhly [122] (p. 20) has put this in a succinct fashion: "the importance of the Dark Age, then, must be that it created the conditions that made possible the growth of this distinctly Greek political organization". To this extent, the stressed ecological conditions that engendered deurbanization and the formation of small isolated communities precipitated the rise of the polis and the Greek city-states. We need to realize, therefore, that perhaps scarcity of resources can also have productive outcomes which otherwise under bountiful conditions might not have occurred. Stanislawski [123] (p. 18) has suggested that instead of seeing the Greek Dark Ages as a period of darkness it should be seen as one of enlightenment with contributions such as: the first use of stone-walled agricultural terraces, the use of chicken eggs in domestic diet, the beginning of the spread of alphabetic writing, the spread of iron, the general use of olive as food, and the first use of waterproof plaster.

Systemic reorganization occurred, and the lengthy duration of the Dark Age is one that we need to note. The fact that it is of such a long duration underscores the length of time required for ecological recovery to take place, and the immensity of the degradation that occurred. What followed in the recovery phase, however, was a Dark Age-conditioned social-cultural and political lifestyle that formed the basis of Western civilization as we know it today.

Given that Dark Ages in world history are significant moments signaling system crisis and system reorganization; the final phase of the Bronze Age crisis led to ecological recovery, certain political-economic realignments and reorganization, and the transition to a new working metal: iron. The Dark Age crisis was *system transformative* for it led to fundamental social system changes evolving to a set of new patterns [25,26,96].

The adoption of iron brought to an end centuries of bronze use that was in the control of palace economies and elites. Gordon Childe [63] has suggested that cheap iron with its wide availability provided the opportunities for agriculture, industry, and even warfare with the adoption of iron as the base metal. With trade route disruption and copper scarcity, the adoption of iron use spread further, especially among the communities in Greece that was isolated as a consequent of Dark Age conditions, for iron was available locally. It led to the development of local iron producing industries [51]. The low cost of iron because it was available locally facilitated its widespread use in agriculture and industry [63,124]. Cultivation was made easier with iron plowshares in heavy clay soils. This enabled the rural communities to participate further in the economy beyond subsistence, and in maintaining a class of miners, smelters, and metal smiths fabricating the iron implements to reproduce material life. Such an explanation is also supported by Heichelheim [125] and Polanyi [126], who have suggested that the widespread adoption iron was the result of the opportunity for rural communities in south Russia, Italy, North Africa, Spain, Gaul, Germany, and Eurasia to work the heavy soils with iron implements, thus increasing their production levels. Production increases can be seen by the fluctuations in grain prices according to Heichelheim [125]. The consequence of such transformation is that the urban elites in the Near East who in the past controlled the grain and other commodities trade suffered losses as a consequence of changing prices, and the falling demand for copper, tin and bronze, which they also controlled.

As a result of the above, the social structures were transformed with the formation of different regional centers in the periphery and in the Mediterranean. The opportunity for the farmers to farm in heavy clay soils utilizing cheap iron implements also provided the conditions for economic and system expansion following the end of the Dark Age where in the past these areas were not as productive. It enabled economic expansion, and the move into newer areas for agriculture as by this time some of the older settled areas were ecologically degraded and overworked.

In addition, at the social system level, the Dark Age crisis thus usher forth the dissociation of high value commodities away from the control of the palace/state, for by the end of the Dark Age, the command palace economies were in the Eastern Mediterranean were dissolved. What emerged was the continued differentiation of commercial/economic structures from the political structures [126]. Instead of bureaucratic palace centered trade, we see the development of mercantile city-states where merchant enterprise replaced the palace-controlled exchange. With this transformation, new forms of political powers and structures emerged. We have the emergence of a new political structure, the city-state (polis) in the Aegean, and the continuation of empire type political structures where the rule was via direct political and military control.

The new political structure, the polis, as a social organization and political concept emerged in 8th century Greece [114,115]. It was, as Morris [115] (p. 752) has stated, unique among ancient states for "its citizen body was actually the state". The rise of such a state form was a consequence of the collapse of the aristocratic society during the Greek Dark Ages. Other factors also precipitated its formation. Deurbanization and the loss of population in the urban areas resulting in the development of isolated communities during the Dark Age engendered the structural conditions for the development of the polis. In addition, with the scarcity of resources and the abundance of poverty leading to less hierarchical social structures, the groundwork for the development of the polis was also put into place. The polis thus was one where all authority was divested to the community unlike previous political forms in Mycenaean Greece. Force, therefore was located in the citizen body as a whole, and thus there was little need for a standing military. Individual natural rights were not sanctioned by a higher power and the highest authority was the polis, *i.e.*, the community. Such a political structure found expression in the Aegean. However, in other parts of the Near East, divine kingship was maintained with some minor modifications. According to Childe [63], Assyria, Babylonia, and Egypt continued as Bronze Age states.

Recovery returned around 700 BC with social systems expanding and growing in complexity again. Expansion came first in the form of colonization by the Greeks in two phases. Between 775 and 675 BC such expansion was for agricultural purposes, where the soils and lands of Greece which were degraded after centuries of erosion and intensive cultivation could no longer produce to meet the needs of the population. The excessive population mostly comprised of poor peasants who were turned into tenant farmers (*hectemores*) with debts that were increasing, and thus forced to swell the cities. With the state of the degraded environment in Greece, with the exception of Boetia, Attica, and Sparta where internal colonization was still possible with some fertile agricultural land left, expansion of the system came with migration to other arenas such as Italy, Sicily, Southern France, and West Asia. Growth in this case comes from a colonization process that was extensive in nature, and a consequence of the ecological crisis of the Dark Age that has just ended. Following the success of the agricultural colonization strategies with surplus generation, a second round of colonization from 675 BC to 600 BC followed, mainly focusing on commercial activities. With this phase of colonization, trade routes were further fixed and strengthened. Wealth for the colonial cities was derived from agricultural exports, trade and production. Other growth poles of the system then were Egypt, Persia and Phoenicia, and as Braudel [58] (p. 225) puts it, the Mediterranean never became a "Greek Lake." With these different centers, no polity ever gained control of the Mediterranean. It was only the arrival of Rome that the Mediterranean became a Roman sea. The growing rise of Rome and the demise of Greece did not interrupt the continuous degradation of the environment [9]. Forests were removed in Northern Africa and almost everywhere Roman rule was established. Mines were dug in Spain, with cities, roads, and production facilities established within the Roman Empire. Crisis emerged again 700 years later, around AD 400 with similar trends and tendencies in terms of ecological and socioeconomic variables like that of the Dark Ages that occurred during the Bronze Age. This time the collapses were not Mesopotamia, Harappa, Mycenaean Greece, Crete or the Hittite Empire, but it was the Western portion of the Roman Empire and the system of the Iron Age.

8. System Crisis in Bronze Age East Asia

The archaeological record reveals the presence of human communities from the Korean peninsula throughout Manchuria, the Gulf of Bohai in East Asia, the Yangtze river, and the islands of Japan as early as the Paleolithic period in 3000 BC [127]. Beyond indication of the dispersal of human communities across East Asia, there is evidence to substantiate the claim that contacts between these communities predate Dark Age periods [127,128]. In particular, one can surmise that due to geographic proximity the peoples of present-day China and Korea have an early history of economic and cultural exchanges. Early in the prehistory of East Asia, one observes contacts among Yemaek, Mongol, Manchu, Han, and other Northern tribes in the Korean Peninsula [129]. In addition, Chinese records indicate the habitation of local tribes such as the Puyo, the Okcho, the Yemaek and the I-Lou in Korea [129]. Furthermore, during China's Shang Dynasty (1600 BC–1046 BC), the Chinese settled at Lolang near modern day Pyongyang. In fact, around 1200 BC, in Northwest Korea, a state was founded under Chinese rule. The Chinese presence was further solidified on the Korean peninsula in 109 BC with an invasion that established four commandeered centers of Chinese administration at Nangnang [130].

8.1. Socioeconomic and Political Connections

Archeological research of the region reveals, five major trade routes in East Asia well in use prior to the 6th century BC: (1) the North route from Siberia; (2) the Korean route via its peninsula and across the Tsushima or Korean Strait; (3) the Jiangsu and Zhejiang route across the East China Sea to Kyushu; (4) the Taiwan and Fujian route via Ryukyu Islands to Kyushu; and (5) and the south sea route from the South Pacific via South China Sea islands to Manchuria [127,131–133]. These major trade routes physically illustrate the ability of human communities early in East Asian pre-history to engage in cultural and economic exchanges. Moreover, the presence of minor routes, or sub routes, through the Korean peninsula also further provided the linkages that connected the Asian mainland to the Japanese islands [127].

Although different explanations have surfaced regarding the flow of goods and people in East Asia, from an archeological perspective the South China Sea was a major route for cultural and economic exchanges [128]. This interaction is observed in the similarities exhibited in jade jewelry, lacquerware, agricultural cultivation, construction, and crops along the East China Sea route [128]. The cultural and economic linkages between China and Japan become evident in the presence of *ge*-shaped large earthenware pots, *yinwen* pottery, circularly-arranged tribal houses and mound-shaped graves in Japan [128]. Scholars note the similarity between items found in Japan and those in the lower Yangtze basin. Further back in history to the New Stone Age, some 7000 years ago, excavation at the Hemudu site in Eastern China revealed the presence of wooden oars and clay boat models [128]. During the same period, similar sites in the Zhoushan Islands off the coast of Zhejiang Province also reveal the ability of ancient communities to travel via waterways [134]. The route from the lower Yangtze basin via the East China Sea to Korea and Japan, was the preferred course of travel in the Late Shengwen period, during the tenth century BC, and became more popular during Japan's late Jomon and Yayoi periods (1500 BC–AD 500) [128]. Agricultural exchanges were not solely confined to seed or crops, but archeologists argue that the origins of the Japanese stone ax, ploughshares, hoes, and crescent-shaped harvesting knives can be traced to the Yangtze basin [128]. Physical evidence suggests that interactions between human communities in East Asia continued and intensified throughout the Bronze Age, and into the Iron Age.

Early iron use in China can be traced back to the Shang period (1766 BC–1122 BC) in a comparable sense to the utilization of iron in the West. Specifically, meteoritic iron was utilized, and was occasionally used in later periods [135–139]. The diffusion of iron throughout East Asia can be tied to the smelting of ore in China's Southern provinces. In this regard, Huang Zhanyue [140,141] provides evidence, along with persuasive arguments, that the smelting of iron in China began in the south and spreads to the Korean peninsula and Japan. Wagner [142–144] reviews Huang Zhanyue's [140] evidence and, coupled with other archeological data, has suggested that iron artifacts can be dated to as early as the Zhou Dynasty, and specific pieces may yet prove that the use of iron can be traced to earlier periods [145].

Chinese involvement on the Korean peninsula from the Shang to the early Han Dynastic period (1600 BC–AD 9) undoubtedly came to affect the political, cultural, and economic life of its inhabitants. China's influence was intensified as successive Chinese kingdoms emerged and sought to expand their political control. For example, during the early Han dynasty Chinese commandeering centers were established within the present-day geographical territory of Korea. Of these commandeering centers Lolang formed the core of Chinese colonial administration during the Han period [142–146]. Lolang not only served an important political role, but also proved to be an important economic point whereby Chinese goods could be distributed to points across Korea and Japan. As the Chinese administration center on the peninsula, Lolang "was in essence a Chinese city where the governor, officials, merchants, and Chinese colonists lived. Their way of life in general can be surmised from the investigation of remains unearthed at T'osong-ni, the site of Lolang administrative center near modern Pyongyang. The variety of burial objects found in their wooden and brickwork tombs attest to the lavish life style of these Chinese officials, merchants, and colonial overlords in Lolang's capital" [146] (p. 14).

Political administration by China of portions of the Korean peninsula influenced not only those populations under direct Chinese rule, but communities further South and East who were exposed and drawn to Chinese culture. The commandeering centers on the Korean peninsula brought China closer and "ultimately (created) a new China-oriented elite class" [146] (p. 14). Within areas outside of Chinese control the absorption of Chinese culture by local populations led to increased economic and cultural exchanges. As the most economically and culturally advanced society in East Asia, China attracted "neighboring states, which coveted the highly advanced Chinese culture" [146] (p. 14).

Although portions of the Korean peninsula were not under direct Chinese political control, the Chinese influence "is apparent from the fact that for the most part the leaders of the . . . states in the Southern half of the peninsula willingly accepted the grants of office and rank, official seals, and ceremonial attire that constituted . . . tokens of their submission to Lolang's (and China's) authority" [146] (p. 14). The availability of natural and human resources on the Korean peninsula made the area economically attractive to the Chinese who "were able to command the labor services of the native population they governed, for (enterprises such as) the large-scale cutting of timber. It is known, too, that iron ore deposits in the Southeast corner of the peninsula were supplied to Lolang" [146] (p. 14).

Although Japan is geographically disconnected from the Asian mainland by the sea, its islands were once connected extensions of the mainland when Ice Age sea levels fell [127]. Historically, land bridges at one point served to link the Chinese mainland to Japan. Early in East Asia's prehistory, initial habitation, cultural growth, and production were closely tied to the Asian mainland. Once the land bridge disappeared, the sea not only separated them, but also provided a method of transportation. According to Wagner [142] (p. 35), the sea makes "it logical that Japan was continuously influenced by mainland cultures since (East Asian prehistory)." Beyond analyses of early maritime trade, archeologists have unearthed ancient East Asian trading links following the discovery of meteorological tools [127]. Further substantiating the connections between human communities, Sima Qian, in the *Records of the Gran Historian*, writes about the use of horse-drawn war chariots introduced from the West and spread through Central Asia during the Shang Dynasty (1600 BC–1046 BC).

8.2. East Asia During the Final Phase of the Bronze Age Crisis

While the early Chinese Shang Dynasty (1600 BC–1046 BC) experienced a political and territorial expansion, the late period is characterized by fragmentation. Entering the Zhou period (770 BC–256 BC) warring factions in China sought to consolidate their power. In the "Annals of Zhou", written by Sima Qian (*ca.* 100 BC), the struggle between Shang and Zhou is well documented. East Asia has undergone periods where political, social, and economic life is disrupted. Although the duration and the precipitating agent of these disruptions requires some specificity, several general points, as related to trade, can be made about a Dark Age period in East Asia. First, previously established trade networks were disrupted. The disruption of trade networks leads to a decrease in economic and cultural contacts between human communities in the region. Secondly, the decrease in trade linkages creates a climate were local products are looked to as substitutes for previously imported foreign objects. However, the imprint of previous exchanges still manifests itself physically in the products that are manufactured, but they increasingly take on a local character. Third, as a result of linkages being disrupted cultural exchanges are also impacted and facilitate the re-emergence of local practices. During the Dark Age period in East Asia, this leads to indigenous or local practices being re-embraced. The archeological record, coupled with the reading of historical documents, indicates that in East Asia the disruption of cultural exchanges leads to a re-emergence of indigenous or local practices. Although we observe the presence of indigenous practices, and locally produced goods, during the pre-Dark Age period, Chinese objects and practices are predominant. As early as the Shang Dynastic period, there is a strong Chinese influence in Korea and Japan as evident in tools and objects traceable to the Yangtze basin. After a successful period of economic and cultural growth in East Asia, problems in the Chinese mainland lead to a decrease in trade exchanges. Internal crisis in China comes to impact the Korean peninsula and Japan.

Additionally, political fragmentation is an important aspect of Dark Ages. The chaos, disorder, and disruptions that characterize these periods in human history are in many respects the result of political strife and struggles over power. During the breakup of Shang rule around the 11th century BC, political power in China was transferred to local warlords. These local warlords were not content to share the Shang Empire. This resulted in the political disunity of China that carried into the Zhou period (770 BC–256 BC). The perpetual military incursions by competing warlords in China extended into the Korean peninsula. Although historically the peninsula has always had a Chinese presence, the move toward greater Korean autonomy from China can be traced to the disorder and wars that engulfed the Asian mainland during subsequent periods. The goal of expansion brought the Chinese Empire to Korea, but it also resulted in disastrous wars and ineffective expeditions.

In addition to the political fragmentation and wars, population loss and dispersal also characterizes Dark Age periods. There is evidence that a cooling period during Japan's late Jomon period (2000 BC–1000 BC) led to a significant depopulation and a downsizing of large settlement areas [147]. Additionally, one can surmise that the constant fighting not only resulted in population losses from combat, but war was also responsible for food shortages and emigration. In particular, border areas suffered mass departures as people attempted to flee war torn areas.

The human imprint in East Asia is clear in the earth because of the industrious husbandry of hundreds of generations, in the degradation of forests, in the eroded and impoverished lands, and in the barren unproductiveness of formerly fertile and populous terrain, all of which attest to prolonged human abuse. Environmental degradation, climate change, and social upheavals all serve as precipitating agents that led to a contraction of the social system and created an opportunity for a restructuring. In subsequent years, that extend into the Chinese Zhou Dynastic period (770 BC–200 BC), core-periphery relations are transformed, political boundaries are reconfigured, trade relations are intensified, and cultural practices are impacted.

8.3. Climate Fluctuations in Bronze Age East Asia

Preceding the rise of China's first dynastic period (2100 BC–1600 BC), prehistory reveals the formation of settlements during the Neolithic Age. Advances in agriculture led to settled communities and the beginnings of city-states. Prehistoric culture in East Asia is characterized by the development of systems of social stratification and the accompanying cultural objects typical of agricultural societies. Communities in East Asia succumb to the first phase of the Bronze Age crisis (2200 BC–1700 BC) during this Neolithic period as fluctuations in climate led to floods, drought and a disruption of settled agriculture. The collapse of Yueshi Culture in East Asia exemplifies the dependent character of social development [148]. Human communities do not develop separately from the limitations, setbacks, or good disposition of nature.

From the Neolithic Age onward, climate fluctuations in East Asia have been documented [149–151]. Historically, these have led to the collapse and flourishing of cultures, population growth and decline, and have halted and supported social development. Related to climate fluctuations, desertification cycles, and decreasing biological diversity over the last 5000 years in the region have been cited as a reason for the emergence and downfall of past empires [152]. Militaristic incursions by outsiders and internal socio-political conditions have long been explanations for the settlement patterns observe in East Asian history. However, it is important to incorporate the observations made in recent paleoclimactic studies that document changes in the physical environment of the Bronze Age world [153].

In the midst of the final phase of the Bronze Age crisis (1200 BC–700 BC), the archaeological record reveals a relocation of human communities [154]. The Western Zhou period (1046 BC–771 BC) in China is characterized by forays into neighboring communities, infighting amongst royals, and peasant uprisings. Although socio-economic issues partly explain the dissatisfaction amongst groups in Zhou society, and the impetus for migration, climate aridity around 1150 BC also pushed people out of former political and economic centers [155]. Written records of the time also corroborate the relocation of peoples. During the Bronze Age crisis, the change in climate led to excessive flooding. From the 16th century BC till 771 BC, several floods struck the Asian mainland that led to migration [156]. Historically, cultural adaptation is observed as a response to challenging social and environmental conditions. However, as demonstrated in East Asia, peoples also sought resource rich areas to recreate lifestyles no longer supported by the current physical environment.

Suggestive of the ebb and flow of history and the cyclical nature of climate, in subsequent periods, human communities are again equipped with the necessary physical conditions to support a sedentary lifestyle. Moving pass the final phase of the Bronze Age crisis and into the Zhou Dynastic period (1046 BC–256 BC), the climate begins to stabilize in the later periods of the Chou Dynasty [154]. Ultimately, an improved environment contributed to the bourgeoning of Bronze Age culture in East Asia and the prosperity observed during the Han Dynasty (202 BC–AD 220). Although climate and other exogenous factors impact social development, it is understood that human activities also contribute to the trajectory of communities.

8.4. Climate and System Crisis

It is clear from our theoretically informed historical narrative that the drivers that caused system crisis have their origins in Culture's relations with the natural environment. Included in this equation are the changes in climate. The late Bronze Age system crisis that impacted both West and East—even though at this point in time, from 1200 BC to 700 BC, there were no systemic connections in terms of trade, cultural exchanges, and socioeconomic relations between the two regions—suggest to us that climate is an important driver that impacted on the social formations in both regions consequently leading to changing socioeconomic and political demise and transitions.

Is climate the factor that is "determinate in the last instance", to borrow a phrase from structural Marxism, for our understanding and explanation of system crisis and transformation? We suggest it is certainly a factor, amongst others, that deserves attention. For the discipline of the social sciences, this might be quite unsettling as the discipline's *raison d'être* is to discover the socioeconomic and political factors that determine the social evolution of human social formations. Our historically informed theoretical framework has demarcated the factors that engender system transformations and crisis. By considering social system evolution over the *long historical time and geographic space*, we have been able to discriminate the relative impacts the various socioeconomic and political factors along with other natural environmental and climatic factors that determine the trajectory of system evolution. The fact that the late Bronze Age crisis showed parallel outcomes between the East and West when systemic trade, cultural and political connections were not in existence (or minimal at best) at this time period in world history, it has enabled us to pinpoint that climate should be considered and further research may reveal it is a principal driver that caused the cacophony of socioeconomic and political change that followed.

8.5. Concluding Remarks

While acknowledging that systemic crisis and transformation feature anthropogenic causal agents, our research makes clear that a more comprehensive study of system transformations must include Nature and the climate. Study of past system crises can inform our understanding of contemporary trends and the potential adjustments that await modern society. To the latter, despite the negative imagery the adjective "Dark" connotes, Dark Ages provide an opportunity for human creativity and ingenuity to surface in response to changing social and physical conditions. The creativity has a lasting impact that, even when conditions do change, endures and comes to alter and influence the character of subsequent exchanges between peoples.

This study also demonstrates that human history can be broadened beyond an anthropocentric discussion of peoples' circumstances, to include an examination of the impact human activities have on the Earth. Tangible remnants of the past remain, such as historical texts and artifacts, which provide vivid evidence of humankind's reach and the ability of human communities to interact directly and indirectly through large expanses as part of a Eurasian global economy. Climate changes and environmental degradation in the ancient world ominously shadows the present and speaks to the humanocentric conduct evident in ecological relations to this day.

Historically, a pattern is observed where periods characterized by prosperity, growth, consumption and materialism, are then followed by a "Dark" epoch that ushers in wars, disease, political instability, economic decline, and a curbing of previous consumptive habits. The social and physical circumstances necessary for unimpeded growth are disrupted allowing for the opportunity for Nature to recoup its losses [9]. This research illustrates that human populations, in interacting with each other and their environments, attempt to accommodate social, political, economic, and cultural activities to very specific environmental conditions. As a result, *the climate* and the ecological landscape are not only central to our analysis, that they should be recognized as the factors that inform the structure of economic and cultural practices that, in turn, conduce the trajectory of social systems.

Acknowledgments: The authors wish to thank Megan Pullin and Dasha Mikhailova who helped along the way.

Author Contributions: The research is the outcome of a collaboration between the authors that involved discussions on design, an extensive review of the literature, analysis, and several drafts of the paper. All authors have read and approved the final manuscript.

Conflicts of Interest: The authors declare no conflict of interest.

References

1. Marx, K. *Early Writings*; Vintage Books: New York, NY, USA, 1963.
2. Catton, W.; Dunlap, R. Environmental Sociology: A New Paradigm. *Am. Sociol.* **1978**, *13*, 41–49.
3. O'Connor, J. Capitalism, Nature, Socialism: A Theoretical Introduction. *Capital. Nat. Social.* **1988**, *1*, 11–38. [CrossRef]
4. O'Connor, J. The Conditions of Production and the Production of Conditions. In *Natural Causes: Essays in Ecological Marxism*; Guilford: New York, NY, USA, 1998; pp. 135–178.
5. Schnaiberg, A. *The Environment from Surplus to Scarcity*; Oxford University Press: New York, NY, USA, 1980.
6. Foster, J.B. Marx's Theory of Metabolic Rift: Classical Foundations for Environmental Sociology. *Am. J. Sociol.* **1999**, *105*, 366–405. [CrossRef]
7. Foster, J.B. *The Ecological Revolution: Making Peace with the Planet*; Monthly Review Press: New York, NY, USA, 2009.
8. Foster, J.B. *The Vulnerable Planet: A Short Economic History of the Environment*; Monthly Review Press: New York, NY, USA, 1999.
9. Chew, S. *World Ecological Degradation: Accumulation, Urbanization, and Deforestation*; AltaMira Press/Rowman and Littlefield Publishers: Lanham, MD, USA, 2001.
10. Chew, S. *The Recurring Dark Ages: Ecological Stress, Climate Changes and System Transformation*; AltaMira Press/Rowman and Littlefield Publishers: Lanham, MD, USA, 2007.

11. Chew, S. *Ecological Futures: What History Can Teach Us*; AltaMira Press/Rowman and Littlefield Publishers: Lanham, MD, USA, 2008.

12. Flotz, R.C. Does Nature Have Historical Agency? World History and Environmental History and How Histories Can Help to Save the Planet. *Hist. Teach.* **2003**, *37*, 9–28. [CrossRef]

13. Foster, J.B. Transcending the Metabolic Rift: A Theory of Crises in the Capitalist World Ecology. *J. Peasant Stud.* **2011**, *38*, 1–46.

14. Moore, J.W. Environmental Crises and the Metabolic Rift in World-Historical Perspective. *Organ. Environ.* **2000**, *13*, 123–157. [CrossRef]

15. Foster, J.B. *Marx's Ecology: Materialism and Nature*; Monthly Review Press: New York, NY, USA, 2000.

16. Foster, J. The Modern World-System as Environmental History? Ecology and the Rise of Capitalism. *Theory Soc.* **2003**, *32*, 307–377.

17. Foster, J. Madeira, Sugar, & the Conquest of Nature in the First Sixteenth Century, Part I: From Island of Timber to Sugar Revolution, 1420–1506. *Rev. J. Fernand Braudel Center* **2009**, *32*, 345–390.

18. Foster, J. Amsterdam is Standing on Norway, Part I: The Alchemy of Capital, Empire, and Nature in the Diaspora of Silver, 1545–1648. *J. Agrar. Chang.* **2010**, *10*, 35–71.

19. Foster, J. Amsterdam is Standing on Norway, Part II: The Global North Atlantic in the Ecological Revolution of the Seventeenth Century. *J. Agrar. Chang.* **2010**, *10*, 188–227.

20. Moore, J. The End of the Road? Agricultural Revolutions in the Capitalist World-Ecology, 1450–2010. *J. Agrar. Chang.* **2010**, *10*, 389–413. [CrossRef]

21. Moore, J. Madeira, Sugar, & the Conquest of Nature in the First Sixteenth Century, Part II: From Regional Crisis to Commodity Frontier, 1506–1530. *Rev. J. Fernand Braudel Center* **2010**, *33*. [CrossRef]

22. Moore, J. This Lofty Mountain of Silver Could Conquer the Whole World: Potosí and the Political Ecology of Underdevelopment, 1545–1800. *J. Philos. Econ.* **2010**, *4*, 58–103.

23. Moore, J. Ecology, Capital, and the Nature of Our Times. *J. World Syst. Anal.* **2010**, *17*, 108–147.

24. Moore, J. *Capitalism in the Web of Life: Ecology and the Accumulation of Capital*; Verso: New York, NY, USA, 2015.

25. Chew, S.C. For Nature: Deep Greening World Systems Analysis for the Twenty-First Century. *J. World Syst. Res.* **2015**, *3*, 381–402. [CrossRef]

26. Chew, S. Ecological Relations and the Decline of Civilizations in the Bronze Age World System: Mesopotamia and Harrapa 2500 BC—1700 BC. In *Ecology and the World System*; Goldfrank, W., Ed.; Greenwood Press: Greenwich, CT, USA, 1999.

27. Roberts, J.T.; Grimes, P. World-Systems Theory and the Environment: A New Synthesis. In *Sociological Theory and the Environment*; Dunlap, R., Ed.; Rowman and Littlefield: Lanham, MD, USA, 2002; pp. 167–196.

28. Friedman, H. What on Earth is the Modern World-System? Food Getting and Territory in the Modern Era and Beyond. *J. World Syst. Res.* **2000**, *6*, 480–515. [CrossRef]

29. Wallerstein, I. *Kondratieff Up or Down. Review II*; Spring: New York, NY, USA, 1979.

30. Wallerstein, I. *The End of the World as We Know It*; The University of Minnesota Press: Minneapolis, MN, USA, 1999.

31. Habermas, J. *Toward a Rational Society: Student Protest, Science and Politics*; Beacon Press: Boston, MA, USA, 1970.

32. Frank, A.G. Transitional Ideological Modes: Feudalism, Capitalism, and Socialism. *Crit. Anthropol.* **1991**, *11*, 171–188. [CrossRef]

33. Frank, A.G.; Gills, B.K. World System Cycles, Crises, and Hegemonial Shifts. *Review* **1992**, *15*, 621–688.

34. Frank, A.G. Bronze Age World System Cycles. *Curr. Anthropol.* **1993**, *34*, 383–429. [CrossRef]

35. Frank, A.G., Ed.; *The World System: 500 or 5000 Years*; Routledge: New York, NY, USA, 1993.

36. Frank, A.G. *ReOrient: Global Economy in the Asian Age*; University of California Press: Berkeley, CA, USA, 1998.

37. Modelski, G. *World Cities*; Faros: Washington, DC, USA, 2003.

38. Modelski, G.; Thompson, W. The Evolutionary Pulse of the World System: Hinterland Incursion and Migrations 4000 BC to AD 1500. In *World System Theory in Practice*; Kardulias, N., Ed.; Rowman and Littlefield: Lanham, MD, USA, 1999; pp. 241–274.

39. Thompson, W. C-Waves, Center-Hinterland Contact and Regime Change in the Ancient Near East: Early Impacts of Globalization. In Proceedings of the International Studies Association Annual Meetings, Los Angeles, CA, USA, 23–26 March 2000.

40. Thompson, W. Trade Pulsations, Collapse, and Reorientation in the Ancient World. In Proceedings of the International Studies Association Annual Meetings, Chicago, IL, USA, 24–27 March 2001.

41. Chase-Dunn, C.; Hall, T. *Rise and Demise: Comparing World-Systems*; Westview Press: Boulder, CO, USA, 1997.

42. Wallerstein, I. *The Modern World-System Vols 1–3*; Academic Press: New York, NY, USA, 1974.

43. Wallerstein, I. The West, Capitalism, and the Modern World System. *Review* **1992**, *15*, 561–620.

44. Beaujard, P. The Indian Ocean in Eurasian and African World-Systems Before the Sixteenth Century. *J. World Hist.* **2005**, *16*, 411–465. [CrossRef]

45. Beaujard, P. Evolution and Temporal Delimitations of Possible Bronze Age World-Systems in Western Asia, Africa and the Mediterranean. In *Interweaving Worlds: Systemic Interactions in Eurasia 7th to 1st Millennia BC*; Wilkinson, T., Ed.; Oxbow Press: Oxford, UK, 2009.

46. Beaujard, P. From Three Possible Iron-Age World-Systems to a Single Afro-Eurasian World-System. *J. World Hist.* **2010**, *21*, 1–43. [CrossRef]

47. Steinberg, T. Down to Earth: Nature, Agency, and Power in History. *Am. Hist. Rev.* **2002**, *107*, 798–820. [CrossRef]

48. Wilkinson, T.J. *Town and Country in SouthEastern Anatolia*; Oriental Institute: Chicago, IL, USA, 1990.

49. Desborough, V.R. *The Greek Dark Ages*; Ernest Benn: London, UK, 1972.

50. Snodgrass, A.M. *The Dark Age of Greec*; Edinburgh University Press: Edinburgh, UK, 1971.

51. Snodgrass, A.M. *Archaic Greece*; MacMillan: London, UK, 1980.

52. Snodgrass, A.M. The Coming of the Iron Age in Greece: Europe's Earliest Bronze/Iron Transition. In *Bronze-Iron Age Transition in Europe*; BAR International Series No. 483; Sorensen, M.L., Thomas, R., Eds.; Archeopress: Oxford, UK, 1989.

53. Braudel, F. *The Mediterranean and the Mediterranean World in the Age of Philip II*; Fontana: London, UK, 1972; Volume 1.

54. Braudel, F. *The Structure of Everyday Life*; Harper and Row: New York, NY, USA, 1981; Volume 1.

55. Braudel, F. *The Wheels of Commerce*; Harper and Row: New York, NY, USA, 1982; Volume 2.

56. Braudel, F. *The Perspective of the World*; Harper and Row: New York, NY, USA, 1984; Volume 3.

57. Braudel, F. *The Identity of France*; Harper and Row: New York, NY, USA, 1989; Volume 1–2.

58. Braudel, F. *Memory and the Mediterranean*; Alfred Knopf: New York, NY, USA, 2001.

59. Kristiansen, K. The Emergence of the European World System in the Bronze Age: Divergence, Convergence, and Social Evolution During the First and Second Millennia BC in Europe. *Sheff. Archaeol. Monogr.* **1993**, *6*, 7–30.

60. Kristiansen, K. *Europe before History*; Cambridge University Press: Cambridge, UK, 1998.

61. Kristiansen, K.; Larsson, T. *The Rise of Bronze Age Society Travels, Transmission, and Transformation*; Cambridge University Press: Cambridge, UK, 2005.

62. Kohl, P. The Ancient Economy, Transferable Technologies, and the Bronze Age World System: A View from the NorthEastern Frontiers of the Ancient Near East. In *Centre and Periphery in the Ancient World*; Rowlands, M., Mogens, L., Kristian, K., Eds.; Cambridge University Press: Cambridge, UK, 1987.

63. Childe, G. *What Happened in History*; Penguin: Harmondsworth, UK, 1942.

64. Childe, G. The Urban Revolution. *Town Plan. Rev.* **1950**, *21*, 3–17. [CrossRef]

65. Bell, B. The Dark Ages in Ancient History I: The First Dark Age in Egypt. *Am. J. Archaeol.* **1971**, *75*, 1–20. [CrossRef]

66. Bell, B. Climate and History of Egypt. *Am. J. Archaeol.* **1975**, *79*, 223–279. [CrossRef]

67. Perlin, J. *A Forest Journey*; Harvard University Press: Cambridge, UK, 1989.

68. Williams, M. *Deforesting the Earth: From Prehistory to Global Crisis*; University of Chicago Press: Chicago, IL, USA, 2003.

69. Rowton, M.B. The Woodlands of Ancient Western Asia. *J. Near East. Stud.* **1967**, *26*, 261–277. [CrossRef]

70. Wilcox, G.H. Timber and Tress: Ancient Exploitation in the Middle East. In *Trees and Timber in Mesopotamia*; Bulletin on Sumerian Agriculture 6; Postgate, J.N., Powell, M.A., Eds.; Cambridge University: Cambridge, UK, 1992.

71. Lal, B.B. *The Earliest Civilization of South Asia: Rise, Maturity, and Decline*; Aryan Books International: New Delhi, India, 1997.

72. Behre, K.E. Some Reflections on Anthropogenic Indicators and the Record of Prehistoric Occupation Phases in Pollen Diagrams. In *Man's Role in the Shaping of the Eastern Mediterranean Landscape*; Bottema, S., Entjesbieburg, G., van Zeist, W., Eds.; Balkema: Rotterdam, The Netherlands, 1990.

73. Neumann, J.; Parpola, S. Climate Change and 11–10th Century Eclipse of Assyria and Babylonia. *J. Near East. Stud.* **1987**, *46*, 161–182. [CrossRef]

74. Weiss, H.; Bradley, R. Archaeology: What Drives Societal Collapse? *Science* **2001**, *26*, 609–610. [CrossRef] [PubMed]

75. Bentaleb, I.; Caratini, C.; Fontungne, M.; Morzadec-Kerfourn, M.; Pascal, J.; Tissot, C. Monsoon Regime Variations During the Late Holocene in SouthWestern India. In *Third Millennium BC Climate Change and Old World Collapse*; Dalfes, H., Ed.; Springer-Verlag: Heidelberg, Germany, 1997.

76. Ratnagar, S. *Encounters: The Westerly Trade of the Harappan Civilization*; Oxford University Press: New Delhi, India, 1981.

77. Fagan, B. *The Long Summer*; Basic Books: New York, NY, USA, 2004.

78. Issar, A. Climate Change and History during the Holocene in the Eastern Mediterranean Region. In *Water, Environment, and Society in Times of Climatic Change*; Issar, A., Brown, N., Eds.; Kluwer: Hague, The Netherlands, 1998.

79. Fairbridge, R.O.; Erol, O.; Karaca, M.; Yilmaz, Y. Background to Mid-Holocene Climate Change in Anatolia and Adjacent Regions. In *Third Millenium BC Climate Change and Old World Collapse*; Dalfes, H., Ed.; Springer-Verlag: Heidelberg, Germany, 1997.

80. Agrawal, D.P.; Sood, R.K. Ecological Factors and the Harappan Civilization. In *Harappan Civilization: Contemporary Perspective*; Possehl, G., Ed.; Oxford University Press: New Delhi, India, 1982; pp. 223–231.

81. Possehl, G. *Ancient Cities of the Indus*; Vikas: New Delhi, India, 1979.

82. Possehl, G. The Harappan Civilization: A Contemporary Perspective. In *Harappan Civilization*; Possehl, G., Ed.; Oxford University Press: New Delhi, India, 1982.

83. Enzel, Y.; Ely, L.L.; Mishra, S.; Ramesh, R.; Amit, R.; Lazar, B.; Rajaguru, S.N.; Baker, V.R.; Sander, A. High Resolution Holocene Environmental Changes in the Thar Desert, Northwest India. *Science* **1999**, *2*, 125–128. [CrossRef]

84. Matthews, R. Zebu: Harbinger of Doom in Bronze Age Western Asia. *Antiquity* **2002**, *76*, 438–446. [CrossRef]

85. Hiebert, F. Bronze Age Central Eurasian Cultures in their Steppe and Desert Environments. In *Environmental Disaster and the Archaeology of Human Response*; Bawden, G., Reycraft, R., Eds.; University of New Mexico Press: Albuquerque, NM, USA, 2000.

86. Krementski, C. The Late Holocene Environmental and Climate Shift in Russia and Surrounding Lands. In *Third Millenium BC Climate Change and Old World Collapse*; Dalfes, H., Ed.; Springer-Verlag: Heidelberg, Germany, 1997.

87. Mellink, M. The Early Bronze Age in Western Anatolia: Aegean and Asiatic Correlations. In *End of the Early Bronze Age in the Aegean*; Cadogan, G., Ed.; Brill: Leiden, The Netherlands, 1986.

88. Butzer, K.W. Socio Political Discontinuity in the Near East c. 2200 BCE. Scenarios from Palestine and Egypt. In *Third Millenium BC Climate Change and Old World Collapse*; Dalfes, H., Ed.; Springer-Verlag: Heidelberg, Germany, 1997.

89. Harrison, T. Shifting Patterns of Settlement in the Highlands of Central Jordan during the Early Bronze Age. *Bull. Am. Sch. Orient. Res.* **1997**, *306*, 1–38. [CrossRef]

90. Jameson, M.; Runnels, C.; Van Andel, T.; Munn, M. *A Greek Countryside: The Southern Argolid from Prehistory to the Present Day*; Stanford University Press: Stanford, CA, USA, 1994.

91. Watrous, L.V. Review of the Aegean Prehistory III: Crete from Earliest Prehistory through the Protopalatial Period. *Am. J. Archaeol.* **1994**, *98*, 695–753. [CrossRef]

92. McGovern, P. Central TransJordan in Late Bronze Age and Early Iron Ages: An Alternative Hypothesis of Socioeconomic Collapse. In *Studies in the History and Archaeology of Jordan 3*; Hadidi, A., Ed.; Routledge and Kegan Pail: London, UK, 1987.

93. Bovarski, E. First Intermediate Period Private Tombs. In *Encyclopedia of Ancient Egypt*; Bard, K., Ed.; Routledge: New York, NY, USA, 1998.

94. Brinkman, J.A. Ur: The Kassite Period and the Period of Assyrian Kings. *Orientalis* **1968**, *38*, 310–348.

95. Knapp, A.B. Thalassocracies in the Bronze Age Eastern Mediterranean Trade: Making and Breaking a Myth World. *World Archaeol.* **1993**, *24*, 332–347. [CrossRef]

96. Sheratt, S. The Growth of the Mediterranean Economy in the Early First Millennium BC. *World Archaeol.* **1993**, *24*, 361–378. [CrossRef]

97. Sheratt, S. Sea Peoples and the Economic Structure of the late 2nd Millennium in the Eastern Mediterranean. In *Mediterranean Peoples in Transition: 13th to Early 10th Century BCE*; Gitin, S., Mazar, A., Sternleds, E., Eds.; Jerusalem Archaeological Society: Jerusalem, Palestine, 1998; pp. 292–313.

98. Sheratt, S. Circulation of Metals and the End of the Bronze Age in the Eastern Mediterranean. In *Metals Make the World Go Round*; Pare, C., Ed.; Oxbow Books: Oxford, UK, 2000; pp. 82–98.

99. Toynbee, A.J. *A Study of History IV and V*; Oxford University Press: Oxford, UK, 1939.

100. Harding, A., Ed.; *Climatic Change in Later Prehistory*; Edinburgh University Press: Edinburgh, UK, 1982.

101. Carter, V.; Dale, T. *Topsoil and Civilization*; University of Oklahoma Press: Norman, France, 1974.

102. Zangger, E. The Environmental Setting. In *Sandy Pylos*; Davis, J., Ed.; University of Texas Press: Austin, TX, USA, 1998.

103. Runnels, C.N. Environmental Degradation in Ancient Greece. *Sci. Am.* **1995**, *272*, 96–99. [CrossRef]

104. Van Andel, T.; Runnels, C.; Pope, O. Five Thousand Years of Land Use and Abuse in the Southern Argolid Greece. *Hesperia* **1986**, *55*, 103–128. [CrossRef]

105. Blintiff, J. Erosion in the Mediterranean Lands: Reconsideration of Pattern, Process, and Methodology. In *Past and Present Soil Erosion*; Bell, M., Boardman, J., Eds.; Oxbow: Orford, UK, 1992.

106. Marinatos, S. The Volcanic Eruption of Minoan Crete. *Antiquity* **1939**, *13*, 425–439.

107. Chadwick, J. *The Mycenaean World*; Cambridge University Press: New York, NY, USA, 1976.

108. Warren, P.M. Minoan Palaces. *Sci. Am.* **1985**, *253*, 94–103. [CrossRef]

109. Carpenter, R. *Discontinuity in Greek Civilization*; Cambridge University Press: Cambridge, UK, 1968.

110. Drews, R. *The End of the Bronze Age*; Princeton University Press: Princeton, NJ, USA, 1993.

111. Lamb, H.R. Carpenter's Discontinuity in Greek Civilization. *Antiquity* **1967**, *41*, 33–34.

112. Bryson, R.A.; Lamb, H.H.; Donley, D.L. Drought and the Decline of Mycenae. *Antiquity* **1974**, *48*, 46–50.

113. Bryson, R.A.; Padoch, C. On Climates of History. *J. Interdiscip. Hist.* **1980**, *10*, 583–597. [CrossRef]

114. Morris, I. *Burial and Ancient Society: The Rise of the Greek City-State*; Cambridge University Press: Cambridge, UK, 1987.

115. Morris, I. Tomb Cult and the Greek Renaissance: The Past in the Present 8th Century BC. *Antiquity* **1988**, *62*, 750–761.

116. Morris, I. *Archaeology as Cultural History: Words and Things in Iron Age Greece*; Blackwell: Malden, MA, USA, 2000.

117. Whitley, J. *Style and Society in Dark Age Greece: The Changing Face of Pre-Literate Society 1100—700 BC*; Cambridge University Press: Cambridge, UK, 1991.

118. Harrison, A.B.; Nigel, S. After the Palace: The Early History of Messinia. In *Sandy Pylos*; Davis, J., Ed.; University of Texas Press: Austin, TX, USA, 1998.

119. Small, D. Handmade Burnished Ware and Prehistoric Aegean Economics. *J. Mediterr. Archaeol.* **1990**, *3*, 3–25. [CrossRef]

120. Rutter, J. Cultural Novelties in the Post Palatial Aegean World: Indices of Vitality or Decline. In *The Crisis Years: The 12th Century BC*; Ward, W., Joukowsky, M., Eds.; Kendall Hunt: Dubuque, IA, USA, 1989.

121. Deger-Jalkotzy, S. The Last Mycenaeans and their Successors Updated. In *Mediterranean Peoples in Transition: Thirteenth to Early Tenth Centuries BCE: In Honor of Professor Trude Dothan*; Gitin, S., Mazar, A., Stern, E., Eds.; Israel Exploration Society: Jerusalem, Palestine, 1998; pp. 114–128.

122. Muhly, J.D. The role of the Sea People of Cyprus during the LC III Period. In *Cyprus at the Close of the Late Bronze Age*; Karageorghis, V., Muhly, J., Eds.; Zavallis: Nicosia, Cyprus, 1984.

123. Stanislawski, D. Dark Age Contributions to the Mediterranean Way of Life. *Ann. Assoc. Am. Geogr.* **1973**, *63*, 397–410. [CrossRef]

124. McNeill, J.R.; McNeill, W. *The Human Web*; Norton: New York, NY, USA, 2003.

125. Heichelheim, F.M. *An Ancient Economic History*; A.W. Sijthoff: Leiden, The Netherlands, 1968; Volume 1.

126. Polanyi, K. *The Livelihood of Man*; Academic Press: New York, NY, USA, 1977.

127. Imamura, K. *Prehistoric Japan: New Perspectives on Insular East Asia*; University of Hawaii Press: Honolulu, HI, USA, 1996.

128. Zhimin, A. Effect of Prehistoric Cultures of the Lower Yangtze River on Ancient Japan. *Kaogu* **1984**, *5*, 439–558.

129. De Bary, T.W., Lee, P.H., Eds.; *Sources of Korean Tradition*; Columbia University Press: New York, NY, USA, 1997.

130. McCune, S. *Korea's Heritage: A Regional and Social Geography*; C.E. Tuttle: Tokyo, Japan, 1956.

131. Tongko, L. Study on Eastern Movement of Japanese. *Kaogu* **1971**, *5*, 439–558.

132. Matsumura, Y. Temporal Distribution of Cereal Remains of Asian Neolithic. *Kikan Koukogaku* **1991**, *37*, 33–35.

133. Takakura, M. The Route of the Arrival of Rice. *Kikan Koukogaku* **1991**, *37*, 40–45.

134. An, Z.M. Chinese Prehistoric Agriculture. *J. Archaeol.* **1988**, *28*, 589–593.

135. Gettens, R.J.; Clarke, R.S.; Chase, W.T. Two Early Chinese Bronze Weapons with Meteoritic Iron Blades. In *Freer Gallery of Art, Occasional Papers 4: 1–11*; Freer Gallery of Art: Washington, DC, USA, 1971.

136. Li, Z. The development of iron and steel technology in ancient China. *Kaogu Xue Bao* **1975**, *2*, 1–22.

137. Li, Z. Studies on the Iron Blade of a Shang Dynasty Bronze Yüeh-axe Unearthed at Kao-ch'eng, Hupei, China. *Kaogu Xue Bao* **1976**, *2*, 17–34.

138. Yuan, J.; Zhang, X. Shang-period Tombs Excavated in Pinggu County, Beijing. *Beijing Munic. Cult. Relics Off.* **1977**, *11*, 1–8.

139. Hua, J. Meteoritic Iron, Meteoritic Iron Artifacts, and the Invention of Iron-Smelting. *KJSW* **1982**, *9*, 17–22.

140. Huang, Z. On the Problem of the First Smelting of Iron and Use of Iron Implements in China. *Wen Wu* **1976**, *8*, 62–70.

141. Huang, Z. Ancient Culture of the Lower Yangtze River and Ancient Japan. *Kaogu* **1990**, *4*, 375–384.

142. Wagner, D.B. The Dating of the Chu Graves of Changsha: The Earliest Iron Artifacts in China? *Acta Orient.* **1987**, *48*, 111–156.

143. Wagner, D.B. *Iron and Steel in Ancient China*; E.J. Brill: New York, NY, USA, 1993.

144. Wagner, D.B. *The State and Iron Industry in Han China*; NIAS: Copenhagen, Denmark, 2001.

145. Zou, H. Stone-chamber Tumulus Sites at Wufengshan in Wuxian County, Jiangsu. *Nianjian* **1984**, *1*, 105–106.

146. Eckert, C.J.; Lee, K.; Lew, Y.; Robinson, M.; Wagner, E. *Korea, Old and New: A History*; Korea Institute: Seoul, Korea, 1990.

147. Habu, J.; Hall, M.E. Climate Change, Human Impacts on the Landscape, and Subsistence Specialization: Historical Ecology and Changes in Jomon Hunter-Gatherer Lifeways. In *The Historical Ecology of Small Scale Economies*; Victor, D.T., Waggoner, J., Eds.; University of Florida Press: Gainesville, FL, USA, 2012.

148. Guo, Y.; Mo, D.; Mao, L.; Wang, S.; Li, S. Settlement Distribution and its Relationship with Environmental Changes from the Neolithic to Shang-Zhou Dynasties in Northern Shandong, China. *J. Geogr. Sci.* **2013**, *23*, 679–694. [CrossRef]

149. Yasuda, Y. Climate Change and the Origin and Development of Rice Cultivation in the Yangtze River Basin, China. *Ambio* **2008**, *2008*, 502–506. [CrossRef]

150. Mao, L.J.; Mo, D.W.; Zhou, K.S.; Guo, W.M.; Jia, Y.F.; Yang, J.H.; Deng, H.; Shi, C.X.; Jia, J.Y. Rare Earth Elements and the Environmental Significance of the Dark Brown Soil in Liyang Plain, Hunan Province, China. *Acta Sci. Circumst.* **2009**, *29*, 1561–1568.

151. Guo, Y.; Mo, D.; Mao, L.; Jin, Y.; Guo, W.; Mudie, P.J. Settlement Distribution and its Relationship with Environmental Changes from the Paleolithic to Shang-Zhou period in Liyang Plain, China. *Quat. Int.* **2014**, *321*, 29–36. [CrossRef]

152. Wang, X.; Chen, F.; Zhang, J.; Yang, Y.; Li, J.; Hasi, E.; Zhang, C.; Xia, D. Climate, Desertification, and the Rise and Collapse of China's Historical Dynasties. *Hum. Ecol. Interdiscip. J.* **2010**, *38*, 157–172. [CrossRef]

153. Huang, C.C.; Su, H. Climate Change and Zhou Relocations in Early Chinese History. *J. Hist. Geogr.* **2009**, *35*, 297–310. [CrossRef]

154. Ouyang, X.; Davis, R.L. *Historical Records of the Five Dynasties*; Columbia University Press: New York, NY, USA, 2004.

155. Huang, C.C.; Zhao, S.; Pang, J.; Zhou, Q.; Chen, S.; Li, P.; Mao, L.; Ding, M. Climatic Aridity and the Relocations of the Zhou Culture in the Southern Loess Plateau of China. *Clim. Chang.* **2003**, *61*, 361–378. [CrossRef]

156. Chen, Y.; James, S.; Shu, G.; Irina, O.; Albert, K. Socio-economic Impacts on Flooding: A 4000-Year History of the Yellow River, China. *J. Hum. Environ.* **2012**, *41*, 682–698. [CrossRef] [PubMed]

sustainability

MDPI

Article

Trans-Boundary Haze Pollution in Southeast Asia: Sustainability through Plural Environmental Governance

Md Saidul Islam [1,*], Yap Hui Pei [2] and Shrutika Mangharam [1]

[1]   Division of Sociology, Nanyang Technological University, 14 Nanyang Drive, Singapore 637332;
     shrutika.mangharam@gmail.com
[2]   Division of Psychology, Nanyang Technological University, 14 Nanyang Drive, Singapore 637332;
     HYAP002@e.ntu.edu.sg
*   Correspondence: msaidul@ntu.edu.sg; Tel.: +65-6592-1519

Academic Editor: Marc A. Rosen
Received: 29 February 2016; Accepted: 13 May 2016; Published: 21 May 2016

Abstract: Recurrent haze in Southeast Asian countries including Singapore is largely attributable to rampant forest fires in Indonesia due to, for example, extensive slash-and-burn (S & B) culture. Drawing on the "treadmill of production" and environmental governance approach, we examine causes and consequences of this culture. We found that, despite some perceived benefits, its environmental consequences include deforestation, soil erosion and degradation, global warming, threats to biodiversity, and trans-boundary haze pollution, while the societal consequences comprise regional tension, health risks, economic and productivity losses, as well as food insecurity. We propose sustainability through a plural coexistence framework of governance for targeting S & B that incorporates strategies of incentives, education and community resource management.

Keywords: slash-and-burn; environmental governance; haze; Indonesia; plural coexistence; global warming; Singapore

1. Introduction

The world's rapidly growing population has been a long-standing cause of concern amongst both economists and environmentalists alike. There is an increasing demand for agricultural and urban spaces to sustain the ever-multiplying demographics. However, due to limited availability of space, the trend of clearing forests to make way for cultivable land has been gaining popularity [1]. One of the most perturbing methods of clearing forests is the utilization of Slash-and-Burn (S & B). The S & B method involves the felling of trees and plants, followed by setting fire to the designated area. Owing to this method's high efficiency and low cost, it has been adopted in a number of developing nations. However, the employment of S & B is not without dire consequences, the most serious of which are trans-national repercussions on the environment, economy and society [2].

Although prevalent across the globe, the practice of S & B is particularly rampant in Indonesia [1,2]. Consequently, neighboring Southeast Asian countries such as Singapore, Malaysia, Brunei and Thailand are negatively affected by S & B techniques in Indonesia [3]. In fact, the trans-boundary haze pollution due to forest fires has become significantly more evident in the recent past, with the extent of air pollution rising to record-high levels. In 1997, for example, due to haze pollution, Singapore recorded a Pollution Standards Index (PSI) level of 226, which rocketed to a reading of 401 in the mid-2013 bout of haze, as reported by BBC News (21 June 2013). These figures demand a deeper analysis of the practice of S & B and the consequences it has, not only on the country in which it is practiced, but also on neighboring nations that are affected by it.

This paper investigates the technique of S & B in a comprehensive manner—studying the reasons for employing S & B, the resulting effects of forest fires, and corrective measures to control the issue. It utilizes the treadmill of production theory to assess the extent to which S & B depletes resources from the environment and simultaneously produces wastes that are harmful to it [4]. First, it analyzes the various factors that encourage the use of S & B in Indonesia. Besides simply being a cheap and efficient method of forest clearing, S & B is also employed to facilitate peatland drainage, logging and establishment of oil palm plantations. Furthermore, weak governance in Indonesia allows for certain groups to exploit common natural resources at the cost of the environment and other sections of society. The paper then proceeds to identify the environmental, societal and economic repercussions, both direct and indirect, of S & B on the countries affected by it. Finally, the paper suggests certain measures that address the concerns surrounding S & B. The implementations of these national and trans-national recommendations would greatly diminish the dangerous impacts of S & B on the affected countries.

2. Framework

2.1. Treadmill of Production

Economic production is the vehicle on which contemporary capitalist societies run. As a result of continuous, unchecked production, a self-sustaining process called the treadmill of production occurs. The treadmill of production theory, a strand of Neo-Marxist understanding of capitalism's relationship with the environment, argues that the continuous race of production through a continuous enhancement of productive forces and practices (S & B in our case) and the need for its continued consumption create a critical interchange of "withdrawals" (extraction of resources from the environment) and "additions" (what is returned to the environment in the form of pollution and garbage). These cycles of withdrawals and additions can disorganize the biospheric systems [4].

The treadmill of production model was further elaborated to incorporate the impacts of production not just on ecological elements, but also on social and economic ones [5]. In the case of S & B, the greatest ecological withdrawal is deforestation, which results in a series of subsequent withdrawals from the environment. These occur in the form of soil erosion and degradation, global warming and climate change and threats to biodiversity. As part of the process of S & B, air pollution in the form of haze is added to the environment, resulting in a plethora of other environmental, political, social, and economic concerns, such as regional tension, health risks, economic and productivity losses and food security issues. Thus, by extracting valuable resources from the eco-system, and contributing hazardous pollutants back to it, S & B is a practice that runs on the treadmill of production.

2.2. Environmental Governance

Environmental governance refers to interventions and regulations that impact the environment. It encompasses mutually beneficial actions and decisions made by the state, communities, corporations and nongovernmental organizations. Hence, these interventions can take the form of international treaties, national policies or local legislation to preserve the quality of the environment, while simultaneously ensuring the well-being of society and the growth of the economy [6]. The environmental governance can be used to recommend certain interventions to mitigate and reduce the impacts of S & B. For our paper, we have used four over-arching themes of environmental governance.

First, with the increasing interconnectedness of today's world, natural resource depletion and waste production spread across geopolitical boundaries. Capital is directed towards countries that have more lenient environmental standards, due to which, resources in these countries are exploited until another country provides easier conditions for production. This "race to the bottom" leaves countries with destroyed natural systems and deep socioeconomic inequalities [6] (p. 300). However, globalization can also aid in the restoration of such nations. With the help of the free flow of information, better technology and the support of transnational environmental institutions, policy initiatives can be established to implement and preserve safe environmental standards.

Second, it is contended that there is a shift towards environmental governance on a "subnational level" [6] (p. 302). The decentralization of governance ensures efficient community-based resource management by those who are more knowledgeable about them, as well as concentrated efforts to protect these resources [6].

The third theme of environmental governance is market- and agent-focused instruments, which aim to favor environmentally sound practices through calculated incentives and costs. These instruments include taxes, subsidies, market incentives and certifications, amongst many other measures, that mobilize individuals to support operations that are the least harmful to the environment. Finally, scholars suggest that, since the repercussions of environmental problems are felt at the local, national, and transnational levels, there needs to be multi-level governance to address these issues [6].

3. Contributing Factors behind S & B Culture

Despite the availability of other, more sustainable alternative, methods of clearing forests (for example, slash-and-mulch, which clears forests by slashing and subsequently planting crops in the mulch, and improved fallow, whereby the land is left fallow to restore fertility), S & B is still rampant across the world. This is due to several factors that make it the most efficient method to implement, which are discussed in this section.

3.1. Perceived Relative Benefits

One of the main reasons why S & B is selected as the method of forest clearing is the perceived economic and environmental benefit of the practice. For example, S & B is often thought to be the most efficient and cost-effective method of clearing land. It is also believed to enhance soil nutrients, balance soil pH levels and soil structure, as well as reduce aluminium presence. Besides these benefits, S & B is viewed as advantageous because it prevents growth of weeds and incidence of pests and diseases [2,3,7–10].

In comparison, however, the alternatives to S & B are perceived as more expensive and as resulting in fewer benefits. Burning assists in the production of ash fertilizer and also aids the eradication of pests and diseases, without which there would be lower crop output or late crops, and higher labor costs. As a result, income would be reduced and poverty would increase [3]. However, it is important to note that these benefits appear to outweigh the costs since they only take into consideration short-term benefits and costs. Greater, long-term costs, both ecological and social, are often ignored, resulting in a misguided perception that S & B is more advantageous than it is harmful [2,11].

3.2. Logging

Logging refers to the extraction of timber from forests. The result of logging often acts as a catalyst to forest fires. For instance, logging leaves behind easily combustible litter on the forest floor. It also leaves behind an open canopy, which then creates drier conditions and permits the growth of extra-combustible vegetation beneath, thus increasing the risk of fire [12–14]. Furthermore, when trees in dense tropical forests are felled, the intertwined roots and vines uproot other trees as well, exacerbating the extent of the aftermath of logging [15]. Logged forests are much more susceptible to fires in comparison to unlogged forests since logging translates into more forests burned and more crown fires [16]. On the other hand, unlogged forests experience less damage, and therefore, are only susceptible to small-intensity surface fires. Evidence of this can be seen in the case of Kalimantan, where 97% of logged forest and peat were destroyed by fire, as compared to 11%–17% of unlogged forest [16].

Another related factor that encourages logging, and subsequently fires, is the construction of roads and irrigation canals, which pave the way for further illegal logging. With this infrastructure in place, people can now access the once-inaccessible forests and peatlands to obtain and transport timber illegally, or even to develop the land for economic purposes. Indeed, as a result, illegal logging rates rose by 44% from 1997 to 2000 [15].

3.3. Oil Palm Plantations

Oil palm is a valuable cash crop for its oil is used as fuel for vehicles, as cooking oil and in cosmetics [15]. Indonesia's climate and soil are suitable for the growth of oil palm, which contribute to its high output and whole-year harvest schedule [13]. Farmers are compelled to grow oil palm to reap the most economic benefits possible from shrinking farm acreage [3]. Forest areas burnt during the 1980s' fires expedited their development into plantations via burning [16]. Since fire is the cheapest and fastest means of clearing land, it was found that 80% of the forest fires were deliberately ignited by plantation companies, and the other 20% by farmers [3,13]. Rapid development in the oil palm sector during the 1990s led Indonesia to expand oil palm plantations to become the world's biggest producer of palm oil, producing 51% of worldwide yields [13,16]. Between 1990–1997, land designated for oil palm plantations doubled to 2.5 million hectares and was projected to increase to 5.5 million hectares by 2000 [17]. With the issuance of Presidential Decree no. 80/1999 in July 1999, 2.8 million hectares of peatlands were targeted for conversion into cash crop estates, the majority being oil palm estates. Logging and oil palm processes can interact [17]. For instance, forestry companies tend to be interested not only in logging but also in the oil palm sector. As a result, logged forest areas are usually converted into oil palm plantations via S & B.

3.4. Government Corruption and Weakness

Many plantation companies from Indonesia, Malaysia and Singapore establish and maintain political connections with Indonesian government officials to receive concessions and face fewer red-tape barriers, such as attaining necessary certification and rights for clearing land more easily and quickly. Government officials are also motivated towards corruption and encouraging S & B due to low pay, a desire for side-line benefits or high cost and difficulty of monitoring and enforcing laws [10,13]. For instance, former Minister of Trade and Industry, Bob Hasan, has been known to channel funds from public avenues such as the Reforestation Fund for private businesses. In addition, 60 million hectares of forests are concentrated in the hands of about 500 companies with logging rights; Barito Pacific Group alone has access to 5.5 million hectares of forest and owns the largest pulp mills worldwide [17]. When interests collide, company representatives settle them with administrative officials under the table [13]. The Ministry of Environment also has weakened authority due to lack of branches in provincial regions. Although an agency called BAPEDAL (Badan Pengendalian Dampak Lingkungan) was set up to counter this issue, it has not shown success. Provincial officials also do not necessarily adhere to state policies on S & B and in fact often disregard them for private interests [17].

As a result, state policies serve little or no disincentive against rule breaking. Perpetrators do not fear punishment and continue violating the rules, establishing a norm of rule-breaking which influences others to do the same, making punishment of rule breakers difficult and inducing officials to either overlook or even aid rule breaking. For instance, Presidential Decree Keppres no. 32/1990 and Indonesian Government Regulation no. 26/2008 curbs the establishing of oil palm plantations on peat extending more than three meters underground, yet a quarter of plantation companies continue violating the rule. In addition, Duta Palma, an Indonesian plantation company, escaped investigation despite extensive history of illegal S & B due to relations with the Indonesian military [13]. Moreover, despite suspending land clearing licences, most companies responsible for the 1997–1998 forest fires continued illegal S & B, even pushing the blame onto one another or to accidents [16].

4. Consequences

While on one hand it appears to be an advantageous practice, S & B has severe consequences, on the other. These can be understood in terms of the treadmill of production theory, with intensive withdrawals of natural resources from the environment, along with large-scale additions to it. This section analyzes the effects of the withdrawals and additions caused by S & B on the environment, society and economy.

4.1. Withdrawals: Deforestation

The most obvious consequence of S & B is the large-scale removal of forests or deforestation [11,15]. Deforestation incurs heavy environmental costs, including soil erosion and degradation, water pollution, desertification, global warming and climate change, vulnerability to natural disasters such as floods, and threats to biodiversity [15]. The impacts last for a long time even after the area is replanted. Subsequent trees and plants growing in deforested areas may store less carbon than before [7]. Less water permeates the soil after deforestation, reducing the rate of replenishing groundwater. Fewer plant roots store sulphur, causing more sulphate ions to enter the atmosphere and fall as acid rain, damaging vegetation, land and marine life [15]. The impacts of the large-scale deforestation that occurs as a result of S & B are far-reaching and long lasting, as described below.

4.1.1. Soil Erosion and Degradation

A direct consequence of deforestation is the increased rate of surface runoff, which speeds up soil erosion and degradation. Soil erosion refers to movement of soil particles via wind or water from one location to another [15]. As a result, soil nutrient levels and density structure are permanently altered, thereby degrading soil productivity. Furthermore, with S & B, forest canopy is opened, exposing soil directly to weather elements such as wind, rain and sunlight. This increases the ease with which soil dries up and is blown or washed away. Soil temperatures and acidity levels are also affected. Higher soil surface temperatures expedite nitrogen loss into the air as well as biomass decomposition [9]. Higher temperatures and soil acidity increases phosphorus sorption, further exacerbating the limited availability of soil phosphorus and negating subsequent effectiveness of adding more fertilizer to increase phosphorus availability [18–20].

Beyond a certain extent of soil erosion, when topsoil productivity decreases by at least 10%, desertification occurs. Deserts or dust bowls are created or expanded as a result. Eroded soil particles get washed into water bodies and cause water pollution via eutrophication or clogging of rivers, lakes and streams. Soil particles may contain herbicide and pesticide remnants, which may be consumed by marine life and possibly kill them [15,19].

4.1.2. Global Warming and Climate Change

Through S & B, trees and plants that absorb and store carbon are cleared out faster than they can grow back [15]. This reduces forests' capacity to absorb human carbon emissions, leading to substantial release of greenhouse gases such as carbon dioxide and methane which enhances the greenhouse effect and accelerates global warming, changes in precipitation, and climate change [2,11,15,21]. Indonesia ranks highest in carbon dioxide emissions from peatland degradation, approximating 900 million tons annually, as draining large areas of peatland causes the peat to decompose into carbon dioxide [22].

Large-scale deforestation can alter regional weather and even climate. With reduced forest canopy, this results in higher ground temperatures and lower humidity [23]. Coupled with reduced transpiration of plants, local rainfall decreases. This makes forests drier and more susceptible to fire [15]. Beyond thirty years, the local climate may change irreversibly such that forests can no longer return or be sustained and may be substituted by less diverse tropical grassland [15,21].

In a vicious positive feedback loop, climate change threatens forests further by increasing susceptibility to insect and pest species that kill more trees. Thus, forest fires are more likely to recur with greater frequency and intensity than before [15]. Climate change also exacerbates peatland degradation by inducing thawing of peatland usually under permafrost conditions during higher temperatures in warmer seasons, causing peat decomposition into large amounts of carbon dioxide and methane.

4.1.3. Threats to Biodiversity

About half the world's known species are housed in tropical forests alone [15]. In fact, Indonesian rainforests have been hailed as biodiversity hotspots [2]. The destruction of vegetation and habitats of native creatures by S & B threatens their livelihood and survival, and pushes them towards the brink of extinction [11,15]—for instance, exclusive orang-utan communities [2,10]. Species that cannot withstand sudden changes in environment, fires or high temperatures, or those that require very specific conditions for survival, are especially likely to be affected [14,23]. They may be unable to withstand prolonged lack of food and water, escape from the fires, or migrate to new homes [14]. Burning also destroys seeds and roots of vegetation, which impedes regeneration [9].

Thus, biodiversity in post-S & B habitats tend to be substantially lower than pre-S & B habitats [23]. With species endangerment or extinction, decreased diversity of genetic resources lowers species' adaptability in response to changing environments, which, in turn, lowers their likelihood of survival further. Thus, a vicious downward spiral is created. Significant loss of many undiscovered plant and animal species which possess medicinal and healing properties, or other attributes that contribute to much-needed products and services, may result [2]. Potential for raising food production and for developing more hardy and nutritious species of crops and animals is also impeded [15]. Moreover, S & B facilitates intrusion of invasive species such as bracken [9,23]. It was also found that post-burning, easily managed weeds (such as wide-leaf annuals) tended to be replaced by harmful perennials [24].

4.1.4. Peatland Drainage

Indonesia has the biggest area of peatlands worldwide, approximating 27 million hectares [12]. Peatlands store water, absorb atmospheric carbon dioxide and house diverse species of plants and animals, including the endangered Sumatran tiger and orang-utan. The problem arises when forestry and plantation corporations drain peatlands for growing oil palm and logging valuable timber. The incidence of this is ever-increasing, with at least half of all the new, projected oil palm plantations being established in peatlands [13]. Another cause for concern is the poor planning of trans-migration programmes, which led to the further degradation of peatlands. Forests were cleared to construct a 4,400 kilometres canal network. The canal was built to assist crop irrigation and soil drainage during dry and rainy seasons, respectively; however, it also drained excessive peatland moisture into the sea, resulting in low water tables which kill vegetation and reduce capacity to absorb water. As a result of this, the peatlands have dried up and become susceptible to fires during the dry season, as evidenced in the 1997 forest fires. Moreover, peat fires seethe underground for years and reignite during dry conditions [12,14].

4.2. Additions: Trans-Boundary Haze Pollution

Haze refers to "a high concentration of particulate matter" [16] (p. 70). S & B creates forest fire emissions that are transported by wind and rain to other countries [16] and can be exacerbated by dry weather or drought from the El-Nino Southern Oscillation (ENSO) [12]. In the Indonesian forest fires of 1997–1998, haze affected not only Indonesia but also neighboring countries such as Malaysia, Thailand, and Singapore [12]. In Kuching, Sarawak in Malaysia, the Air Pollution Index (API), registered an all-time high of 849 [16]. Concentrations of sulphur dioxide (SO_2), carbon dioxide (CO_2), methane (CH_4) and particulate matter (PM_{10}) exceeded baseline concentrations by at least ten times, five times, two times and twenty times, respectively [7]. In Singapore, haze in 1994 and 1997 from forest fires resulted in prolonged high levels of PM_{10} at 150-180μg m^{-3} [7] as well as a fifty percent spike in carbon monoxide (CO) concentrations [25]. Haze results in negative outcomes for the environment. It inhibits photosynthesis, reducing forests' ability to absorb carbon, which worsens global warming [7]. Furthermore, it has greater risks on society and the economy.

4.2.1. Regional Tension

Haze issues have led to political tension between Indonesia and its neighbors such as Malaysia and Singapore. For instance, while Malaysia and Singapore alleged to help Indonesia fight against its forest fires, Indonesia was also censured for its persistent lack of improvement in instituting fire control and measures. In turn, Indonesia held trans-national firms responsible for unrestrained illegal logging, which left its forests vulnerable to destructive blazes. As reported in popular Dailies, in 2013, Agung Laksono, the in-charge Minister, in response to the haze episode, then criticized Singaporeans for being immature and childish, rousing widespread anger. The Indonesian president, Susilo Bambang Yudhoyono, had to express remorse on behalf of Indonesia to right the repercussions faced by neighboring countries.

4.2.2. Health Risks

The health of approximately seventy-five million people is affected by haze each year [13]. Haze contains $PM_{2.5}$ that contains toxic trace metals such as copper and chromium; inhalation can result in cancer, for every 1 in 200 people [12]. Each $10\mu g\ m^{-3}$ increase in particulate matter is associated with increased lung cancer risk by 8% [12]. $PM_{2.5}$ particles are also miniscule enough to penetrate the lungs deeply, increasing risk of respiratory-related diseases such as bronchitis and asthma. Indeed, inhalation accounts for 70% of $PM_{2.5}$ in the lungs [12]. All in all, haze is associated with respiratory disease, associated hospital admissions, risk of cancer, eye conditions, as well as death [10,12,14].

People residing or working in haze-affected areas, such as fire-fighters and plantation workers, are especially prone to health risks. It was found that concentrations of $PM_{2.5}$, trace metal and nitrated polycyclic aromatic hydrocarbons (PAHs) were highest in areas nearest to peat fires, such as Sumatra and Kalimantan, with severe health consequences. Even outside of Indonesia, the health impacts of haze are strongly felt. For example, in Singapore, the 1997 haze saw a 12% increase in respiratory illnesses and a 19% rise in occurrence of asthma [1,26]. During the same period in Malaysia, the number of respiratory patients increased from 250 per day to 800 per day [26,27]. Simultaneously, there was a huge increase in occurrence of asthma, bronchitis and conjunctivitis across Malaysia [27,28]. The total cost of health damage in Malaysia was approximately RM 129 million during the 1997 haze [27]. Peat fires that smoulder emit especially high amounts of $PM_{2.5}$ [12]. S & B produces gases such as CO and hydrocarbons that contribute to ozone formation [29]. Ozone pollution can cause lung damage and inflammation, and respiratory diseases. Ozone is also the main constituent of smog, which increases eye and throat discomfort as well as the risk of illness [14].

4.2.3. Economic Tensions

Haze pollution and health risks have various ripple effects including hampering economic productivity in affected Southeast Asian countries, especially Singapore and Malaysia [10,13]. More people fell ill due to haze, amounting to heftier medical fees and work absenteeism, which translated to work productivity loss. Additional impacts were seen in the form of declining tourism and recreation in haze-affected areas, which affected performance of businesses [10]. Schools and businesses were shut; flights were delayed or cancelled [10,30]. Kalimantan even experienced lack of food and water [14].

For Singapore, losses incurred from the 1997 haze amounted to US$163.5–US$286.2 million. Greatest loss occurred in the tourism sector, amounting to US$136.6–US$210.5 million. Recreation suffered due to poor scenery and visibility, amounting to costs of US$23.2–US$71.2 million. Health losses amounted to US$3.8–US$4.5 million. Businesses, especially retail and food-and-beverage sectors, suffered as most people stayed indoors during the haze and did not leave their homes longer than necessary [10]. Table 1 summarizes the total damage costs in Singapore due to 1997 haze [30] (p. 182). Indonesia also incurred heavy losses amounting to US$20.1 billion, approximately 50% of its government income in 1997. Its tourism sector declined since tourist hotspots were affected by fire and haze [2]. For instance, most of Kutai National Park in East Kalimantan was burnt [17]. Also in 1997, an Indonesian air flight carrying 234 people on board crashed due to poor visibility from haze, and remains to-date the deadliest aviation disaster in Indonesian history [17].

These calculated losses are likely to be below the actual true costs since not all costs can be fully taken into consideration [10]. The economic costs that countries incur comprise of private costs that are usually taken into account, such as damages and loss of economic goods and services [2]. However, social costs or negative externalities such as loss of forests and corresponding ecosystem services are usually overlooked.

Table 1. Summary of the total damage costs in Singapore due to the 1997 Haze.

Impacts of Haze Damages	Upper Bound Estimation (US$)	Lower Bound Estimation (US$)
Health damage (cost of illness, loss of earnings or productivity, preventive expenditures etc.)	4,517,629	3,776,708
Loss to tourism	210,449,067	136,577,290
Loss in visibility and views	71,137,941	23,057,133
Loss in recreation activities	94,170	94,170
Damage costs per person	95.39	54.50
Damage costs per household	369.90	211.31
% of 1996 Gross Domestic Product (GDP)	0.32	0.18

4.2.4. Food Security Issues

Food security is the state whereby most or all people in a population can get healthy food on a daily basis. S & B relates to food security via net primary productivity (NPP) and the role of producers (usually trees and plants). NPP refers to "the rate at which producers use photosynthesis to produce and store chemical energy, minus the rate at which they use some of this stored chemical energy through aerobic respiration" [15] (p. 61). In other words, only biomass stored in producers, represented as NPP, is available as nutrients for consumers; NPP is thus the limiting factor for survival. Housing huge quantities and species of producers, tropical rain forests are very high in NPP. In S & B, NPP decreases significantly which translates into decreased nutrients available for consumption and use.

NPP is affected by soil productivity. With decreased soil phosphorus availability, duration of yearly harvests may be reduced and the soil becomes less fertile over time, especially if S & B episodes recur [20]. This translates into inadequate and unstable food supplies, threatening the food security and livelihood of farmers, their families and businesses [11]. Reduced soil phosphorus availability is further compounded by crop harvesting which clears away plant material that constitutes sources of phosphorus, as well as erosion in agricultural systems and deforested areas.

5. Sustainability through Plural Environmental Governance

The complex nature of the issue needs integrated environmental efforts which we call "plural environmental governance" (Figure 1). It involves, among other initiatives, intervention based on globalization, decentralized environmental governance, market and agent focused instruments of environmental governance, and cross-scale environmental governance. These have been expanded below.

Figure 1. Plural environmental governance

5.1. Intervention Based on Globalization

To monitor S & B activity, research on haze prevention, techniques to spot burning and to interpret patterns of fire is required, yet it has not been sufficiently addressed and included in preventive costs. Incentives such as international aid for governments to invest in research are warranted. Countries affected by haze from Indonesia's forest fires can offer Indonesia aid equivalent to the maximum damage each of them incurred [10]. Singapore also provides Malaysia and Indonesia satellite data to aid in haze research, courtesy of the Center for Remote Imaging, Sensing and Processing (CRISP) at the National University of Singapore. Another suggestion is debt-for-nature swaps or conservation concessions, whereby countries receive financial aid or have their debts waived in return for preserving forests and natural resources [15]. Research can be conducted on means of determining economic value of ecosystem goods and services, as well as cost-benefit analysis, so that economic products can be optimally priced to include social costs, and to maximize land use among competing economic activities. Research on ENSO permits understanding of its characteristics and patterns of occurrence, which influence effectiveness of S & B policies. Research on potential techniques includes remote sensing, air quality modelling, and Geographical Information Systems (GIS). Adopting a combination of these techniques in parallel, coupled with land ownership records, help in regulating S & B activities as they can help pinpoint perpetrators and hold them responsible. However, they require consistent monitoring and precise, accurate data. Furthermore, adopting techniques in isolation may not depict actual situations completely and accurately [10].

5.2. Decentralized Environmental Governance

5.2.1. Community Resource Management

Forests are ideal for management by local communities, due to their clear boundaries, making it possible to determine rights of access and to monitor usage. For instance, illegal or inappropriate S & B is easily spotted, hence the perpetrator is more likely to be caught and punished. In communities where people have known each other for long, complex interpersonal relationships are established, which facilitate development of shared community norms and expectations. Violation for personal gain would cause the perpetrator to risk heavier losses such as losing respect and trust or being ostracized, which could threaten his future survival within the group. As a result, individuals are motivated to refrain from unacceptable S & B. Moreover, local communities are likely to have adequate knowledge of forest resources and to be highly dependent on forest resources for survival. Hence, they are concerned about its overexploitation; this culminates in a participatory style of creating rules agreed upon by everyone, such that rules are perceived as fair and adhered to voluntarily [18,31]. Complex relationships, coupled with shared norms and adequate knowledge of forest resources, facilitate knowledge transmission throughout social networks via word-of-mouth, which is perceived as more credible and persuasive, thereby more influential in decisions involving S & B. Key strengths of community resource management lie not only in its potential to induce voluntary compliance to S & B regulations but also to enhance the spread of and perceived efficacy of S & B alternatives, thus increasing adoption rates.

5.2.2. Education

People can be educated on the long-term outcomes of S & B, ways of regulating S & B activities and emissions, as well as fire-free alternatives [10,15,32]. Such information can be disseminated via formal channels such as national media, or via informal channels such as word-of-mouth from community members [11]. To be effective, it is important to customize information to make it understandable, credible, personally relevant, motivating, and attention capturing for intended parties. Examples include emphasis on on-going losses and costs incurred from S & B, keeping information simple, direct and relevant, using striking and tangible images of S & B costs to evoke moderate fear, and pairing them with strategies to reduce the fear. These strategies would include adopting alternatives, strategies to increase perceived self-capability of executing alternatives by emphasizing the ease of grasping new technology, and strategies to ease the transit to alternatives by lowering costs and providing on-site guidance [18,33]. Elicit public commitment to phasing out S & B by signing a statement to do so, broadcasting the names of participating individuals and corporations on national television (with prior consent sought), and so on. This reduces likelihood of detraction as they now are motivated to uphold a positive public image of walking the talk and to maintain self-esteem [33]. Note that eventual adoption of S & B alternatives can be influenced not just by individual attributes e.g., level of education, but also by farm characteristics such as size and type of crop grown, and by institutional factors such as land ownership policies [11].

5.3. Market- and Agent-Focused Instruments of Environmental Governance

5.3.1. Incentives and Rewards

Perpetrators are usually aware of costs of no-burning and benefits of S & B accruing to self, but not the costs of S & B and benefits of no-burning that accrue to society. As a result, they perceive benefits of S & B as overriding its costs [11]. Incentives in the form of regulations, taxes, rewards, and so on serve to correct this misguided perception. The purpose is to increase perceived costs and reduce perceived benefits of S & B, as well as to increase perceived benefits of S & B alternatives [18]. Note that S & B alternatives should address not only environmental needs and concerns but also that of parties involved [3].

Rewards and assistance can be provided to parties that comply with S & B regulations or those that are willing to incorporate fire-free alternatives. For instance, rural communities can receive funding or other rewards if bigger-than-permitted fires have not occurred in the vicinity in any particular year, providing them with an impetus to control fires and to report violations. Companies, especially smaller ones, can receive government funds and subsidies in areas of technology adoption, training and consultation in forestry management [10]. An international fund can be set up to help farmers in developing countries adopt more sustainable fire-free alternatives for land clearing and agriculture [15]. In addition, alternative employment may be offered to farmers to pull them out of poverty [2].

5.3.2. Regulations and Policies

At present, a complete ban on burning is not feasible as it can increase farmers' poverty. In addition, it is not possible to monitor every single violation since farmers are likely to burn smaller areas at a time that are not so easily detected. The initial ban on burning in 1984 had to be renewed in 1997 due to lack of adherence [3]. Regulating S & B is more feasible in mitigating its negative environmental impact than complete bans. To be effective, perceived probability of being penalized and severity of penalties need to be sufficiently high [18]. Thus, laws and regulations pertaining to S & B need to be consistently and rigorously implemented. An example could be imposition of strict conditions for granting forestry licenses. Another suggestion is to build fire-fighting capability in advance, adopt the newest technology to minimize emissions, pay a deposit in advance to cover potential future costs of pollution and buy insurance [10]. Perpetrators can also be made liable to foot damages in the event of loss from fire, regardless of extenuating circumstances, to increase adoption of precautionary measures.

Authorities would also need to punish illegal logging severely [15]. A tax on land clearing, proportionate to acreage of land owned, can be made mandatory on forestry and plantation companies, to cover costs of land clearing undertaken by a central state agency; this would lower companies' need or inclination to resort to S & B, as S & B would then constitute an additional cost. For this tax scheme to work, the state agency in charge needs to be highly responsive to requests and to clear land efficiently [2]. Corruption within the government needs to be stamped out. The Indonesia Corruption Watch investigates cases of corruption whereby government officials have illegal connections with forestry and plantation companies, or enjoy private benefits from such collusions, and prosecutes perpetrators [13]. To systematically eradicate corruption at all levels, from top management to provincial branches, greater transparency of government rules and operations, as well as efficient and effective communication and cooperation among different agencies and levels of government, are required.

To emphasize costs of S & B, given that losses usually matter more than equivalent gains, the media can publicly blacklist identities of companies that violate regulations; the government can release actual costs of S & B, breaking it down into subcomponents such as private and social costs [10].

Burning of smaller areas one at a time can be regulated [3]. During dry seasons or impending drought, burning can be prohibited, requiring farmers and companies to adopt fire-free alternatives such as grinding and mulching [10]. Incentives to do so require that parties do not incur associated losses such as smaller yields. Therefore, farmers' views need to be adopted to understand more clearly perceived barriers and costs to adopting fire-free alternatives. In addition, consistent, rigorous monitoring and enforcement tends to be more feasible for company operations but less so for individual farmers [3].

6. Conclusions

S & B is a complex phenomenon, with multiple interacting factors and consequences that vary across people, situations and time [18]. We have used the treadmill of production theory to unpack the causes and consequences of this practice and proposed a plural environmental governance model to formulate potential solutions. As discussed, potential interventions warrant multi-faceted, multi-disciplinary approaches adopted in parallel, which underlies the essence of a plural coexistence framework. Scholars delineate a community-based forest program that incorporates all three strategies of incentives, education and community resource management [33]. The program aims to resolve issues of S & B in Indonesia while providing participants with employment and access to forest assets to help them rise above poverty. Various stakeholders such as communities and national and international Non-Governmental Organizations (NGOs) collaborate to ensure its long-term feasibility. Under this program, participants apply for licenses to manage the forests, which are certified by the Forest Stewardship Council (FSC). Participants also undergo training on knowledge and skills in forestry management and the FSC-certified wood market, with emphasis on the ecological value of forests. Rules regarding eligibility for the program and production of FSC-certified wood are specified, such as the allowed maximum width and number of trees to be cut. With that, each group of farmers allocated to a plot of forest decides among themselves the specific areas to cut and submits the decision to a local cooperative for compiling the harvest schedule. Upon wood production, farmers receive partial payment, with the remaining payment pending receipt after sales. Participants also receive additional income in the form of dividends. Throughout the various processes of license application, training, and wood production, participants engage in much social interaction with other community members, which fosters a sense of belonging, collective security, as well as responsibility towards the group. Such a system has the potential to draw more people in to expand its scope of influence because of the embedding of incentives and education within the context of the community, which targets many S & B factors in parallel.

Acknowledgments: This research was supported by Nanyang Technological University's Undergraduate Research Experience on Campus (URECA) program and a Tier-1 grant from the Ministry of Education, Singapore. The authors would like to thank the reviewers for their helpful comments.

Author Contributions: Md Saidul Islam conceptualized and designed the research. Yap Hui Pei and Shrutika Mangharam wrote the paper and analysed the data. All authors have read and approved the manuscript.

Conflicts of Interest: The authors declare no conflict of interest.

References

1. Sastry, N. Forest Fires, Air Pollution and Mortality in Southeast Asia. *Demography* **2002**, *39*, 1–23. [CrossRef] [PubMed]
2. Varma, A. The economics of slash and burn: A case study of the 1997–1998 Indonesian forest fires. *Ecol. Econ.* **2003**, *46*, 159–171. [CrossRef]
3. Ketterings, Q.M.; Wibowo, T.T.; van Noordwijk, M.; Penot, E. Farmers' perspectives on slash-and-burn as a land clearing method for small-scale rubber producers in Sepunggur, Jambi Province, Sumatra, Indonesia. *For. Ecol. Manage.* **1999**, *120*, 157–169.
4. Schnaiberg, A. *Environment: From Surplus to Scarcity*; Oxford University Press: New York, NY, USA, 1980.
5. Schnaiberg, A.; Pellow, D.N.; Weinberg, A. The Treadmill of production and the environmental state. *Environ. State Press.* **2002**, *10*, 15–32.
6. Lemos, M.C.; Agrawal, A. Environmental governance. *Annu. Rev. Environ. Resour.* **2006**, *31*, 297–325. [CrossRef]
7. Davies, S.J.; Unam, L. Smoke-Haze from the 1997 Indonesian forest fires: Effects on pollution levels, local climate, atmospheric CO_2 concentrations, and tree photosynthesis. *For. Ecol. Manage.* **1999**, *124*, 137–144. [CrossRef]
8. Kato, M.D.S.; Kato, O.R.; Denich, M.; Vlek, P.L. Fire-Free alternatives to slash-and-burn for shifting cultivation in the Eastern Amazon region: The role of fertilizers. *Field Crops Res.* **1999**, *62*, 225–237. [CrossRef]
9. Kleinman, P.; Pimentel, D.; Bryant, R.B. The ecological sustainability of slash-and-burn agriculture. *Agric. Ecosyst. Environ.* **1995**, *52*, 235–249. [CrossRef]
10. Quah, E. Transboundary pollution in Southeast Asia: The Indonesian fires. *World Dev.* **2002**, *30*, 429–441. [CrossRef]
11. Schuck, E.C.; Nganje, W.; Yantio, D. The role of land tenure and extension education in the adoption of slash and burn agriculture. *Ecol. Econ.* **2002**, *43*, 61–70. [CrossRef]
12. Betha, R.; Pradani, M.; Lestari, P.; Joshi, U.M.; Reid, J.S.; Balasubramanian, R. Chemical speciation of trace metals emitted from Indonesian peat fires for health risk assessment. *Atmos. Res.* **2013**, *122*, 571–578. [CrossRef]
13. Varkkey, H. Patronage Politics as a driver of economic regionalisation: The Indonesian oil palm sector and transboundary haze. *Asia Pac. Viewp.* **2012**, *53*, 314–329. [CrossRef]
14. Brown, N. Out of control: Fires and forestry in Indonesia. *Trends Ecol. Evol.* **1998**, *13*, 41. [CrossRef]
15. Miller, G.; Spoolman, S. *Living in the Environment: Principles, Connections, and Solutions*; Nelson Education: Scarborough, ON, Canada, 2011.
16. Gellert, P.K. A brief history and analysis of Indonesia's forest fire crisis. *Indonesia* **1998**, 63–85. [CrossRef]
17. Cotton, J. The "haze" over Southeast Asia: Challenging the ASEAN mode of regional engagement. *Pac. Aff.* **1999**, 331–351. [CrossRef]
18. Gardner, G.; Stern, P. *Environmental Problems and Human Behavior*; Allyn & Bacon: Boston, MA, USA, 1996.
19. Ketterings, Q.M.; van Noordwijk, M.; Bigham, J.M. Soil phosphorus availability after slash-and-burn fires of different intensities in rubber agroforests in Sumatra, Indonesia. *Agric. Ecosyst. Environ.* **2002**, *92*, 37–48. [CrossRef]
20. Tinker, P.B.; Ingram, J.S.; Struwe, S. Effects of slash-and-burn agriculture and deforestation on climate change. *Agric. Ecosyst. Environ.* **1996**, *58*, 13–22. [CrossRef]
21. Skinner, B.J.M.; Skinner, B.B.J.; Murck, B. *The Blue Planet: An Introduction to Earth System Science*; Wiley: Hoboken, NJ, USA, 2011.
22. Cleary, D.F.; Mooers, A.Ø.; Eichhorn, K.A.; Van Tol, J.; De Jong, R.; Menken, S.B. Diversity and community composition of butterflies and odonates in an Enso-Induced fire affected habitat mosaic: A case study from East Kalimantan, Indonesia. *Oikos* **2004**, *105*, 426–448. [CrossRef]
23. Pound, B. Alternatives to Slash and Burn Agriculture in the Ichilo-Sara Region of Bolivia. Available online: http://r4d.dfid.gov.uk/PDF/Outputs/NatResSys/R6165FTR.pdf (accessed on 19 May 2016).
24. Mukherjee, P.; Viswanathan, S. Contributions to CO concentrations from biomass burning and traffic during haze episodes in Singapore. *Atmos. Environ.* **2001**, *35*, 715–725. [CrossRef]

25. Lumpur, K. Biregional Workshop on Health Impacts of Haze-Related Air Pollution. World Health Organization. Available online: http://iris.wpro.who.int/bitstream/handle/10665.1/5982/RS_98_GE_17_MAA_eng.pdf (accessed on 19 May 2016).

26. Afroz, R.; Hassan, M.N.; Ibrahim, N.A. Review of air pollution and health impacts in Malaysia. *Environ. Res.* **2003**, *92*, 71–77. [CrossRef]

27. Hassan, M.N.; Choo, W.Y.; Afroz, R.; Rahman, M.M.; Theng, L.C. Estimation of health damage cost for 1997-haze episode in Malaysia using the Ostro model. In Proceedings of the Malaysian Science and Technology Congress (MSTC), Ipoh, Perak, Malaysia, 13–15 November 2000.

28. Kita, K.; Fujiwara, M.; Kawakami, S. Total ozone increase associated with forest fires over the Indonesian region and its relation to the El Nino-Southern oscillation. *Atmos. Environ.* **2000**, *34*, 2681–2690. [CrossRef]

29. Nasrul, B. Program of community empowerment prevents forest fires in Indonesian peat land. *Procedia Environ. Sci.* **2013**, *17*, 129–134.

30. Quah, E.; Johnston, D. Forest fires and environmental haze in Southeast Asia: Using the 'Stakeholder' approach to assign costs and responsibilities. *J. Environ. Manage.* **2001**, *63*, 181–191. [CrossRef] [PubMed]

31. Feder, G.; Slade, R. The acquisition of information and the adoption of new technology. *Am. J. Agric. Econ.* **1984**, *66*, 312–320. [CrossRef]

32. Clayton, S.; Myers, G. *Conservation Psychology: Understanding and Promoting Human Care for Nature;* Wiley-Blackwell: Hoboken, NJ, USA, 2009.

33. Harada, K. Certification of a community-based forest enterprise for improving institutional management and household income: A case from Southeast Sulawesi, Indonesia. *Small-Scale For.* **2014**, *13*, 47–64. [CrossRef]

sustainability

MDPI

Article

The Making of Sustainable Urban Development: A Synthesis Framework

Hui-Ting Tang and Yuh-Ming Lee *

Institute of Natural Resources Management, National Taipei University, 151 University Road, San Shia District, New Taipei City 23741, Taiwan; s810075101@webmail.ntpu.edu.tw
* Correspondence: yml@mail.ntpu.edu.tw; Tel.: +886-2-8674-1111 (ext. 67333)

Academic Editor: Md Saidul Islam
Received: 29 February 2016; Accepted: 14 May 2016; Published: 19 May 2016

Abstract: In a time of rapid climate change and environmental degradation, planning and building an ecologically sustainable environment have become imperative. In particular, urban settlements, as a densely populated built environment, are the center of attention. This study aims to build a clear and concise synthesis of sustainable urban development not only to serve as an essential reference for decision and policy makers, but also encourage more strategically organized sustainability efforts. The extensive similarities between environmental planning and a policy-making/decision-making/problem-solving process will be carefully examined to confirm the fundamental need to build a synthesis. Major global urban sustainability rankings/standards will be presented, discussed, and integrated to produce a holistic synthesis with ten themes and three dimensions. The study will assemble disparate information across time, space, and disciplines to guide and to facilitate sustainable urban development in which both environmental concerns and human wellbeing are addressed.

Keywords: sustainable urban development; synthesis framework; environmental planning

---

1. Introduction

1.1. Challenges of Climate Change and Environmental Degredation: Cities on the Front Line

The climate change we are facing now is of large scale and high speed, unprecedented and unseen in the past. It occurs across national borders and geographical boundaries and has already taken its toll on humankind. The Fourth Assessment Report (AR4) compiled by the Intergovernmental Panel on Climate Change (IPCC) makes a shocking but truthful observation: global average surface temperatures have increased by about 0.74 °C over the past one hundred years (between 1906 and 2005) and 2005 and 1998 were the two warmest years in the instrumental global surface air temperature record since 1850 [1]. In the Fifth Assessment Report (AR5) released in 2013, new atmospheric temperature measurements are used and the IPCC goes further to "show an estimated warming of 0.85 °C (1.5 °F) since 1880 with the fastest rate of warming in the Arctic" [2].

Several different scenarios of the 21st century global temperatures and greenhouse gases (GHGs) concentrations have been described in the AR5, and it has been projected that "global surface temperature increases will exceed 1.5 °C and keep rising beyond 2100 in all scenarios except the lowest-emission scenario" [2]. The speed of global warming is picking up and, without cooperative measures from around the world to limit GHGs emissions, "in the scenarios with higher rates of emissions, warming is likely to exceed 2 °C by 2100, and could even exceed 4 °C" [2]. Also noted by the IPCC is that rising sea levels are a particularly serious outcome of global warming. Worldwide sea level is expected to increase by 8–88 cm during the 21st century [3].

As temperatures increase, more floods, droughts, diseases, famines, and wars will follow, creating millions of dislocated people and destroying ecosystems. According to a team of health and climate scientists from the World Health Organization (WHO) and the University of Wisconsin at Madison, global warming and climate change will not only threaten our health in the future, but also cause more than 150,000 deaths and five million illnesses every year. This number is estimated to double by 2030 [4]. Once the 2 °C threshold of temperature increase is passed, the balance of ecosystem will be thrown off, food and water safety will be compromised, and extreme weather events will strike. Ultimately comes the extinction of all species.

The United Nations Framework Convention on Climate Change (UNFCCC) defines climate change as "a change of climate which is attributed directly or indirectly to human activity that alters the composition of the global atmosphere and which is in addition to natural climate variability observed over comparable time periods" [5]. It is stated clearly that climate change here refers only to the type originating from human causes. After decades of careful observation and examination, the scientific community has reached a consensus, concluding that climate change is indeed happening and various human activities are to take the majority of the blame.

Such a claim is supported by the United Nations Human Settlements Programme (UN-HABITAT), which has identified that, since cities are heavily populated and concentrated with human activities like manufacturing and consumption, they produce nearly 60%–70% of the total GHGs emissions. However, cities all over the world take up in total only 2% of the land [6]. It should also be noted that cities not only rank as the most prominent GHGs sources, but also "concentrate disproportional parts of the economy, resource consumption and the decision-making power in most countries" [7]. A staggeringly high proportion of "75% of the global economic production takes place in urban areas" [7]. Not surprisingly, it comes with a price: cities consume 75% of the planet's resources, generate a comparable percentage of waste, including air pollution, solid waste, and toxic effluents [8] and use 67% of the total global energy consumption [7]. From this perspective, it follows that urbanization, or to be more specific, urban development is indeed strongly associated with environmental degradation.

1.2. Reserch Rationale and Objective

Around the globe, rapid urbanization has created immense burdens on public infrastructure, such as transit systems and utility facilities. It has also produced a highly stressed and strained ecosystem. The ongoing reciprocal action between climate change and urbanization further complicates the situation and has greatly threatened the global natural environment, economic development, social stability, and human wellbeing. Combining highly concentrated population and economic assets, urban settlements are truly the places where human impacts on the environment are most extensive, persistent, and focused. Accordingly, it is well recognized that careful planning of the environment of urban settlements will be the crucial step to securing a sustainable future.

The fact that cities are the places where a large portion of economic activity and consumption take place means that human impacts on the environment will be the most intense. Contrariwise, environmental impacts on human society will be the most visible. In short, cities are fundamental to climate change management efforts. This compelling fact defines the scope of argument in this study. Namely, the focus of environmental planning presented and discussed below will be perceived from the perspective of urban settlements. We aim to address the challenge of sustainable urban development by means of offering a concise synthesis framework. It will register all concerns in environmental issues and human wellbeing. It will not only serve as a fundamental reference for decision and policy makers, but also encourage more strategically organized efforts in sustainable environmental planning.

In this study, Section 1 presents an overview of the problems of climate change and environmental degradation originating from human activities. The rest of the study is organized as follows. Section 2 begins with an investigation of the role of cities both in creating and in addressing the issues of sustainable urban development. Then comes an extensive review of a range of concepts and approaches developed by a variety of bodies, showing their focuses and areas of overlap and divergence. Section 3 moves on to argue that environmental planning is a form of policymaking which in its turn is a type of problem-solving that requires a clear formulation of agenda, hence the need of a clear synthesis framework. The framework is finally offered by Section 4, which merges together all the approaches considered through an integrative methodology in order to give them consistency and exhaustiveness. Section 5 in conclusion indicates directions for future applications and research.

2. Overview of Urban Planning and Sustainable Urban Development

2.1. The Role of Urban Planning in Sustainable Urban Development

The world's first cities can be dated back to 3500 B.C. It is generally agreed by scholars that the Uruk Cluster in Mesopotamia is humanity's first great urban center and city [9]. It was located 150 miles south of the modern-day Baghdad and, ever since its establishment, cities have come in to existence all over the world. Early into the 21st century, cities have started to appear with greater frequency. According to statistics calculated by Global Health Observatory, a program run by the WHO, as of 2010, more than half of all people live in urban areas. It is projected that by 2030, six out of every 10 will live in cities, and by 2050, this proportion will increase to seven out of 10 [10].

Although cities cover only a trivial percentage of the land, they are densely populated and create a high volume of economic activities. It has been observed by the International Bank for Reconstruction and Development that "by enabling density—the concentration of people and economic activities in a small geographic space—cities have helped transform economies for many centuries" [11] (p. 1). Statistically, 50% of world gross domestic product (GDP) is generated on just 1.5% of the world's land, practically all of it in cities [12]. Cities are characterized by high population densities and prosperous human activities and, as stated above, around 70% of GHGs released into the atmosphere are attributed to urban residents [6]. It is hoped by focusing on the planning of cities, the most prominent GHGs sources, we might facilitate sustainability and improve human comfort and development at the same time.

Urban planning is defined as "the planning and designing of buildings, roads, and services in a town" [13]. In "urban planning," we deal with two concepts: "urban environment" and "planning." Even though the first term is frequently used, it does not mean that it has a universally-agreed-upon definition. In fact, as to what an urban area stands for or what it is comprised of, we still do not have a consensus [14]. In most countries, whether a settlement or population should be classified as rural or urban often depends on its population number, density, physical characteristics, or administrative functions [15]. The International Council for Science proposes a synthesized definition to call urban environment "the natural, built and institutional elements that determine the physical, mental and social health and wellbeing of people who live in cities and towns" [16] (p. 8). As for "planning," if used in a city or business context, it usually refers to "the establishment of goals, policies, and procedures for a social or economic unit" [17]. From the discussion above, we can see that "cities have thus been planned from the beginning, enabling new settlements, economic specialization, and cultural expression" [11].

Urbanization is most evident in the context of cities where the majority of global population resides and therefore brings about the most significant environmental impact. It can be reasonably inferred that human activity is the principal driving force of various kinds of environmental problems. With so many residents and properties in them, cities are also the most vulnerable in the face of extreme weather events or other climate-related impacts. For example, compared with rural areas, big cities will encounter a more rapid temperature increase because of the heat island effect. According to the United States Environmental Protection Agency, any city with one million people or more can be 1–3 °C warmer than surrounding areas in terms of the annual mean air temperature [18]. This shall increase or aggregate health problems. From this point of view, cities are indeed both the victimizers and the victims. To solve this cyclical problem, cities, or urban settlements, should be carefully planned from the beginning.

2.2. Different Views and Aspects of Sustainable Urban Development

To address the challenges of climate change and environmental degradation, more holistic planning of urban development has become our immediate priority. In 1973, the United Nations Environmental Progamme (UNEP) declared 5 June of every year as the World Environment Day to promote global environmental awareness of the importance of taking prompt action to protect and to preserve the Earth. Its theme in 2005 was "Green Cities" and used the slogan "Plan for the Planet!" Starting from 1986, the UN-HABITAT nominated the first Monday of every October as the World Habitat Day. Every year, in the commemoration of this day, a specific topic on urban environment and development is celebrated, such as "Planning Our Urban Future" in 2009, "Better City, Better Life" in 2010, and "Cities and Climate Change" in 2011. They all call for cities around the world to alleviate pressure on the ecosystem and to ensure the quality and security of our living environment.

The aim of sustainable urban development has emerged and spawned numerous urban settlement theories, including the "Healthy City", "Sustainable City", "Low-Carbon City", "Transit-Oriented City", "Compact City", "Smart City", "Green City", and "Livable City". These theories may come with different concerns in different areas, but they all share one central idea and ultimate goal: achieving maximum development with minimum resource consumption and environmental impact to ensure the well-being of both humans and the Earth.

Investigating the relationship between humans and environment has always attracted considerable attention. The concept of the "Healthy City" is used in the field of public sanitation and city design; it emphasizes how policies can influence human health. It originated in the mid-19th century and its modern incarnation appeared in the Initiative on Healthy Cities and Villages advocated by the WHO in 1986 [19]. At the time, 11 cities were initially chosen to participate in the project and follow the principle of "Health for All (HTA)" [20]. The WHO points out that many factors, including society, economy, and environment, influence human health, so planning a Healthy City not only involves public health protection, but also requires efforts in political, economic, and social arenas [21]. A Healthy City will bring many benefits, such as "a clean, safe physical environment of high quality", "the meeting of basic needs for all the city's people", and "an ecosystem that is stable now and sustainable in the long term" [20].

The WHO initiated the Healthy Cities Project in 1990 and 47 countries participated in it during the first stage. At the time, the WHO drew up 53 Healthy Cities Indicators as initial references and continued to collected relevant data and statistics. After meticulous study and analysis, the 53 indicators were condensed into 32 and were classified in four categories: Health Indicators, Health Service Indicators, Environmental Indicators, and Socio-economic Indicators [22].

Since the urban environment comprises a wide range of elements and its form of planning is varied, the "Sustainable City" has become a major trend in many countries. It takes environmental impacts into consideration during the design phase of city planning and encourages residents to actively reduce their energy and water consumption and to limit their emissions of GHGs and other pollutants. In 2002, the International Environmental Technology Centre of the UNEP and the Environment Protection Authority of Victoria in Australia collaborated to hold an international expert panel in Melbourne. From it, the Melbourne Principles for Sustainable Cities were developed. The vision promoted by the principles is to "create environmentally healthy, vibrant and sustainable cities where people respect one another and nature, to the benefit of all" [23]. Rather than a fixed framework, the principles are designed to be flexible enough to be adopted by any cities and they provide a starting point for decision-makers on the journey towards sustainability, assisting government officials in understanding the implications of decisions taken at a broad strategic level [23].

The "Low-Carbon City" is sometimes referred to as the "Low-Emission City". To confront the issue of ever-increasing GHGs emissions, the UNEP and the UNFCCC have been advocating Adaption and Mitigation: the former addresses the adverse effects of climate change, responds to the impacts of existing climate change, and improves resilience against future impacts [24]; the latter refers to reduction or prevention of GHGs emissions. For example, "mitigation can mean using new technologies and renewable energies, making older equipment more energy efficient, or changing management practices or consumer behavior" [25].

A Low-Carbon City uses mitigation strategies in urban planning with the aim of enlisting efforts from not only the public and private sectors but the whole community significantly to cut down emissions. In *Global Report on Human Settlements 2009—Planning Sustainable Cities: Policy Directions* published by the UN-HABITAT, it has been strongly advocated that "the key objective of the trend towards 'carbon neutral' cities is to ensure that every home, neighbourhood and business is carbon neutral. Carbon neutral cities are able to reduce their ecological footprint through energy efficiency and by replacing fossil fuels" [26] (p. 149). From this statement, we can reasonably infer that "low carbon," or the ultimate "carbon neutral", has become the goal of all sustainable urban development. Such awareness and action are essential if the world is to shift to "post-carbon cities" [27]. The World Bank launched the Low-Carbon Livable Cities Initiative in September 2013 and planned to help 300 large cities in developing countries to transition into low-carbon settlements in the next four years. Assistance will come in the form of planning and financing and necessary assistance will be promptly provided.

Along the same lines, several programs have been in place to help cities reach the goal of carbon emissions reduction or carbon neutrality. Examples include the Cities for Climate Change program by the Local Governments for Sustainability, the Clinton Foundation's C-40 Climate Change Initiative, Architecture 2030, and the UN-HABITAT's Cities for Climate Change Initiative. These programs all stress the importance of reducing energy use wherever and whenever possible, especially in the building and transportation sectors. Since transport creates the primary form of any city, it is frequently regarded as the most fundamental infrastructure for a city [28] and naturally should be the focus of any urban sustainability efforts.

One of the dominant features of modern cities is high density. Those in developing countries often have much higher density than those in developed countries. If vehicles in these confined spaces are not controlled in numbers, or have poorly-maintained fossil fuel engines, serious air pollution will surely follow. Therefore, cities have to rigorously monitor and manage such emission sources [29]. Transit-Oriented Development (TOD) has the potential to address this issue. TOD represents a neighborhood incorporating a mélange of land uses centered around a transit station [30]. Within a short walking distance from the core, usually in ten minutes, residents can easily access all kinds of daily services, such as retail stores, offices, and residential quarters. The function and importance of TODs are emphasized as follows [31] (p. 2):

> "the location, mix, and configuration of land uses in TODs are designed to encourage convenient alternatives to the auto, to provide a model of efficient land utilization, to better serve the needs of [...] diverse households, and to create more identifiable, livable communities".

TOD can not only reduce car use per capita by 50%, but save households about 20% of their income because they can manage with average one fewer car, or even none [32]. It also enables low carbon housing. For instance, in the United States, shifting 60% of new growth to compact/high-density patterns will reduce CO_2 emissions by as much as 85 million metric tons annually by 2030 [33]. Compared with traditional community development, TOD expands facets of economy, comfort, and environment. As identified by Belzer and Autler [34], measures of livability which relate to TOD include reduction of gasoline consumption, increased walkability and access to public transportation, decreased traffic congestion, positive health outcomes, and more convenient access to services, activities, and public spaces. Cities, or the built environment, are all too often the most prominent

GHGs sources. In other words, they are the key to success of any efforts towards emission reduction. TOD illustrates that, in urban development, environmental concerns and human interest can be balanced at the same time under the common goal of sustainability for all.

The "Compact City" also strives for TOD and plans for roads, streets, and neighborhood networks that promote walkability and are convenient for all users. It is high in density and social diversity, emphasizing the optimal provision of infrastructures in cities of small and medium size and advocating local production and consumption. In it, economic and social activities often overlap and community development is focused on the neighborhood. Therefore, energy and space efficiency can be greatly enhanced.

In *Urban Patterns for A Green Economy—Leveraging Density*, a report published by the UN-Habitat in 2012, five Ds that characterize a compact city are proposed: Density, Diversity, Design, Destination, and Distance to Transit. The five Ds show the importance of making good decisions on locations, urban structures, and street networks in order to weave an urban fabric conducive to walking, cycling, and public transit [35]. With similar ideals in mind, the Institute for Transportation and Development Policy (ITDP) released a report named *Europe's Vibrant New Low Car(bon) Communities*. It puts forth eight principles for smart urban growth, or a smart city: promote walking, prioritize bicycle networks, create dense street networks, support high-quality transit, plan for mixed land use, match density with transit capacity, create compact regions, and regulate parking and local road use [36]. In the report, the ITDP emphasizes the importance of walking, cycling, and quality public transportation and believes that the key to emission reduction is to cut back on the use of vehicles that burn fossil fuels. The belief is actually summed up in the title of the report, *Europe's Vibrant New Low Car(bon) Communities*, as the word "car" is plainly stated.

In literature on the subject on sustainable urban settlements, the notion of "greenness" has also become influential in recent years. It is frequently presented as "greening" or "green", and can be found in various city rankings, such as the European Green Capital Award (EGCA) and the Green City Index. The EGCA aims to "reward cities which are making efforts to improve the urban environment and move towards healthier and sustainable living areas" [37]. Siemens AG and the Economist Intelligence Unit (EIU) collaborate to survey cities in more than 120 countries. The focal geographical regions cover Europe, Latin America, United States, Canada, Asia, and Africa. Cities are assessed and compared in terms of environmental performance. The final evaluation results will be compiled and presented as the Green City Index, showing weaknesses and strengths of each region and each city. The Green City Index is targeted to measure and to rate the environmental performance of cities, "touching on a wide range of environmental areas, from environmental governance and water consumption to waste management and greenhouse gas emissions" [38] (p. 4).

As urban settlements represent a built environment with various man-made architectural structures, the concept of greenness is also embodied in contemporary building standards. Both homes and commercial buildings use large amounts of energy for heating, cooling, cooking, and management of waste. Attempts to rein in such energy use and its subsequent GHGs emissions from fossil fuel combustion have led to an increase of green building standards that promote better occupant comfort and lower environmental impacts at the same time. In general, a green building aims to be responsible to the environment during its entire life cycle and to increase its energy efficiency at different stages, including siting, design, construction, operation, maintenance, renovation, and demolition. It requires close cooperation among design teams, architects, engineers, and clients [39]. Compared with traditional ones, green buildings expand concerns of economy, utility, durability, and comfort [40]. Around the world, "incentives or requirements for buildings to meet green-building standards have been used in some cities as part of a move towards carbon neutrality" [26] (p. 41).

The notion of "livability" is also highly noteworthy. It is sometimes presented as "liveability" or "livable/liveable", and appears in numerous documents from both public and private sector organizations. For efforts made by public organizations, the most recent and significant one is the Better Life Initiative, which is the culmination of research results published by the Organisation for

Economic Co-operation and Development (OECD) in 2011. The OECD has put in more than a decade of work and has subsequently assembled internationally comparable measures of well-being, called the Better Life Index. The Index is one of the core products from the Initiative. It invites users to compare well-being across countries according to the following 11 topics: community, education, environment, civic engagement, health, housing, income, jobs, life satisfaction, safety and work-life balance [41]. The OECD's goal lies in "developing statistics to capture aspects of life that matter to people and that shape the quality of their lives" [41] (p. 1).

For notions deployed by private sector organizations, there are also numerous examples. By way of example, the EIU runs a global survey of livability entitled the "EIU Liveability Ranking". It states that livability "assesses which locations around the world provide the best or the worst living conditions" [42] (p. 1). Mercer, a global leading human resources consultancy, publishes "Quality-of-Living Reports" that rank cities in terms of quality factors including political/social/economic environment, medical/health considerations, and education [43]. Monocle, a global affairs magazine, holds its annual "Quality of Life Survey", previously named "The Most Liveable Cities Index". It rates the "components and forces that make a city not simply attractive or wealthy but truly liveable" [44] and announces every year its top 25 livable cities in the world, based on "statistics collected on population, international flights, crime, sunshine, tolerance, unemployment rate, upcoming developments, electric car charging points, culture, bookshops, green space, street life, and dinner on a Sunday" [45].

Presented above are some of the most notable and most frequently-cited global livability rankings. Through them, we may get a fuller understanding of what livability is. However, livability "does not come packaged in a single accepted definition" [46] because the concept has constantly been associated with an abundance of social characteristics and physical aspects. It involves not only elements of the daily physical environment but ideals of placemaking. From this point of view, "a livable community is one that has affordable and appropriate housing, supportive community features and services, and adequate mobility options, which together facilitate personal independence and the engagement of residents in civic and social life" [47].

3. Environmental Planning as Read in Policy-Making, Decision-Making, and Problem-Solving

Innumerable countries are now faced with the same challenge: how to design and develop sustainable urban settlements. Environmental planning in this sense is much like a problem-solving process. We should also bear in mind that for any environmental planning to be sustainable, it should take into consideration the environmental, social, political, economic, governance, and ethics factors that can influence and determine the relationship between natural systems and human systems. All concerns have to be addressed in balance in order to render well-rounded decisions and polices.

If inspected in fine enough detail, policy-making is in principle strongly similar to decision-making, as interpreted since the 1950s. Harold Lasswell [48] was one of the first to view the overall process of policy-making through the lens of "phases" or "stages." He put forward the following seven "stages" of "the decision process": intelligence, promotion, prescription, invocation, application, termination, and appraisal. Since Lasswell's identification of the seven stages, there has been an abundance of variants to the number and specification of the stages [49–56]. Though the set of stages has been challenged and placed under scrutiny over the years, it remains a firm basis for subsequent study in policy science and policy analysis [57].

Stemming from the work of Lasswell are countless policy-making models. Stokey and Zeckhauser [58] propose a five-step process in which the analysts are charged to determine the underlying problem and objective to be pursued, set out possible alternatives, predict the consequences of each alternative, determine the criteria for measuring the achievement of alternatives, and to indicate the preferred choice of action. Anderson's [59] policy process model has six stages: problem identification, agenda formation, policy formulation, policy adoption, policy implementation, and policy evaluation. Quade [60] lists five elements: problem formulation, searching for alternatives, forecasting the future environment, modeling the impacts of alternatives, and evaluating the alternatives. Hill [61] and Jann and Wegrich [57] also describe five steps: agenda setting, policy formulation, decision-making, implementation, and evaluation.

These kinds of policy-making processes have a strong resemblance to decision-making. A large number of decision-making processes or models have been proposed in literature over the years. In a simplified and generalized sense, a decision-making process can be portrayed as a problem-solving process. After analyzing 25 such decision processes, Mintzberg *et al.* [62] summarizes them into a three-phase model: (1) Identification of issues and goals; (2) Development of alternative solutions; and (3) Selection of alternative [63]. The rational decision-making model [63,64] has often been used as a reference frame when depicting decision-making processes. Though it can be a reductionist version of reality, it can still offer a close impression of the decision-making process via six steps: goal clarification, solution search, solution analysis, solution evaluation, decision, and control [64].

In Figure 1, we can clearly see the similarities among the classic policy-making process, the decision-making process, and the problem-solving process.

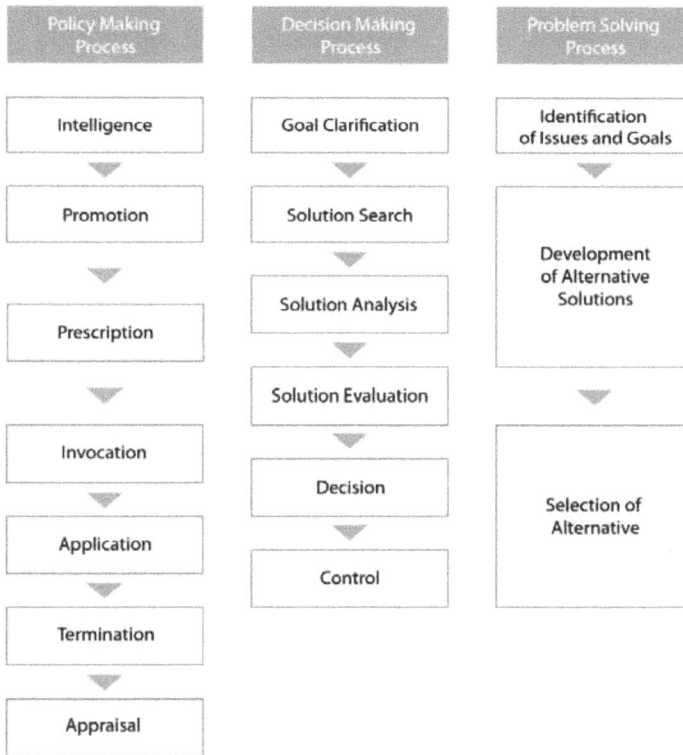

Figure 1. Comparability among policy-making, decision-making, and problem-solving processes.

Whether it is the policy-making, the decision-making, or the problem-solving process, the first and foremost step is to determine areas of concerns based on available information. Any later efforts in subsequent steps can thus be made in a more efficient and directed way. Applying such a concept in the context of sustainable urban development, we can readily infer that a synthesis framework that encompasses the complete range of human and environmental wellbeing will be the top priority (Figure 2). Since the concept of sustainable urban development is multidimensional, efforts made in the field could be diverse, random, and not strategically organized. Such a framework should be able to provide the big picture, that is, to summarize complex issues for supporting policy makers and encouraging more focused efforts.

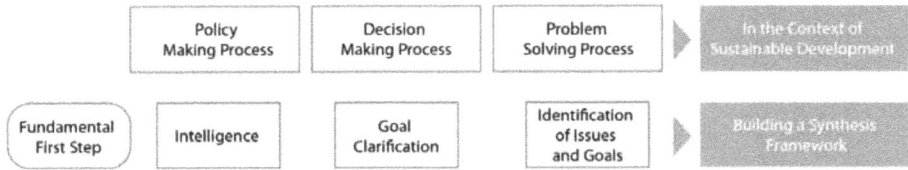

Figure 2. Building a synthesis framework as fundamental first step.

4. A Synthesis Framework with Dimensions and Themes

This section elaborates on how the synthesis framework is constructed and consists of four subsections:

- Subsection 4.1: Defining Sustainable Urban Development
- Subsection 4.2: Global Rankings/Standards of Urban Sustainability: Different Focal Areas
- Subsection 4.3: Global Rankings/Standards of Urban Sustainability: Sorting Indicators into Themes
- Subsection 4.4: Global Rankings/Standards of Urban Sustainability: A Synthesis Framework

It begins with Subsection 4.1, which establishes an original definition of sustainable urban development. An integrative methodology used in this study to build the synthesis framework is introduced in Subsection 4.2, and 10 representative urban sustainability rankings/standards are selected in this subsection. All indicators from the rankings/standards are collected and sorted into new themes in Subsection 4.3 Finally, Subsection 4.4 integrates the new themes into dimensions and produces a synthesis framework.

4.1. Defining Sustainable Urban Development

In the search for a sustainable development pathway, the United Nations World Commission on Environment and Development published in 1987 *Our Common Future*, also known as the *Brundtland Report*. It is considered the starting point of the global discourse on sustainability and defines sustainable development as "development that meets the needs of the present without compromising the ability of future generations to meet their own needs" [65] (p. 37). Sustainable development carries different meanings to different people, subject to their position in societies [66–68]. "It takes on meaning within different political ideologies and programmes underpinned by different kinds of knowledge, values and philosophy" [69] (p. 3). Thus far, there has been no consensus on how such development should be defined or attained.

The concept of sustainable urban development is thus ever-changing and evolving. It is sometimes defined in terms of the economic sustainability of a city, that is, its potential "to reach qualitatively a new level of socio-economic, demographic and technological output which in the long run reinforces the foundations of the urban system" [70]. This way of thinking seeks to continue economic growth and is now regarded as a relatively weaker form of sustainable development. Others may put more emphasis on the social sustainability and base the concept on a broad range of social principles of futurity, equity, and participation, especially involvement of public citizens in the land development process [71]. When viewed alongside environmental concerns, the concept also embodies environmental sustainability, meaning the pursuit of urban form that synthesizes land development and nature preservation and places the protection of natural systems into a state of vital equipoise [72]. In general, countries around the world are called to minimize environmental impact and to improve the social conditions of individuals and the community [73]. In summary, principles of achieving sustainable urban development are generally based on environmental, economic, and social considerations [74–76].

Although current discussions appear to focus more on the environment and economy, cities are still fundamentally human habitats. In contrast to the weaker form of sustainable development, a stronger form "represents a revised form of self-reliant community development which sustains people's livelihoods using appropriate technology" [69] (p. 4). Since cities are for people [77], sustainable cities should be "places where people want to live and work, now and in the future. They meet the diverse needs of existing and future residents, are sensitive to their environment, and contribute to a high quality of life. They are safe and inclusive, well planned, built and run, and offer equality of opportunity and good services for all" [78] (p. 56). It is prescient that human health, wellbeing, safety, security and opportunity will be influenced by the way urban settlements are planned, designed, developed and managed [79]. It should also be noted that social development and economic productivity depend on citizens whose mental and physical needs are satisfied. City inhabitants' comfort hence plays a significant role in sustainable urban development.

Sustainable urban development is indeed a multilayered concept. It synthesizes land development and nature preservation. It also refers to the capacity of nature to support its activities, the vitality of a city as a complex system, and the quality of life of its inhabitants. In other words, sustainable urban development covers many fields of activity such as environmental protection, human development, and inhabitant wellbeing. However, despite all the discussions, no single or agreed meaning has been produced. Taking account of all the concerns stated above, this study proposes to define sustainable urban development as the capacity of any significant human settlements to maintain environmental quality and carrying capacity, to support socio-economic development and management, and to provide sufficient services and livelihoods to all current and future inhabitants. That is, the practicable and full realization of sustainability can only take place in the overlap, or the dynamic, among the three fundamental capacities (Figure 3).

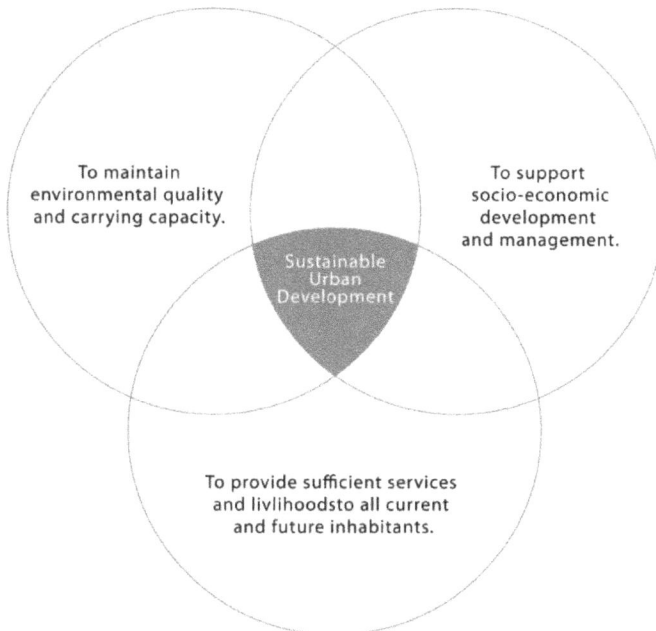

Figure 3. Sustainable urban development—defined as capability in three aspects.

4.2. Global Rankings/Standards of Urban Sustainability: Different Focal Areas

In order to facilitate an improved understanding on the state of, or changes to, urban settlements in relation to better sustainability performance, different sets of frameworks, indicators, and assessment tools have been developed [80,81]. A broad range of urban sustainability indicators has also been in use across different cities and regions, varying in accordance with their particular needs and goals [82,83]. From an initial look, these indicators or rankings/standards appear individually from different sources, leaving the impression that they are proposed as ad-hoc solutions to the emerging environmental challenges. However, each is actually sensibly put together in line with the growing trends in urban environmental planning.

For instance, in recent decades, the concept of New Urbanism, also called Smart Growth or Transit-Oriented Development, has begun to take form. It started out as a reaction against the perceived environmental, economic and social problems of earlier generations of urban planning. New Urbanism advocates "restructuring of public policy and development practices to support the following principles: neighborhoods should be diverse in use and population; communities should be designed for the pedestrian and transit as well as the car; cities and towns should be shaped by physically defined and universally accessible public spaces and community institutions; urban places should be framed by architecture and landscape design that celebrate local history, climate, ecology, and building practice" [84]. From a thoroughgoing critique of the impacts of urbanization, many have also made the case for "walkable, human-scaled neighbourhoods as the building blocks of sustainable communities and regions" [85]. Such conception are materialized into six fundamental features, including a clear neighborhood center that satisfies all residents' daily needs, the five minute walk, a street network in the form of a continuous web, narrow and versatile streets, mixed land use, and special sites for special buildings [85].

These New Urbanist features have in reality been translated into indicators, such as "shift of transport mode" in the Low Carbon Cities Framework, "local transport" in European Green Capital Award, "green transport promotion" in Green City Index, "density" in Sustainable Cities Index, "complete neighborhood/compact city" in Indicators for Sustainability, and "street life" in Quality of Life Survey. Each of these indicators serves as a parameter "which points to, provides information about, and/or describes the state of a phenomenon/environment/area" [86,87]. Indicators have the role of measuring performance. They must be clear, simple, scientifically sound, verifiable, and reproducible [88]. According to the European Evaluation Network for Rural Development [89], an indicator must be SMART: Specific, Measurable, Achievable, Relevant, and Time-related. They help make tangible an otherwise rather abstract concept, that is, in this case, urban sustainability.

"An indicator quantifies and aggregates data that can be measured and monitored to determine whether change is taking place" [90], and change can often bring cost reduction and service improvement outcomes. In Asia, the Green City Index by Siemens AG has projected potential cost savings of US\$2.7 billion from various projects or clean technology deployments in the 22 Asian cities surveyed and "bulk of the estimated savings will be generated from energy consumption and energy efficiency initiatives" [91]. In Denmark, there is the Copenhagen 10-Step Program; the results are also highly positive and can be described in measurable terms. The city has: (1) reduced the number of cars in its center by eliminating parking spaces at a rate of 2–3 percent per year; (2) introduced the City Bike system, allowing anyone to borrow a bike from any one of the 110 bike stands located around the city center for a small refundable coin deposit; and (3) encouraged 34 percent of Copenhageners working in the city to bicycle to their jobs [92].

However, as much as efforts from different parties have been made in applying sustainability indicators, the results can sometimes be mixed and a number of outcomes can even fall short in terms of facilitating sustainability performance [93–95]. It has been contended that an inadequate selection of indicators [80,94] and the lack of consensus on urban sustainability indicators among different approaches [96,97] have been causing confusion and have led to, in some cases, failure to achieve favorable sustainability results. Furthermore, policymakers and city authorities today are faced with a huge array of available urban sustainability rankings/standards and the sheer number and diversity of them can be overwhelming [98]. There are still no pertinent standards or universal methods for selecting urban sustainability indicators [99].

Among various measures currently in use, there has not been any comprehensive framework that can cover the three fundamental capacities that define sustainable urban development: (1) to maintain environmental quality and carrying capacity; (2) to support socio-economic development and management; and (3) to provide sufficient services and livelihoods to all current and future inhabitants. Hence, this study proposes an integrative methodology (Figure 4) to select urban sustainability rankings/standards with different focal areas and integrate them into a synthesis framework that can encompass a complete range of urban sustainability concerns.

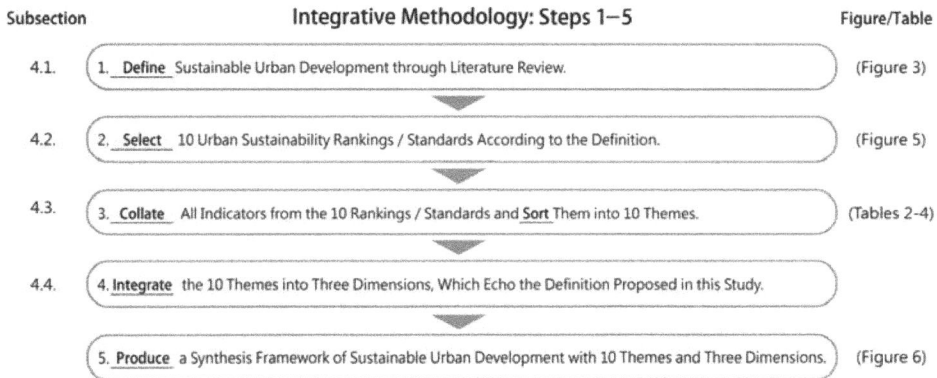

Subsection	Integrative Methodology: Steps 1–5	Figure/Table
4.1.	1. **Define** Sustainable Urban Development through Literature Review.	(Figure 3)
4.2.	2. **Select** 10 Urban Sustainability Rankings / Standards According to the Definition.	(Figure 5)
4.3.	3. **Collate** All Indicators from the 10 Rankings / Standards and **Sort** Them into 10 Themes.	(Tables 2-4)
4.4.	4. **Integrate** the 10 Themes into Three Dimensions, Which Echo the Definition Proposed in this Study.	
	5. **Produce** a Synthesis Framework of Sustainable Urban Development with 10 Themes and Three Dimensions.	(Figure 6)

Figure 4. An integrative methodology of building a synthesis framework of sustainable urban development.

Step 1 of the five-step methodology has been carried out in Subsection 4.1. Now we proceed to Step 2, where we select representative rankings/standards that correspond to the three fundamental capacities of sustainable urban development defined in this study. There are 10 in total (Figure 5) and they are chosen to optimize the purpose of this study.

- *To maintain environmental quality and carrying capacity*: Selected are rankings/standards named with reference to "Low-Carbon" or "Green" (e.g., Low Carbon Cities Framework by the Malaysian Ministry of Energy, Green Technology and Water [100], the European Green Capital Award [101], and the Siemens AG Green City Index [38]). Their main concerns relate to the natural environment with relatively minor considerations of socio-economic issues.
- *To support socio-economic development and management*: Selected are rankings/standards labeled with "Sustainability," or "Health" (e.g., the Sustainable Cities Index of the Australian Conservation Foundation [102], Indicators for Sustainability by Sustainable Cities International [103], and the WHO Healthy Cities Indicators [22]). They usually focus on socio-economic development, public infrastructure, and human health-related statistics. Environmental or ecosystem preservation is of secondary importance.
- *To provide sufficient services and livelihoods to all current and future inhabitants*: Selected are rankings/standards titled as "Livable" or "Life/Living" (e.g., the OECD Better Life Index [104], the EIU Liveability Ranking [42], the Mercer Quality-of-Living Report [105], and the Monocle Quality of Life Survey [45]). Their emphasis on socio-economic and medical services and provision of inhabitant physical and mental wellbeing is strong. Environmental interests are limited.

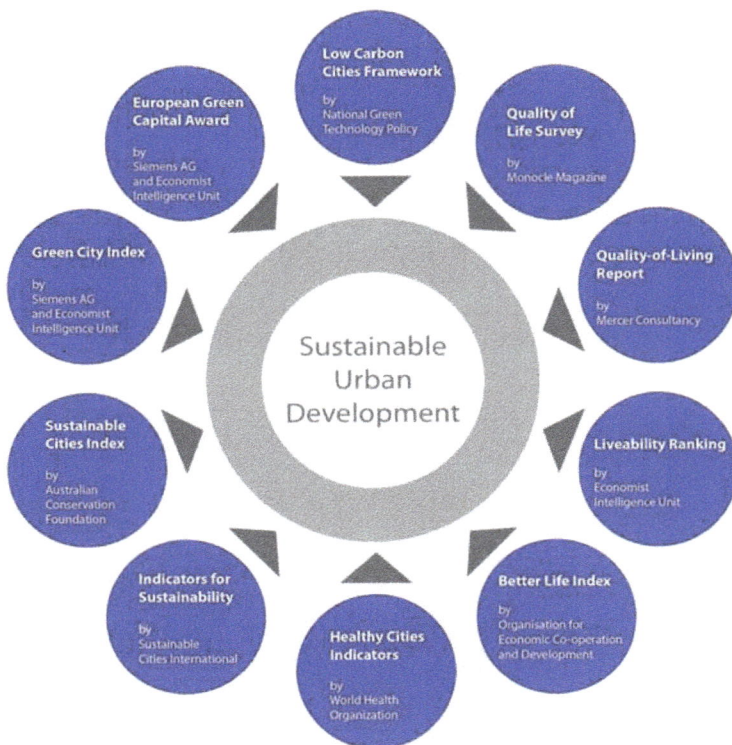

Figure 5. Sustainable urban development—incorporating ten major global urban sustainability rankings/standards with different focal areas.

4.3. Global Rankings/Standards of Urban Sustainability: Sorting Indicators into Themes

Urban sustainability rankings/standards currently in use are composed of indicators that address different concerns. "Indicators are selected to provide information about the functioning of a specific system, for a specific purpose—to support decision-making and management" [90]. The common ground to be found among all these rankings/standards is that they all aim to promote sustainable urban development by aggregating diverse information into focused and applicable knowledge [106].

However, issues covered in sustainable urban development can be innumerable. For example, according to the Division for Sustainable Development of the United Nations (UN-DESA), "urban planning, transport systems, water, sanitation, waste management, disaster risk reduction, access to information, education and capacity-building are all relevant issues to sustainable urban development" [107]. In addition, in 2003, the British Office of the Deputy Prime Minister (ODPM, UK) launched a programme of action called *Sustainable Communities: Building for the future*. In it, the most important requirements of sustainable communities are set out as below [108] (p. 5):

(1)   A flourishing local economy to provide jobs and wealth;
(2)   Strong leadership to respond positively to change;
(3)   Effective engagement and participation by local people, groups and businesses, especially in the planning, design and longterm stewardship of their community, and an active voluntary and community sector;
(4)   A safe and healthy local environment with well-designed public and green space;

(5)   Sufficient size, scale and density, and the right layout to support basic amenities in the neighbourhood and minimise use of resources (including land);

(6)   Good public transport and other transport infrastructure both within the community and linking it to urban, rural and regional centres;

(7)   Buildings—both individually and collectively—that can meet different needs over time, and that minimise the use of resources;

(8)   A well-integrated mix of decent homes of different types and tenures to support a range of household sizes, ages and incomes;

(9)   Good quality local public services, including education and training opportunities, health care and community facilities, especially for leisure;

(10) A diverse, vibrant and creative local culture, encouraging pride in the community and cohesion within it;

(11) A "sense of place";

(12) The right links with the wider regional, national and international community.

As the pace of urbanization continues to accelerate, many cities are faced with "an urgent need for a transition towards a future that maximises their liveability and sustainability" [109]. The notion of urban sustainability becomes increasingly intertwined with livability, which represents "the sum of the factors that add up to a community's quality of life—including the built and natural environments, economic prosperity, social stability and equity, educational opportunity, and cultural, entertainment and recreation possibilities", as defined by the Partners for Livable Communities (PLC) [110]. In short, as put by the British Department for Communities and Local Government (DCLG), a sustainable and livable city should be an environment that is both inviting and enjoyable, where inhabitants would want to live and work now and in the future [111].

From the above discussion, it can be observed that different types of issues embody different concerns. In many cases, the concerns are unbalanced and fails to concurrently address the environmental, socio-economic, and inhabitant wellbeing aspects. Therefore, Table 1 collects issues proposed from multiple sources and summarizes them into ten major themes: (1) Environmental Quality Monitoring; (2) Natural Resource Consumption; (3) Lowering Environmental Impact and Maintaining Carrying Capacity; (4) A Sound Socio-economic Environment; (5) Adequate Infrastructure; (6) Development Strategy Considering Both Human and Natural Environment; (7) Sports, Leisure and Recreation; (8) Consumer Goods and Services; (9) Cultural Diversity and Tolerance; and (10) Sense of Wellbeing and Work-Life Balance.

In the 10 themes above, the first three are considered more environmental, the second three more socio-economic, and the last four more inhabitant wellbeing-oriented. With themes clearly laid out, we continue with Step 3 of the integrative methodology: collate all indicators from the 10 major global urban sustainability rankings/standards (Figure 5) and follow the specified theme coverage in Table 1 to re-arrange all indicators into 10 themes. Tables 2–4 show how these hundreds of indicators are sorted into themes for easy and comprehensive reference.

Table 1. Ten themes and specified theme coverage.

Proposed by	Urban Sustainability Issues	Major Theme Summarized	Theme Coverage
UN-DESA	Water	Natural Resource Consumption	Water quality/consumption, Energy intensity/performance/consumption
ODPM, UK	Requirement #5		
PLC	Natural environment		
UN-DESA	Disaster risk reduction	Environmental Quality Monitoring	Atmospheric and biological environment monitoring, Air pollution monitoring
PLC	Natural environment		
UN-DESA	Transportation systems, Waste management	Lowering Environmental Impact and Maintaining Carrying Capacity	Green transport, Green buildings, Waste, Green space planning and land use, Resource productivity improvement
ODPM, UK	Requirements #4, 5, 6, 7		
PLC	Built environment		
UN-DESA	Education	A Sound Socio-economic Environment	Social stability, Public participation, Education, Housing, Economy
ODPM, UK	Requirements #1, 2, 3, 8, 9		
PLC	Economic prosperity, Social stability and equity, Educational opportunity		
UN-DESA	Transport systems, Water, Sanitation	Adequate Infrastructure	Transportation network, Water/Energy/Telecom infrastructure, Health and medical services
ODPM, UK	Requirement #9		
UN-DESA	Urban planning, Access to information, Capacity-building	Development Strategy Considering Human and Natural Environment	City space planning, Decision-making and action management
ODPM, UK	Requirements #2, 3		
ODPM, UK	Requirement #9	Sport, Leisure and Recreation	(same as title of theme)
PLC	Entertainment and recreation possibilities		
ODPM, UK	Requirement #5	Consumer Goods and Services	(same as title of theme)
ODPM, UK	Requirements #10, 11, 12	Cultural Diversity and Tolerance	(same as title of theme)
PLC	Cultural possibilities		
DCLG, UK	An environment that is both inviting and enjoyable, where inhabitants would want to live and work now and in the future	Sense of Wellbeing and Work-Life Balance	(same as title of theme)

Table 2. Indicators sorted into environmental themes: environmental quality monitoring/natural resource consumption/lowering environmental impact and maintaining carrying capacity.

Themes Ranking/Standard	Natural Resource Consumption	Environmental Quality Monitoring	Lowering Environmental Impact and Maintaining Carrying Capacity
Low Carbon Cities Framework		✓ Urban Greenery and Environmental Quality	✓ Waste ✓ Low Carbon Buildings ✓ Green Transport Infrastructure ✓ Clean Vehicles ✓ Urban Greenery and Environmental Quality
European Green Capital Award	✓ Water Management ✓ Waste Water Treatment ✓ Energy Performance	✓ Ambient Air Quality (PM_{10}, $PM_{2.5}$) ✓ Ambient Air Quality (NO_2) ✓ Climate Change: Mitigation and Adaptation ✓ Nature and Biodiversity	✓ Waste Production and Management ✓ Green Urban Areas Incorporating Sustainable Land Use
Green City Index	✓ Water Consumption ✓ Water ✓ System Leakages ✓ Wastewater Treatment ✓ Water Efficiency and Treatment Policies ✓ Energy Consumption ✓ Energy Intensity ✓ Energy Consumption of Residential Buildings	✓ Particulate Matter ✓ Sulfur Dioxide ✓ Nitrogen Dioxide ✓ CO_2 Emissions ✓ CO_2 Intensity ✓ CO_2 Reduction Strategy ✓ Clean Air Policies ✓ Ozone	✓ Municipal Waste Production ✓ Waste Recycling ✓ Waste Reduction and Policies ✓ Energy-efficient Buildings Standards ✓ Energy Efficient Buildings Initiatives ✓ Use of Non-car Transport ✓ Size of Non-car Transport Network ✓ Green Transport Promotion ✓ Green Land Use Policies ✓ Renewable Energy Consumption ✓ Clean and Efficient Energy Policies

Table 2. *Cont.*

Ranking/Standard Themes	Natural Resource Consumption	Environmental Quality Monitoring	Lowering Environmental Impact and Maintaining Carrying Capacity
Sustainable Cities Index	✓ Water	✓ Air Quality (level of particulate matter) ✓ Climate Change ✓ Ecological Footprint ✓ Biodiversity ✓ Food Production	✓ Green Building
Indicators for Sustainability		✓ Air quality (PM_{10}, $PM_{2.5}$) ✓ Reduce Greenhouse Gases/Energy Efficiency	✓ Waste/Reuse/Recycle ✓ Green Spaces
Healthy Cities Indicators	✓ Water quality ✓ Percentage of Water Pollutants Removed from Total Swage Produced	✓ Atmospheric Pollution (dust fallout) ✓ Atmospheric Pollution (SO_2, NO_2)	✓ Household Waste Collection Quality Index ✓ Household Waste Treatment Quality Index ✓ Relative Surface Area of Green Spaces in the City ✓ Public Access to Green Space
Better Life Index	✓ Water Quality	✓ Air Pollution (PM_{10})	
Liveability Ranking			
Quality-of-Living Report	✓ Sewage	✓ Air Pollution ✓ Climate ✓ Record of Natural Disasters	✓ Waste Disposal
Quality of Life Survey		✓ Population	✓ Electric Car Charging Points ✓ Green Space ✓ Sunshine

Table 3. Indicators sorted into socio-economic themes: a sound socio-economic environment/adequate infrastructure/development strategy considering both human and natural environment.

Themes Ranking/Standard	A Sound Socio-economic Environment	Adequate Infrastructure	Development Strategy Considering Human and Natural Environment
Low Carbon Cities Framework	✓ Community Services	✓ Shift of Transport Mode ✓ Traffic Management ✓ Infrastructure Provision ✓ Energy (infrastructure) ✓ Water Management (infrastructure)	✓ Site Selection ✓ Urban Form
European Green Capital Award		✓ Local Transport	✓ Eco Innovation and Sustainable Employment ✓ Integrated Environmental Management
Green City Index			✓ Green Action Plan ✓ Green Management ✓ Public Participation in Green Policy
Sustainable Cities Index	✓ Employment ✓ Household Repayments ✓ Public Participation ✓ Education	✓ Health ✓ Transport	✓ Density
Indicators for Sustainability	✓ Unemployment Rates/Jobs, Economic Growth ✓ Housing ✓ Quality Public Space ✓ Education	✓ Water quality/Availability ✓ Mobility ✓ Sanitation ✓ Health	✓ Complete Neighborhood/Compact City

Table 3. *Cont.*

Ranking/Standard \ Themes	A Sound Socio-economic Environment	Adequate Infrastructure	Development Strategy Considering Human and Natural Environment
Healthy Cities Indicators	✓ Living Space ✓ Percentage of Population Living in Substandard Accommodation ✓ Estimated Number of Homeless People ✓ Unemployment Rate ✓ Percentage of People Earning Less than the Mean per Capita Income ✓ Percentage of Child Care Places for Pre-school Children ✓ Percentage of All Live Births to Mothers >20; 20–34; 35+ ✓ Abortion Rate in Relation to Total Number of Live Births ✓ Percentage of Disabled Persons Employed	✓ Public Transport ✓ Public Transport Network Cover ✓ Mortality: All Causes ✓ Cause of Death ✓ Low Birth Weight ✓ Existence of a City Health Education Program ✓ Percentage of Children Fully Immunized ✓ Number of Inhabitants per Practicing Primary Health Care Practitioner ✓ Number of Inhabitants Per Nurse ✓ Percentage of Population Covered by Health Insurance ✓ Availability of Primary Health Care Services in Foreign Languages ✓ Number of Health Related Questions Examined by the City Council Every Year	✓ Derelict Industrial Sites ✓ Pedestrian Streets ✓ Cycling in City
Better Life Index	✓ Quality of Support Network ✓ Years in Education ✓ Student Skills ✓ Educational Attainment ✓ Consultation on Rule-making ✓ Voter Turnout ✓ Housing Expenditure ✓ Dwellings with Basic Facilities ✓ Rooms per Person ✓ Household Financial Wealth ✓ Household Net Adjusted Disposable Income ✓ Job Security ✓ Personal Earnings ✓ Long-term Unemployment Rate ✓ Employment Rate ✓ Homicide Rate ✓ Assault Rate	✓ Self-reported Health ✓ Life Expectancy	

Table 3. *Cont.*

Themes Ranking/Standard	A Sound Socio-economic Environment	Adequate Infrastructure	Development Strategy Considering Human and Natural Environment
Liveability Ranking	✓ Prevalence of Petty Crime ✓ Prevalence of Violent Crime ✓ Threat of Terror ✓ Threat of Military Conflict ✓ Threat of Civil Unrest/Conflict ✓ Availability of Private Education ✓ Quality of Private Education ✓ Public Education Indicators ✓ Availability of Good Quality Housing	✓ Availability of Private Healthcare ✓ Quality of Private Healthcare ✓ Availability of Public Healthcare ✓ Quality of Public Healthcare ✓ Availability of Over-the-counter Drugs ✓ General Healthcare Indicators ✓ Quality of Road Network ✓ Quality of Public Transport ✓ Quality of International Links ✓ Quality of Energy Provision ✓ Quality of Water Provision ✓ Quality of Telecommunications	
Quality-of-Living Report	✓ Political Stability ✓ Crime, Law Enforcement ✓ Currency Exchange Regulations ✓ Banking Services ✓ Standards and Availability of International Schools ✓ Rental Housing ✓ Household Appliances ✓ Furniture ✓ Maintenance Services	✓ Medical Supplies and Services ✓ Infectious Diseases ✓ Electricity (public services) ✓ Water (public services) ✓ Public Transportation ✓ Traffic Congestion	
Quality of Life Survey	✓ Crime ✓ Unemployment Rate	✓ International Flights	✓ Upcoming Developments

Table 4. Indicators sorted into inhabitant wellbeing-oriented themes: sports, leisure and recreation/consumer goods and services/cultural diversity and tolerance/sense of wellbeing and work-life balance.

Ranking/Standard \ Themes	Sports, Leisure and Recreation	Consumer Goods and Services	Cultural Diversity and Tolerance	Sense of Wellbeing and Work-Life Balance
Low Carbon Cities Framework				
European Green Capital Award				
Green City Index				
Sustainable Cities Index				✓ Subjective Wellbeing
Indicators for Sustainability				
Healthy Cities Indicators	✓ Sport and Leisure			
Better Life Index				✓ Life Satisfaction ✓ Time Devoted to Leisure and Personal Care ✓ Employees Working Very Long Hours
Liveability Ranking	✓ Sporting Availability	✓ Food and Drink ✓ Consumer Goods and Services	✓ Cultural Availability ✓ Level of Corruption ✓ Social or Religious Restrictions ✓ Level of Censorship	
Quality-of-Living Report	✓ Theatres ✓ Cinemas ✓ Sports and Leisure	✓ Restaurants ✓ Availability of Food/Daily Consumption Items ✓ Cars	✓ Media Availability and Censorship ✓ Limitations on Personal Freedom	
Quality of Life Survey	✓ Street Life ✓ Bookshops	✓ Dinner on a Sunday	✓ Tolerance ✓ Culture	

4.4. Global Rankings/Standards of Urban Sustainability: A Synthesis Framework

Steps 1–3 of the integrative methodology have been completed in Section 4.1, Section 4.2 and Section 4.3:

- Step 1: Define sustainable urban development through literature review.
- Step 2: In line with the definition developed in this study, select 10 global rankings/standards of urban sustainability.
- Step 3: Collate all indicators from the 10 rankings/standards and sort them into 10 themes according to the specified theme coverage established in this study: (1) Natural Resource Consumption; (2) Environmental Quality Monitoring; (3) Lowering Environmental Impact and Maintaining Carrying Capacity; (4) A Sound Socio-economic Environment; (5) Adequate Infrastructure; (6) Development Strategy Considering Both Human and Natural Environment; (7) Sports, Leisure and Recreation; (8) Consumer Goods and Services; (9) Cultural Diversity and Tolerance; and (10) Sense of Wellbeing and Work-Life Balance.

Now, this subsection further proposes the following as the final two steps:

- Step 4: Integrate the 10 themes into three dimensions: (1) Environmental Quality and Carrying Capacity (Themes 1–3); (2) Environmental Management and Development Strategy (Themes 4–6); and (3) Lifestyles of Sustainability (Themes 7–10). The three dimensions correspond directly to the definition of sustainable urban development proposed in this study: to maintain environmental quality and carrying capacity, to support socio-economic development and management, and to provide sufficient services and livelihoods to all current and future residents.
- Step 5: Produce a synthesis framework of sustainable urban development (Figure 6).

Sustainable development has often been identified as composed of economic, social, and environmental goals [112]. However, "a paradigm that does not have a central focus on human health and wellbeing may fail to recognize the critical systemic relationships involved and thus the opportunities for identification of strategies that generate cobenefits" [113]. In other words, a focus primarily on environment or economy may risk excluding inhabitants' comfort or wellbeing from the benefits of sustainable development of cities. To ensure successful and sustainable urban environmental planning, an interwoven approach that addresses concerns in natural environment and resources, infrastructure and socio-economic development, and inhabitants' wellbeing should be adopted. The three aspects must receive equal attention and importance.

The framework proposed in this study addresses exactly the three aspects. It is a synthesis of existing approaches, incorporating the 10 rankings/standards advocated by international and regional organizations. It overcomes the heterogeneity of a myriad of indicators currently offered and addresses a certain confusion surrounding the topic of sustainable urban development. Careful reference to all the three dimensions and the 10 themes of the framework will enable environmental planning that exemplifies a balanced intersection among various sustainability goals. Indicators from multiple urban sustainability approaches are collated and sorted into specific themes for ease of quick reference and possible selection. More indicators can be added or removed in accordance with emerging needs or gained experience and that allows policy and decision makers to customize their best practices in individual cases. For any existing environmental planning policies or programs, the synthesis framework with all its components can also serve as a checklist to assess the policy strengths and weaknesses.

Figure 6. Sustainable urban development—a synthesis framework with three dimensions and ten themes.

5. Conclusions

Since many countries are moving into a fast-growing and transforming stage, there is global dialogue and consensus that urbanization will continue to bring about compelling global and local changes. To adapt and respond to changes, the study has hence collected major global urban sustainability rankings/standards and provided a newly devised synthesis framework of sustainable urban development with 10 themes and three dimensions. In summary, it has:

- enabled the idea of sustainability in various urban settlement theories to be explored through a review of current notions in literature;
- approached the multifaceted concept of sustainable urban development from the perspectives of policy-making, decision-making, and problem-solving processes to establish the essentiality of developing a synthesis framework;
- re-organized and integrated major global urban sustainability rankings/standards into newly and clearly defined dimensions and themes under a concise framework to help identify a more holistic approach to realizing the goal of livable, ecological, and sustainable cities; and
- devised a synthesis framework that is globally encompassing and adaptive for any cities to use in their policy-and-decision-making processes towards a sustainable future.

This project contributes to the ongoing discussion of urban sustainability. To facilitate truly sustainable urban development, we first inspected the evolution of views on human–environment relations in urban settlements theories to examine the interconnectedness between human societies and ecosystems. An original definition of sustainable urban development is offered, bringing a clearer understanding of this multidimensional phenomenon. We then examined the high degree of similarities between environmental planning and a decision process (including policy-making, decision-making, and problem-solving) to confirm the need to build a synthesis framework. Through the procedures explained earlier, we have established a synthesis framework based on integration of current approaches and concepts. The framework has managed to put some order in a broad and partly inconsistent literature. As underlying guidance, it will provide the conceptual and practical scaffolding for creating new policies and encourage more strategically organized efforts in sustainable environmental planning.

Acknowledgments: The authors would like to thank the reviewers for their thoughtful review and valuable comments. This research is financially supported by the Ministry of Science and Technology, Taiwan, under the Grant number of MOST104-2621-M-305-001.

Author Contributions: Both authors have made substantial contributions in conceptualizing the research design, reviewing and analyzing extensive literature, and developing and delivering the final research results. The final version has been approved by all authors.

Conflicts of Interest: The authors declare no conflict of interest.

References

1. Intergovernmental Panel on Climate Change. *Climate Change 2007: Synthesis Report—An Assessment of Intergovernmental Panel on Climate Change*; Intergovernmental Panel on Climate Change: Valencia, Spain, 2007.
2. Center for Climate and Energy Solutions. IPCC AR5 Working Group I Highlights. Available online: http://www.c2es.org/science-impacts/ipcc-summaries/fifth-assessment-report-working-group-1 (accessed on 11 December 2015).
3. Watson, R.T. Climate Change 2001. In Proceedings of the Resumed Sixth Conference of Parties to the United Nations Framework Convention on Climate Change, Bonn, Germany, 19 July 2001.
4. Olympians. Deaths from Global Warming Expected to Rise as Earth Changes, Scientists Say. Available online: http://www.theolympian.com/2009/05/18/854028/deaths-from-global-warming-expected.html (accessed on 20 November 2015).
5. United Nations Framework Convention on Climate Change. United Nations Framework Convention on Climate Change. Available online: http://unfccc.int/essential_background/convention/background/items/2536.php (accessed on 25 November 2015).
6. United Nations Human Settlements Programme. Cities and Climate Change: Global Report on Human Settlements 2011. Available online: http://unhabitat.org/books/cities-and-climate-change-global-report-on-human-settlements-2011/ (accessed on 3 October 2015).
7. United Nations University, Institute for the Advance Study of Sustainability. Sustainable Urban Future—Cities and Climate Change. Available online: http://urban.ias.unu.edu/index.php/cities-and-climate-change/ (accessed on 29 April 2015).
8. Girardet, H. Giant Footprints. *Our Planet* **1996**, *8*, 21–23.
9. Hays, J. Early Man and Ancient History—Mesopotamia (Sumerians, Babylonians, and Assyrians). Available online: http://factsanddetails.com/world/cat56/sub363/item1532.html (accessed on 28 March 2015).
10. World Health Organization and United Nations Human Settlements Programme. Hidden Cities: Unmasking and Overcoming Health Inequities in Urban Settings. Available online: http://www.who.int/kobe_centre/publications/hiddencities_media/p1_who_un_habitat_hidden_cities.pdf (accessed on 6 February 2015).
11. Lall, S.V. *Planning, Connecting, and Financing Cities—Now: Priorities for City Leaders*; International Bank for Reconstruction and Development/World Bank: Washington, DC, USA, 2013.
12. World Bank. *World Development Report: Reshaping Economic Geography*; World Bank: Washington, DC, USA, 2008.
13. Macmillan Dictionary. Urban Planning. Available online: http://www.macmillandictionary.com/us/dictionary/american/urban-planning (accessed on 12 December 2015).
14. Champion, A.G.; Hugo, G.J. *New Forms of Urbanization: Beyond the Urban/Rural Dichotomy*; Ashgate Publishing Limited: Aldershot, UK, 2003.

15. Van dePoel, E.; O'Donnell, O.; Van Doorslaer, E. Urbanization and the spread of diseases of affluence in China. *Econ. Hum. Biol.* **2009**, *7*, 200–216. [CrossRef] [PubMed]

16. International Council for Science. *Report of the ICSU Planning Group on Health and Wellbeing in the Changing Urban Environment: A Systems Analysis Approach*; International Council for Science: Paris, France, 2011.

17. Merriam-Webster. Planning. Available online: http://www.merriam-webster.com/dictionary/planning (accessed on 5 December 2015).

18. United States Environmental Protection Agency. Heat Island Effect. Available online: http://www.epa.gov/heat-islands (accessed on 17 November 2015).

19. Awofeso, N. The Healthy Cities approach—Reflections on a framework for improving global health. *Bull. World Health Organ.* **2003**, *81*, 222–225. [PubMed]

20. World Health Organization Regional Office for Europe. *Twenty Steps for Developing a Healthy Cities Project*, 3rd ed.; World Health Organization Regional Office for Europe: Copenhagen, Denmark, 1997.

21. World Health Organization Regional Office for Europe. Types of Healthy Settings: Healthy Cities. Available online: http://www.who.int/healthy_settings/types/cities/en/ (accessed on 8 November 2015).

22. Webster, P.; Price, C. *Healthy Cities Indicators: Analysis of Data from Cities across Europe*; World Health Organization Regional Office for Europe: Copenhagen, Denmark, 1996.

23. Danish Architecture Center and Cities. 2002 Melbourne Principles: Respect for People and Nature. Available online: http://www.dac.dk/en/dac-cities/sustainable-cities/historic-milestones/2002--melbourne-principles-respect-for-people-and-nature/ (accessed on 6 May 2016).

24. United Nations Framework Convention on Climate Change. Adaptation. Available online: http://unfccc.int/adaptation/items/4159.php (accessed on 20 November 2015).

25. United Nations Environment Programme. Climate Change Mitigation. Available online: http://www.unep.org/climatechange/mitigation/ (accessed on 3 January 2015).

26. United Nations Human Settlements Programme. *Planning Sustainable Cities—Global Report on Human Settlements 2009*; Eearthscan: London, UK, 2009.

27. Lerch, D. *Post Carbon Cities: Planning for Energy and Climate Uncertainty*; Post Carbon: Portland, OR, USA, 2007.

28. Kostoff, S. *The City Shaped*; Thames and Hudson: London, UK, 1991.

29. Jain, S. Smog city to clean city: How did Delhi do it? In *Mumbai Newsline*; The Indian Express: Noida, India, 2004.

30. Calthorpe, P. *The Next American Metropolis: Ecology, Community, and the American Dream*; Princeton Architectural Press: New York, NY, USA, 1993.

31. Calthorpe Associates and Mintier Associates. Transit-Oriented Development Design Guidelines for Sacramento County Planning and Community Development Department. Available online: http://www.per.saccounty.net/PlansandProjectsIn-Progress/Documents/General%20Plan%202030/GP%20Elements/TOD%20Guidelines.pdf (accessed on 6 April 2015).

32. Cervero, R. *Effects of TOD on Housing, Parking and Travel, Transit Cooperative Research Program Report 128*; Federal Transit Administration: Washington, DC, USA, 2008.

33. Ewing, R.H.; Bartholomew, K.; Winkelman, S.; Walters, J.; Chen, D. *Growing Cooler: The Evidence on Urban Development and Climate Change*; Urban Land Institute: Washington, DC, USA, 2007.

34. Belzer, D.; Autler, G. Transit Oriented Development: Moving from Rhetoric to Reality. Discussion Paper Prepared for the Brookings Institution Center on Urban and Metropolitan Policy and the Great American Station Foundation. Available online: http://www.ocs.polito.it/biblioteca/mobilita/TOD.pdf (accessed on 2 April 2015).

35. United Nations Human Settlements Programme. *Urban Patterns for a Green Economy—Leveraging Density*; UNON Publishing Services Section: Nairobi, Kenya, 2012.

36. Foletta, N.; Field, S. Europe's Vibrant New Low Car(bon) Communities. Available online: http://esci-ksp.org/wp/wp-content/uploads/2011/11/Europe%E2%80%99s-Vibrant-New-Low-Carbon-Communities.pdf (accessed on 7 December 2015).

37. European Green Capital Award (EGCA). About EGCA. Available online: http://ec.europa.eu/environment/europeangreencapital/about-the-award/index.html (accessed on 5 April 2015).

38. Siemens, A.G. *European Green City Index: Assessing the Environmental Impact of Europe's Major Cities*; Corporate Communications and Government Affairs, Siemens AG: Munich, Germany, 2009.

39. Ji, Y.; Plainiotis, S. *Design for Sustainability*; China Architecture and Building Press: Beijing, China, 2006.
40. United States Environmental Protection Agency. Green Building Basic Information. Available online: https://archive.epa.gov/greenbuilding/web/html/about.html (accessed on 5 April 2015).
41. Organization for Economic Co-operation and Development. Executive Summary. Available online: http://www.oecdbetterlifeindex.org/media/bli/documents/BLI_executive_summary_2014.pdf (accessed on 12 December 2015).
42. Economist Intelligence Unit. A Summary of the Liveability Ranking and Overview. Available online: http://pages.eiu.com/rs/eiu2/images/Liveability_rankings_2014.pdf (accessed on 28 November 2015).
43. Mercer. Quality-of-Living Rankings Spotlight Emerging Cities. Available online: http://www.mercer.com.tw/insights/view/2014/quality-of-living-rankings-spotlight-emerging-cities.html (accessed on 28 November 2015).
44. Willis, D. What's a Livable International City? Available online: http://www.lyonalacarte.com/?What-s-a-livable-international (accessed on 6 February 2015).
45. Adelaide Capital City Committee. Monocle Quality of Life Survey. Available online: http://capcity.adelaide.sa.gov.au/cities/global-city-rankings/monocle-quality-of-life-survey/ (accessed on 15 December 2015).
46. Godschalk, D. Land use planning challenges: Coping with conflicts in visions of sustainable development and livable communities. *J. Am. Plan. Assoc.* **2004**, *70*, 1–9. [CrossRef]
47. American Association of Retired Persons. Beyond 50.05—A Report to the Nation on Livable Communities: Creating Environments for Successful Aging. Available online: http://assets.aarp.org/rgcenter/il/beyond_50_communities.pdf (assessed on 5 April 2016).
48. Lasswell, H.D. *The Decision Process: Seven Categories of Functional Analysis*; University of Maryland Press: College Park, MD, USA, 1956.
49. Mack, R. *Planning and Uncertainty*; John Wiley: New York, NY, USA, 1971.
50. Rose, R. Comparing public policy: An overview. *Eur. J. Polit. Res.* **1973**, *1*, 67–94. [CrossRef]
51. Brewer, G.D. The policy science emerge: To nature and structure a discipline. *Policy Sci.* **1974**, *5*, 239–244. [CrossRef]
52. Anderson, J. *Public Policy Making*; Praeger Publishing: New York, NY, USA, 1975.
53. Jenkins, W. *Policy Analysis: A Political and Organizational Perspective*; Martin Robertson: Oxford, UK, 1978.
54. May, J.V.; Wildavsky, A.B. *The Policy Cycle*; Sage Publications: Beverly Hills, CA, USA, 1978.
55. Brewer, G.; Deleon, P. *The Foundations of Policy Analysis*; Dorsey Press: Homewood, IL, USA, 1983.
56. Hogwood, B.W.; Gunn, L. *Policy Analysis for the Real World*; Oxford University Press: Oxford, UK, 1984.
57. Jann, W.; Wegrich, K. Theories of the policy cycle. In *Handbook of Public Policy Analysis*; Fischer, F., Miller, G.J., Sidney, M.S., Eds.; CRC Press: Boca Raton, FL, USA, 2007; pp. 43–62.
58. Stokey, E.; Zeckhauser, R. *A Primer for Policy Analysis*; W.W. Norton: New York, NY, USA, 1978.
59. Anderson, J.E.; Brady, D.W.; Bullock, C.S., III; Stewart, J., Jr. *Public Policy and Politics in America*, 2nd ed.; Brooks/Cole: Monterey, CA, USA, 1984.
60. Quade, E.S. *Analysis for Public Decisions*, 2nd ed.; Elsevier: New York, NY, USA, 1982.
61. Hill, M. *The Public Policy Process*; Pearson Education Limited: Harlow, UK, 1997.
62. Mintzberg, H.; Raisinghani, D.; Théorêt, A. The structure of "unstructured" decision processes. *Adm. Sci. Q.* **1976**, *21*, 246–275. [CrossRef]
63. Rasmussen, J.; Brehmer, B.; Leplat, J. *Distributed Decision Making: Cognitive Models for Cooperative Work*; John Wiley & Sons: West Sussex, UK, 1991.
64. Badke-Schaub, P.; Gehrlicher, A. Patterns of decisions in design: Leaps, loops, cycles, sequences and metaprocesses. In Proceedings of the International Conference on Engineering Design (ICED), Stockholm, Denmark, 19–20 August 2003.
65. World Commission on Environment and Development. *Our Common Future—Brundtland Commission*; Oxford University Press: Oxford, UK, 1987.
66. Organisation for Economic Co-operation and Development. What is Sustainable Development? *OECD Environ. Sustain. Dev.* **2008**, *18*, 16–31.
67. Robert, K.W.; Parris, T.M.; Leiserowitz, A.A. What is sustainable development? Goals, indicators, values, and practice. *Environ. Sci. Policy Sustain. Dev.* **2005**, *47*, 8–21. [CrossRef]
68. Robinson, J. Squaring the circle? Some thoughts on the idea of sustainable development. *Ecol. Econ.* **2004**, *48*, 369–384. [CrossRef]

69. Huckle, J. Realizing sustainability in changing times. In *Education for Sustainability*; Huckle, J., Sterling, S., Eds.; Earthscan: London, UK, 1996.

70. Ewers, H.; Nijkamp, P. Urban sustainability. In *Urban Sustainability*; Nijkamp, P., Ed.; Avebury: Gower House, UK, 1990; pp. 8–10.

71. Friends of the Earth. *Planning for the Planet: Sustainable Development Policies for Local and Strategic Plans*; Friends of the Earth: London, UK, 1994.

72. Lyle, J.T. *Regenerative Design for Sustainable Development*; John Wiley & Sons, Inc.: New York, NY, USA, 1994.

73. United Nations. *Rio Declaration on Environment and Development*; United Nations Department of Economic and Social Affairs: Rio de Janeiro, Brazil, 1992.

74. Basiago, A.D. The search for the sustainable city in 20th century urban planning. *Environmentalist* **1996**, *16*, 135–155. [CrossRef]

75. Haughton, G. Developing sustainable urban development models. *Cities* **1997**, *14*, 189–195. [CrossRef]

76. Haughton, G.; Hunter, C. *Sustainable Cities*; Taylor and Francis: London, UK, 2003.

77. Gehl, J. *Cities for People*; Island Press: Washington, DC, USA, 2010.

78. Office of the Deputy Prime Minister. *Sustainable Communities: People Places and Prosperity*; Office of the Deputy Prime Minister: London, UK, 2005.

79. McMichael, A.J. The urban environment and health in a world of increasing globalization: Issues for developing countries. *Bull. World Health Organ.* **2000**, *78*, 1117–1126. [PubMed]

80. Briassoulis, H. Sustainable development and its indicators: Through a (planner's) glass darkly. *J. Environ. Plan. Manag.* **2001**, *44*, 409–427. [CrossRef]

81. Davison, F. Planning for performance: Requirements for sustainable development. *Habitat Int.* **1996**, *20*, 445–462. [CrossRef]

82. Brandon, P.S.; Lombardi, P. *Evaluating Sustainable Development in the Built Environment*; Blackwell: Oxford, UK, 2005.

83. Verbruggen, H.; Kuik, O. Indicators of sustainable development: An overview. In *In Search of Indicators of Sustainable Development*; Kuik, O., Verbruggen, H., Eds.; Springer: New York, NY, USA, 1991.

84. Congress for the New Urbanism. Charter of the New Urbanism. Available online: http://cnu.civicactions.net/charter (accessed on 2 April 2016).

85. Duany, A.; Plater-Zyberk, E.; Speck, J. *Suburban Nation: The Rise of Sprawl and the Decline of the American Dream*; North Point Press: New York, NY, USA, 2000.

86. Gabrielsen, P.; Bosch, P. *Environmental Indicators: Typology and Use in Reporting*; European Environment Agency: Copenhagen, Denmark, 2003.

87. Organisation for Economic Co-operation and Development. *OECD Environmental Indicators: Development, Measurement and Use*; OECD: Paris, France, 2003.

88. Mega, V.; Pedersen, J. *Urban Sustainability Indicators*; European Foundation for the Improvement of Living and Working Conditions: Dublin, Ireland, 1998.

89. European Evaluation Network for Rural Development. What is a SMART indicator? Available online: http://enrd.ec.europa.eu/enrd-static/evaluation/faq/en/indicators.html (accessed on 28 April 2016).

90. Food and Agriculture Organization of the United Nations. Pressure-State-Response Framework and Environmental Indicators. Available online: http://www.fao.org/ag/againfo/programmes/en/lead/toolbox/refer/envindi.htm (accessed on 15 October 2014).

91. Content for Engineers (CFE) Media. Siemens Eyeing $2.7 B in Cost Savings from "Clean" Technology Use. Available online: http://m.csemag.com/articlepage/siemens-eyeing-27-b-in-cost-savings-from-clean-technology-use/41644fe0ff90ec11048da8f7c8f4b68e.html (accessed on 5 April 2016).

92. New Urbanism. Pedestrian Cities/Quality of Life. Available online: http://www.newurbanism.org/pedestrian.html (accessed on 3 April 2016).

93. Alshuwaikhat, H.M.; Nkwenti, D.I. Visualizing decision-making: Perspectives on collaborative and participatory approach to sustainable urban planning and management. *Environ. Plan.* **2002**, *29*, 513–531. [CrossRef]

94. Seabrooke, W.; Yeung, C.W.S.; Ma, M.F.F. Implementing sustainable urban development at the operational level (with reference to Hong Kong and Guangzhou). *Habitat Int.* **2004**, *28*, 443–466. [CrossRef]

95. Selman, P. Three decades of environmental planning: What have we really learned? In *Planning Sustainability*; Kenny, M., Meadowcroft, J., Eds.; Routledge: London, UK, 1999.

96. Legrand, N.; Planche, S.; Rabia, F. *Integration d'Indicateurs de Developpement Durable dans un Outil d'Aide a la Decision*; École des Ingénieurs de la Ville de Paris: Paris, France, 2007.

97. Lazzeri, Y.; Planque, B. Elaboration d'indicateurs pour un systeme de suivi-evaluation du developpement durable. Available online: http://www.territoires-rdd.net/recherches/planque/tome1_planque.pdf (accessed on 28 April 2016).

98. Zavadskas, E.; Kaklauskas, A.; Šaparauskas, J.; Kalibatas, D. Vilnius urban sustainability assessment with an emphasis on pollution. *Ekologija* **2007**, *53*, 64–72.

99. Kahn, M.E. *Green Cities: Urban Growth and the Environment*; Brookings Institution Press: Washington, DC, USA, 2006.

100. Kementerian Tenaga, Teknologi Hijau dan Air. Low Carbon Cities—Framework & Assessment System. Available online: http://esci-ksp.org/wp/wp-content/uploads/2012/04/Low-Carbon-Cities-Framework-and-Assessment-System.pdf (accessed on 5 December 2015).

101. European Green Capital Award. Will Your City be the European Green Capital in 2017? Available online: http://ec.europa.eu/environment/europeangreencapital/wp-content/uploads/2013/02/Will-your-city-2017_Web-Copy-F01.pdf (accessed on 25 April 2015).

102. Australian Conservation Foundation. Sustainable Cities Index—Ranking Australia's 20 Largest Cities in 2010. Available online: http://www.acfonline.org.au/sites/default/files/resources/2010_ACF_SCI_Report_Comparative-Table_and_Fact-Sheets.pdf (accessed on 9 September 2015).

103. Sustainable Cities International. Indicators for Sustainability—How Cities are Monitoring and Evaluating Their Success. 2012. Available online: http://www.mayorsinnovation.org/images/uploads/pdf/2_-_International_Case_Studies.pdf (accessed on 9 October 2015).

104. Organization for Economic Co-operation and Development. OECD Better Life Index. Available online: http://www.oecdbetterlifeindex.org/ (accessed on 25 April 2015).

105. Mercer. Newsroom: Vienna Tops Latest Quality of Living Rankings. Available online: http://www.uk.mercer.com/newsroom/2015-quality-of-living-survey.html (accessed on 25 April 2015).

106. Hiremath, R.B.; Balachandra, P.; Kumar, B.; Bansode, S.S.; Murali, J. Indicator-based urban sustainability—A review. *Energy Sustain. Dev.* **2013**, *17*, 555–563. [CrossRef]

107. United Nations Division for Sustainable Development. Sustainable Cities and Human Settlements. Available online: https://sustainabledevelopment.un.org/topics/sustainablecities (accessed on 5 April 2016).

108. Office of the Deputy Prime Minister. *Sustainable Communities: Building for the Future*; Office of the Deputy Prime Minister: London, UK, 2003.

109. JPI Urban Europe. Launching the JPI Urban Europe Strategic Research and Innovation Agenda—Transition towards Sustainable and Liveable Urban Futures. Available online: http://jpi-urbaneurope.eu/downloads/programme-29-30-september-2015/ (accessed on 6 April 2016).

110. Partners for Livable Communities. What is Livability? Available online: http://livable.org/about-us/what-is-livability (assessed on 10 January 2016).

111. Department for Communities and Local Government. *State of the English Cities: Liveability in English Cities*; Technical Report. Department for Communities and Local Government: London, UK, 2006.

112. Adams, W.M. The Future of Sustainability: Re-thinking Environment and Development in the Twenty-first Century. Available online: http://cmsdata.iucn.org/downloads/iucn_future_of_sustainability.pdf (accessed on 8 April 2016).

113. Siri, J.; Capon, A. Global Sustainable Development Report (GSDR) 2015 Brief—Health and Wellbeing in Sustainable Urban Development. Available online: https://sustainabledevelopment.un.org/content/documents/632481-Siri-Health%20and%20Wellbeing%20in%20Sustainable%20Urban%20Development.pdf (accessed on 3 April 2016).

sustainability

MDPI

Article

Sustainability within the Academic EcoHealth Literature: Existing Engagement and Future Prospects

Aryn Lisitza [1] and Gregor Wolbring [2,*]

[1] Cumming School of Medicine, University of Calgary, Calgary, AB T2N4N1, Canada; ablisitz@ucalgary.ca
[2] Department of Community Health Sciences, Cumming School of Medicine, Stream of Community Rehabilitation and Disability Studies, University of Calgary, 3330 Hospital Drive NW, Calgary, AB T2N4N1, Canada
* Correspondence: gwolbrin@ucalgary.ca; Tel.: +1-403-210-7083

Academic Editor: Md Saidul Islam
Received: 11 December 2015; Accepted: 18 February 2016; Published: 25 February 2016

Abstract: In September 2015, 193 Member States of the United Nations agreed on a new sustainable development agenda, which is outlined in the outcome document *Transforming our world: the 2030 Agenda for Sustainable Development*. EcoHealth is an emerging field of academic inquiry and practice that seeks to improve the health and well-being of people, animals, and ecosystems and is informed in part by the principle of sustainability. The purpose of this study is to investigate which sustainability terms and phrases were engaged in the academic EcoHealth literature, and whether the engagement was conceptual or non-conceptual. To fulfill the purpose, we searched four academic databases (EBSCO All, Scopus, Science Direct, and Web of Science) for the term "ecohealth" in the article title, article abstract, or in the title of the journal. Following the search, we generated descriptive quantitative and qualitative data on n = 647 academic EcoHealth articles. We discuss our findings through the document *Transforming our world: the 2030 Agenda for Sustainable Development*. Based on n = 647 articles, our findings suggest that although the academic EcoHealth literature mentions n = 162 sustainability discourse terms and phrases, the vast majority are mentioned in less than 1% of the articles and are not investigated in a conceptual way. We posit that the 2030 Agenda for Sustainable Development gives an opening to the EcoHealth scholars and practitioners to engage more with various sustainability discourses including the 2030 Agenda for Sustainable Development.

Keywords: EcoHealth; sustainability; sustainable development; 2030 Agenda for Sustainable Development

1. Introduction

1.1. Sustainability

According to Gooden, the term "sustainability" first showed up in 1714, in the book *Forest Economy or Guide to Tree Cultivation Conforming with Nature* by Hans Carl von Carlowitz [1]. The term was employed to discuss how the use of timber as a natural resource could be managed for continued long-term use [1]. The 1980 World Conservation Strategy: Living Resource for Sustainable Development report, which was prepared by the International Union for Conservation of Nature and Natural Resources (IUCN), coined the term "sustainable development" amidst the use of other phrases such as "sustainable utilization of species or ecosystems", "to maintain resources sustainably", "sustainable wildlife utilization" and "sustainable system" [2]. In 1987, the World Commission on Environment and Development (WCED) released the report *Our Common Future*, also known as the Brundtland Report after the chair of the Commission [3] with its vision of what sustainable development signifies. The report used the term "sustainable development" 189 times applying sustainability to over 16 areas [3]. The 1987 Brundtland report altered the discourse of sustainable development to encompass three main dimensions: environmental, economic, and social sustainability [4]. The expansion of the topic of sustainable development has enabled it to become a conceptual model, which "encompasses complex changes in society in order to achieve the ends of economic development, environmental protection and social justice" [5]. However, despite the expansion of the topic, the conceptualization of the term "sustainable development" has also been widely debated and criticized [6–12]. Since these two pioneering reports, several discourses encompassing divergent views of sustainability have appeared including sustainable future, sustainable lifestyle, sustainability science, or sustainable consumption, (for articles on these and other aspects of sustainability discourses see [13–30]).

In the latest chapter on the topic of sustainable development, 193 Member States of the United Nations agreed on a new sustainable development agenda, which is outlined in the document *Transforming Our World: the 2030 Agenda for Sustainable Development* [31]. This document seeks to build off of the Millennium Development Goals and to shift the world onto a sustainable and resilient path, while tackling peace, poverty, and equity [32].

EcoHealth is one field of academic inquiry and practice that is informed in part by the principle of sustainability [33,34].

1.2. EcoHealth and Sustainability

EcoHealth is an emerging field [35–37] that seeks to make a positive difference in the health and well-being of people, animals, and the ecosystems [38] by studying the impact that changes in the biological, physical, social, and economic environments have on such health and well-being [39]. The field of EcoHealth relies on transdisciplinarity, participation, and equity as their three methodological pillars [39,40]. In 2012, Charron proposed to expand the three pillars of field of EcoHealth to six principles: systems thinking, transdisciplinary research, participation, gender, social equity, knowledge to action and, finally, sustainability [34]. The first principle, systems thinking, holds that the component parts of a system, in this case humans, animals, and the environment, should not be understood in isolation, but rather they should be understood within the context of the interactions and linkages between each of the components that make up a system and affect other systems [39]. Transdisciplinary research demands an inclusive vision of ecosystem-related health issues encountered within EcoHealth and relies on a common framework of blended concepts and theories taken from multiple disciplines and stakeholders such as researchers, community representatives, and decision-makers [39]. Participation refers to the aim of cooperation and collaboration both within and across the scientific realm, decision-making groups, and the community [39]. Social and gender equity seeks to "address unequal and unfair conditions impinging on the health and well-being of women and other disadvantaged groups in society" [34]. Knowledge-to-action is the idea that knowledge generated by research is then implemented and applied to improve the environment and the health and well-being of humans [39]. Finally, importantly, the last principle is sustainability, which refers to EcoHealth's goal to protect ecosystems and improve degraded environments to maintain the health and well-being of today's people and future generations [34]. The identification of sustainability as one of six key principles is in line with Leung *et al.*'s work that says that the thinking and practice of EcoHealth has been "heavily shaped by the sustainable development movement of the 1980's" and various aspects of the Brundtland Report [33]. Indeed, the principle of sustainability is seen to inform the field of EcoHealth "to make ethical, positive, and lasting changes which are environmentally sound and socially acceptable" [34]. EcoHealth research is seen to contribute to the improvement of people's health while also advancing sustainable development [41]. According to Kingsley *et al.*, "EcoHealth involves research and practice to promote sustainability of individuals, animals and biodiversity by linking the complex interaction of ecosystem, socio-cultural and economic factors" [42]. It is expected that the six principles including systems thinking, transdisciplinary research, participation, sustainability, gender, social equity, and knowledge to action [34] influence each other.

Given the linkage between the field of EcoHealth and the concept of sustainability as well as knowledge that sustainable development and other sustainability terms and phrases are still being debated and critiqued, the purpose of this study is to understand how the academic EcoHealth literature engages with sustainability terms and phrases. We generated descriptive quantitative and qualitative data to identify which sustainability related terms and phrases are present in the academic EcoHealth literature and if the engagement is conceptual or non-conceptual. We discuss the results through the lens of (a) the vision of the 2014 [43] and 2016 [44] EcoHealth conference, (b) the paper *EcoHealth Research in Practice: Innovative Applications of an Ecosystem Approach to Health* which expands on sustainability as a principle for EcoHealth [34], and (c) the recent outcome document *Transforming Our World: The 2030 Agenda for Sustainable Development* [31] of the United Nations.

2. Experimental Section

2.1. Framing Analysis Through Three Lenses

The analytical framework of this paper is a framing analysis, which is typically used to investigate differing interpretations of a topic or an issue [45]. According to Entman, the basis of framing is to "select some aspects of perceived reality, and make them more salient in a communicating text, in such a way as to promote a particular problem definition, causal interpretation, moral evaluation, and/or treatment recommendation for the item described" [46]. Given the diversity of sustainability discourses and concepts, there are numerous ways that sustainability could be framed within the EcoHealth field.

2.2. Data Source

Four academic databases (EBSCO All, (an umbrella database that consists of over 70 other databases), Scopus, Science Direct and Web of Science) were searched on 4 May 2015 for the term "ecohealth" in the article title, article abstract, article keyword, or in the title of journal ("topic" in the case of Web of Science). The article hits were exported as RIS files and imported into Endnote software where all duplicate articles were identified and eliminated. After the elimination of duplicates, a total of n = 647 academic articles were downloaded and imported into the ATLAS-ti7© software [47], a qualitative analysis software, to produce both quantitative and qualitative data.

2.3. Data Analysis

Step 1: We auto-coded for the term "sustain" in all n = 647 articles within ATLAS-ti7©. All terms that contained the word "sustain" were analyzed for the context in which the term "sustain" and all other terms containing the term "sustain" were used. From this, we generated a list of all of the different sustainability terms or phrases that were associated with or containing the term "sustain" throughout the n = 647 articles.

Step 2: In ATLAS-ti7©, we then auto-coded for each of the sustainability terms or phrases that we had generated (in Step 1) from the original "sustain" code in order to obtain quantitative hits.

Step 3 All n = 647 articles were searched for the sustainability terms and phrases that were present in [3,31] using the auto-coding function within ATLAS-ti7©.

Step 4: Tables were generated for the results of Steps 1–3. The hit counts indicate how many times each term was present in the n = 647 articles, while article count indicates the number of the articles that the term appeared in, which did not include the reference section of each article (Section 3.1 and the Supplementary Materials).

Step 5: Of the list of sustainability terms and phrases generated in Steps 1–3, we chose those that were present in more than one percent of the n = 647 articles (excluding the generic terms sustain* or sustainable or sustainability) for qualitative analysis. We analyzed the context in which each of these sustainability terms and phrases were used, focusing on whether or not the phrases and terms were engaged with in a conceptual way. Conceptual engagement for our purposes is seen as the special attention, interest and exploration or questioning of term or phrase that goes beyond a simple mention or fact about the term or phrase (Section 3.2.1).

Step 6: We performed proximity searches for content containing the term "sustain", reflecting the EcoHealth principle of sustainability, and each of the Ecohealth principles: "systems thinking", "transdisciplinary" "participation", "gender", "equity", and "knowledge to action" reflecting the other five EcoHealth principles (Section 3.2.2).

2.4. Limitations

Only articles written in the English language were downloaded, which means the study has excluded viewpoints present in academic literature written in other languages. Articles were drawn from only four academic databases. Given that databases reflect a certain focus, this could have led to a biased collection of EcoHealth articles. Likewise, only articles that had the term "ecohealth" in the article title, abstract, list of keywords or in the title of a journal were downloaded. Other similar discourses and terms such as "ecological health" were not included and this, too, could have led to a selection bias within the EcoHealth discourse. Finally, we investigated whether or not sustainability terms have been engaged with conceptually by looking for the explicit mention of certain terms. It is possible that we could have missed articles that deal with terms conceptually, although in an implicit way.

3. Results

3.1. Quantitative Data

To obtain an overview of how the n = 647 EcoHealth articles engage with sustainability as a topic, we searched the n = 647 articles for words containing "sustain" (Step 2). The sustainability terms and phrases that emerged from Step 2 and that are present in more than one percent of the n = 647 articles are presented in Table 1. Table S1 lists all the sustainability terms and phrases we found in the n = 647 articles and specifies which sustainability terms were present or not in the Brundtland report and the 2015 United Nations outcome document *Transforming Our World: The 2030 Agenda for Sustainable Development* [3,31].

Table 1. The sustainability discourse terms present in more than 1% of the n = 647 articles as well as the number of hit counts and the number of articles that each term appears in, using ATLAS-ti7© software.

Term	Hit Counts (ATLAS-ti7©)	Number of Articles (ATLAS-ti7©) (n = 647)
sustain	2313	305
sustainable	985	207
sustainability	1002	176
unsustainable	40	32
sustainable development	105	28
ecosystem sustainability and sustainable ecosystems	34	20
environmental sustainability I environmental sustainable	23	16
sustainable use	25	15
health and sustainability	26	13
sustainable management	14	11
long-term sustainability I longer-term sustainability	12	10
sustainable health	11	10
ecological sustainability/ecological sustainable	13	10
sustainable solutions	12	9
sustainable future	8	6

Table 1 reveals that excluding the three generic terms (sustain*, reflecting every word that contained sustain; sustainability and sustainable), only n = 12 sustainability related terms and phrases were present in more than one percent of the n = 647 EcoHealth articles that we covered. Of the n = 12 terms, "sustainable development" was the concept that was mentioned the most showing up in n = 28 or 4.32% of the n = 647 articles.

Table S1 (Supplementary Materials) reveals that a further n = 142 sustainability related terms and phrases were present in less than one percent of the n = 647 articles; n = 35 that were present in [3,31] were not present in the n = 647 articles.

3.2. Qualitative Data

To gain further insight in how the n = 647 EcoHealth articles engaged with the sustainability terms and phrases, we looked at the context or way in which the n = 12 terms, that were present in more than one percent of the n = 647 articles, were covered within the academic EcoHealth literature. We focused our analysis on whether or not the terms were engaged with conceptually, meaning the terms were explored or questioned, as opposed to being mentioned without further engagement.

3.2.1. Conceptual Engagement with Sustainability Terms or Phrases

3.2.1.1. Sustainable Development (SD)

n = 28 articles mentioned the term "sustainable development" in the body of their text. Of these n = 28 articles only n = 7 articles [41,48–52] mentioned sustainable development more than twice in the body. Of the n = 28 articles, only n = 8 engaged with the term conceptually. Two of the nine articles talked about the issue of stakeholder participation [41,53]. Another two articles thematized transdisciplinarity [41,54], one of the six principles of EcoHealth. Two articles spoke about water issues [52,53], another on risk discourses [55], and one other on indicators [48]. One article reflected specifically on the 2012 EcoHealth conference in China [56] and two more articles mentioned the usefulness of EcoHealth [41,50]. One article outlined the shortcomings of Rio+20 extensively [50].

In regards to stakeholder participation, Boischio *et al.* argue that sustainable development requires "many forms and areas of knowledge to guide practical actions on the ground" [41], a sentiment that is also reflected by Lam *et al.* when they state that "sustainable development decision-making requires the perspectives of all segments of society" [53]. Boischio *et al.* give voice to Christens *et al.*'s [57] critiques and responses surrounding the application of participatory methods in development and highlight the problem of power distribution [41]. Two articles thematize transdisciplinarity [41,54]. Boischio *et al.* argue that transdisciplinarity might help to achieve multi-stakeholder engagement as the authors see transdisciplinarity as an effort "to create a common vision and language to overcome differences in perspective and priorities" between empirical, normative and technical disciplines [41]. Orozco makes the point that transdisciplinary research can contribute to sustainable development [54] without further thematizing the issue.

Bringing transdisciplinarity and stakeholder participation together, Boischio *et al.* contend that EcoHealth approaches can respond to the call by different groups "for a more pluralistic and transdisciplinary exploration of sustainable development alternatives based on multi-stakeholder participation approaches" [41].

Two articles focused on the topic of water [52,58]. Bunch *et al.* argue that addressing both biophysical and social environments at the same time can improve human health while promoting sustainable development, and that working on water issues "can overcome the missed opportunity to focus on the commonalities between health promotion and sustainable development" [52]. Lipchin focuses on sustainable water management options, arguing that solutions for sustainable development will not be based on more water for more development, but will come "from a new land and water management system that is sensitive to social, cultural and ecological resources". Lipchin contends that their Dead Sea Basin project has to answer many questions, one of which asks how sustainable development plans to provide incentives to promote local forms of environmental security and equitable access to goods and services [58].

One article focused on risk discourses [55]. Rao contends that "one of the key success factors in the path towards sustainable development is the ability to manage transitional risks", "especially in the context of adopting newer technologies and economic upheavals" [55]. For that, he says, it is important to understand risk culture and risk related behavior of the population [55]. Rao presents this idea in his paper *A Conceptual Framework Outlining Key Components of Risk Culture and Their Interrelationships* [55].

As to indicators, Rapport and Singh argue that a "comprehensive system for State of Environment Reporting (SOER) must take into account indicators of stress on ecosystems, indicators of the state of the system (*i.e.,* ecosystem structure and function), and indicators of social response (policy interventions)". In addition, Rapport and Singh quote Agenda 21, Principle 1 to make the point that SOER framework should allow for more positive, harmonious relations with nature [48].

One article by Custer reflected on the 2012 EcoHealth conference in China and its engagement with the sustainable development goals. The article stated that the desire of EcoHealth practitioners to contribute to future Sustainable Development Goals (SDGs) "may best be represented by a similar balance in the consideration of human health and the environment" and that "a symmetry between the ecosystem approach to health and the health approach to ecosystem may best meet the desires and challenges of current and future EcoHealth advocates" [56]. In the article, it is argued further that "the EcoHealth lens, or the ecosystem approach to health, could be applied to issues such as "environmental sustainability and food security; central to future SDG" [56].

One article thematized that "Rio+20 fell short of promoting a balanced integration of the social, economic, and environmental pillars of sustainable development" [50] and that "the way toward the post-2015 SDGs will likely be more effective if it highlights the full gamut of linkages between sustainable development, global environmental change, health, and well-being" [50]. It is argued in the article that a strengthening of knowledge on the linkages between ecosystem processes, anthropogenic changes, socio-economic changes, and human health and well-being is needed, which is provided by the "development of more integrated research, linking together medical, veterinary, natural, economic, and social sciences, as well as working at multiple scales (local, regional, and global scales)" [50]. This is an endeavor that the author saw as being strongly supported by the One Health and EcoHealth initiatives [50]. Indeed, the author argues that outcomes of One Health and EcoHealth research show a way toward more sustainable ecological, economic, and social development outcomes, including global health equity [50]. The author concluded that "health and ecosystems are inextricably linked to all development sectors and that the inter-linkages should be recognized as a cross-sectoral issue within the sustainable development goals [50].

3.2.1.2. Ecosystem Sustainability or Sustainable Ecosystems

n = 16 articles covered the phrase "ecosystem sustainability" or "sustainable ecosystems" once and n = 4 articles covered it twice, none more than twice. All articles cover some context around the concepts. Five articles used the Adaptive Methodology for Ecosystem Sustainability and Health to engage conceptually with ecosystem sustainability [37,52,59–62]. One article argued that "situating human health within a theoretical framework of ecological thinking enables health professionals to see that the determinants of health are components of a complex adaptive system" which, in turn, allows "the potential for ecological thinking and settings approaches to be applied to understand, advance, and indeed maintain both population health and ecosystem sustainability" [63]. One argued that an EcoHealth approach is the most appropriate way to address communicable diseases because the EcoHealth approach "strives for improved human health and well-being, based on sustainable ecosystems, with more equitable development and less poverty" [64]. One article covered the Millennium Ecosystem Assessment [65]. Another article highlighted a special issue of EcoHealth which had "explicit interest in (re)integrating indigenous perspectives on ecosystem sustainability and health" [66].

One article voiced the sentiments that Indigenous peoples have the same goal of understanding the complex interrelationship between human health and ecosystem sustainability [67]. Water was linked to the concept of ecosystem sustainability in three articles. An editorial in the journal EcoHealth *Water, Ecology, and Health* stated that "no environmental issue is so profoundly critical to human health and ecosystem sustainability" as water is [68]. Although the editorial did not cover the term "ecosystem sustainability" beyond its engagement with the linkage between ecology and water [69]. A second article argued that research on the costs of diseases and disabilities that can be attributed to environmental contaminants is "relevant to ecosystem sustainability because environmental, human and economic health are all indicators of sustainability" [70]. The third article argued that the goal of ecosystem sustainability, which is resilience and health for humans and all species, challenges the EcoHealth community to consider all forms of knowledge in order to increase our understanding of complex problems affecting health, ecosystems, and society and to mobilized actions [69]. One article highlighted the need to deal with the relationships between population health and ecosystem sustainability [71]. Finally, one article stated that "the ecosystem and health relationship can be measured by indicators of environmental health-risk exposure, human morbidity or mortality, or human well-being and ecosystem sustainability approaches" [72].

3.2.1.3. Sustainable Use

One article thematized sustainable use in relation to medicinal plants of the Maya, highlighting the decrease in knowledge and its impact on sustainable use, while arguing that "modeling the geographical distribution of a medicinal plant species is a key issue when considering its conservation and "Sustainable use" [73]. Most articles, however, just used the term "sustainable use" as a goal. One article for example made the link that the experience of the ecosystem approach to health is that "a participatory process can directly and indirectly encourage stewardship of resources for sustainable use, empower marginalized groups through knowledge sharing and capacity building, and empower communities to take charge of environmental management actions based on research evidence" [59].

3.2.1.4. Health and Sustainability and Sustainable Health

n = 13 articles covered the phrase "health and sustainability", n = 5 mentioning it more than once. Of the ones mentioning it more than once, Bunch *et al.* makes the case that the concept of resilience bridges health and sustainability [52] and that "focusing on watersheds as a setting for health and sustainability encourages a view of health–water relationships that goes beyond the traditional focus of water management on drinking water supply, sanitation, and contaminants" [52]. At the same time, Bunch *et al.* identifies the "evaluation of the role of watersheds as a place based context in which to govern for both health and sustainability" as a governance challenge [52]. Charron makes the case that in order to respond to the Lancet special commission on the MDGs health and sustainability challenge, more work has to be done on the ecological dimensions of health [37]. Connell *et al.* concluded that "the overarching goals of health and sustainability facilitate collaboration among disciplines" but "that differences arise from how each approach operationalizes systems as variables and indicators" and, therefore, that "the concepts of health and sustainability can be used to study ecosystems and livelihoods at various scales" [74]. Patrick *et al.* outline that the individual competencies inherent to health promotion that are useful for engaging with health and sustainability are "individual behavior, organization and social change, partnership development, advocacy for policy and legislative change, and community engagement" [75]. In their study they performed interviews that revealed the sentiment that the "absence of a comprehensive framework to guide action on health and sustainability" is a main barrier for "incorporating sustainability into healthcare practice" [75]. The interviews revealed further that the local agenda 21, healthy cities, Ottawa Charter for Health Promotion, environments for health, the Climate Change Adaptation: a Framework for Action, and the Social Model of Health were identified as facilitators to the "congruence between health and sustainability goals" [75].

One article simply stated that "ecosystem health is as much about our own health and sustainability as it is about the health and sustainability of the other species with which we coexist, and of the entire system as a whole" [48].

n = 10 articles used the phrase sustainable health, two mentioning the phrase twice and none more than two times. The phrase was mostly used to highlight a goal, although one article stated that Laos goals of modernizing its social systems, people's lifestyles in terms of socioeconomic development, and the conservation of culture and the social and natural environment should be harmonized with the pursuit of sustainable health [76].

Three articles simply stated that EcoHealth promotes "the sustainable health of people, animals, and ecosystems by formally connecting [the] social and ecological determinants of health" [77] (see also [78,79]). One article argued that the International Association for Ecology for Health's focus is "to promote sustainable health through scientific discovery and understanding at the confluence of disciplines" [80]. Harris *et al.* argued that the settings approach to health promotion is one way to ensure sustainable health gains for the elderly, thereby reducing local and global burdens of disease [63]. The settings approach is described as "an ecological model of health promotion that focuses on the whole system or organization as the context for introducing changes that promote health" [63]. Leung *et al.* argued that community engagement strategies are essential in the creation of "sustainable health outcomes that could be replicated across neighborhoods and communities" [33]. Finally Stephen *et al.* argued for the involvement of indigenous perspectives in the discourses that foster sustainable, healthy prospects for future generations [66].

3.2.1.5. Sustainable Solution

Only one article mentioned the phrase sustainable solution more than once. The authors of this article questioned whether highly-pathogenic avian influenza HPAI control policies in Southeast Asia are generally within the confines of solitary disciplines rather than long-term sustainable solutions, which could integrate truly transdisciplinary approaches [81]. The authors call "for more research directed to ecosystem approaches to health management in order to inform development of sustainable solutions that improve the health and livelihoods of communities" and they posit that "clear guidance from EcoHealth research is needed to identify primary areas of investigation that will yield sustainable solutions of high impact on poverty, livestock and human health, and environmental management" [81]. Returning to the concept of sustainable solution(s), one article thematized that "eroded social infrastructure along with market-oriented ideologies may serve to promote short-term interventions over longer term sustainable solutions" [82]. One article argued that the transdisciplinary approach is critical to building sustainable solutions [83] and two others argue that true community involvement is needed for sustainable solutions [49,84]. The idea that the failure to appreciate how complex systems interact has ultimately prevented sustainable solutions from being adopted is being questioned [85]. Lastly, indigenous research is highlighted as an essential activity that needs to take place within each region of the world if sustainable solutions are to be found [86].

3.2.1.6. Sustainable Management

Only one article mentioned sustainable management more than once. This article argued that sustainable management of water resources is seen to be achieved by applying Dublin principles which includes ecological principles, institutional principles, and instrument principles [87]. Another article highlighted the divergent views of stakeholders: "while most, if not all, stakeholders agreed with the need to manage the industry and individual farms in a sustainable way, we found divergent opinions and approaches to sustainability among critics, supporters, and managers of salmon farming, such that there was no shared foundation from which to define measurable criteria or indicators for sustainable management programs" [88].

3.2.1.7. Ecological Sustainability

Ecological sustainability was mentioned in n = 8 articles; however only three articles mentioned what one could conceive as conceptual coverage of ecological sustainability. In one article, it is argued "that enhancement of ecological sustainability will be followed by enhancement of social sustainability" [57]. Another looked at the Millennium ecosystem assessment [89], stating that the role of the health community "in safeguarding ecological sustainability is still a matter for debate" [89] and, lastly, one argued that nexus between Indigenous health and natural ecosystem conservation is a prerequisite for achieving global ecological sustainability [90].

3.2.1.8. Environmental Sustainability or Environmental Sustainable

The phrases "environmental sustainability" or "environmental sustainable" were mentioned in n = 16 articles. Most articles mentioned the phrases to indicate an outcome. One article questioned the "proper role of the government in ensuring environmental sustainability under China's changing systems of private enterprise" [91]. Another argued that although 180 nations signed the Millennium Development Goals declaration, an enhanced focus on the role of the environment appears necessary; the Millennium Ecosystem Assessment concluded that global marine systems including coastal habitats are overharvested and in decline [92] (see also [93] covering Millennium Ecosystem Assessment and environmental sustainability). One article argued that innovative ideas and paradigms, in the "real world" and in research, are needed to address the challenge of how human communities can avoid compromising human health while meeting growing demands on resources and ecosystem services, while at the same time promoting thriving, resilient communities and environmental sustainability [76]. Another article reflected on the coverage of environmental sustainability at the 2012 EcoHealth conference [56]. In one article, it is argued that low adoption of restructuring livestock-keeping methods and strategies may be due to "the focus on a single species outcome rather than an integrated outcome that balances environmental sustainability with community partnership and free choice of economic activities" [81]. Another article argued that "simultaneously and systematically embracing environmental sustainability, transdisciplinarity, social justice and gender equity, as well as stakeholder participation provides a pathway, not only to understand complex problems in public health but also to translate that knowledge into effective policy and action at the local, national and global levels" [33].

3.2.1.9. Long-Term Sustainability

The phrase "long-term sustainability" appeared in n = 10 articles, although it mostly consisted of non-conceptual engagement. Two articles talked about the inability to produce long-term sustainability, although these too are non-conceptual in that there is no further questioning of the material. The first article stated that "at the local scale, the management of ecosystem resources tends not to take (a) sufficient account of the needs for long-term sustainability [93] and the second stated that conventional economic growth is incompatible with long-term sustainability [94].

Two terms have no conceptual engagement.

3.2.1.10. Unsustainable

The term "unsustainable" was mentioned in n = 32 articles. No article engaged with the term "unsustainable" conceptually. The term was used to outline unsustainable practices, which included agriculture (n = 4), (ignoring) local characteristics, harvest (n = 4), human society, human development, food production (n = 2), technical solutions (n = 3), livestock production, management of resources, land use patterns, urban waste management, economic growth (n = 2), increased use of cars, unregulated water use, unsustainable social norms and values, rainforest destruction, hunting practice, healthcare material source use, and unsustainable development.

3.2.1.11. Sustainable Future

The term "sustainable future" was only mentioned as a goal but no article engaged with the concept further to discuss what a sustainable future should be, or who should decide that and how. One article stated that "further integration of health impact assessments with the environmental impact assessment process can provide [a] more meaningful cost-benefit analysis and better decision making for sustainable futures" [95]. One simply stated that "the first years of this century have seen significant advances in integrating the many perspectives on what it will take to achieve a healthy and sustainable future" [93]. Another stated that there are "unique opportunities provided by the continuation of this project in these slum settlements with regard to post-tsunami community development and rebuilding towards a socially and ecologically sustainable future" [96]. One highlighted that "the founding Editorial of EcoHealth encouraged the emerging field to be seen in the context of parallel and complementary efforts", and described the collective endeavor as a "transdisciplinary imperative for a sustainable future" [97]. Finally, one article noted that "EcoHealth 2012 was the latest in a series of reminders—suggesting that revisiting, challenging, and rediscovering long-standing questions are a treasured part of our collective journey, and offer fertile ground for the transformative changes required to realize a healthy, just and sustainable future" [98].

3.2.2. Linking Sustainability to the Other Five Principles of EcoHealth

To deepen our understanding of how the n = 647 academic EcoHealth articles have engaged with the sustainability terms and phrases, we applied a second strategy whereby we investigated how the sustainability terms and phrases were mentioned in relation to the other five principles of EcoHealth as proposed by Charron [34].

3.2.2.1. "Sustain" and "Equity"

n = 23 documents covered "sustain" and "equity" in the same paragraph (we used equity rather than just social equity). Of those, n = 14 articles did not have coverage beyond simply listing each EcoHealth principle.

Of the n = 9 articles which did not simply list the six principles, two articles made the point that the natural environment is related to health based on the eight prerequisites for health—peace, shelter, education, food, income, a stable eco-system, sustainable resources, and social justice and equity—of the Ottawa Charter for Health Promotion [99] (see also [100]). One article argued that prioritizing sustainable watershed management for the improvement of human health fosters, among others, sustainable livelihoods, and equity [101]. Another article questioned the discourses around Rio+20 and argued that participatory approaches that encourage rather than suppress negotiation and debate generate benefits "related to equity, sustainability, democratic accountability, and managing uncertainty" [102]. In a fifth article *Towards a Better Integration of Global Health and Biodiversity in the New Sustainable Development Goals Beyond Rio+20*, it is argued that sustainable development efforts over the last 20 years have not resulted in health equity despite being considered a goal of the global health field, and that the discussions around SDGs offer an opportunity to "show the way toward more sustainable ecological, economic, and social development outcomes including global health equity" [50]. The sixth article posits that "implementation issues encountered when working across disciplines, using participatory approaches, ensuring equity in the process, and building capacity for the sustainability of interventions, may apply more generally across EcoHealth projects [103]. Orozco *et al.* in their article *Development of Transdisciplinarity Among Students Placed with a Sustainability for Health Research Project* highlights that students generate the term "social–ecological balance" to link "concepts such as agricultural sustainability (more agronomy students), social equity (more health education students), and environmental justice (the law student)" [54]. Patrick *et al.* gives voice to Hanlon and Carlisle who "suggest a new ideology; one that emphasizes the rights of global citizens while seeking a sustainable solution to current and future ecological challenges ... a reprioritization of society towards values which promote well-being, health and equity, while reducing inequalities and over-consumption" [75]. Hanlon and Carlisle "infer the need for leadership, systems thinking and social change, which are the essential features of health promotion practice" [75]. Finally, one article outlined that EcoHealth connections that enable a world that supports social and gender equity, ecosystem sustainability, and health for humans and other species allows its participants to be agents of change and "to challenge the dogma of neutral science" [69].

3.2.2.2. "Sustain" and "Knowledge-to-Action"

n = 8 documents covered "sustain" and "knowledge to action" within the same paragraph. n = 6 simply listed all the EcoHealth principles. Aside from Charron's 2012 article that outlines the six principles including knowledge-to-action and how it relates to the other principles [34], Spiegel *et al.* in their article *Barriers and Bridges to Prevention and Control of Dengue: The Need for a Social–Ecological Approach* investigated the effectiveness of dengue fever control program that lacked an integrated approach but included other EcoHealth principles, and they found that it was particularly sustainability that was a challenge in all programs investigated [85].

3.2.2.3. "Sustain" and "Systems Changes"

n = 7 documents covered "sustain" and "systems change" within the same paragraph. Similar to the other principles, n = 5 simply listed all the EcoHealth principles. One article outlined the synergy between the EcoHealth and the One Health movement in regards to EcoHealth principles such as "systems thinking, inter- and trans-disciplinary research and collaborative participation" [79]. One article used "EcoHealth as a transdisciplinary lens" to investigate the linkage between sustainable livelihoods and ecosystem health" [74]. One article applied the Driving forces–Pressures–State–Impact–Response (DPSIR) framework "for integrating social, cultural, and economic aspects of environmental and human health into a single framework" [104].

3.2.2.4. "Sustain" and "Participation"

n = 26 documents covered "sustain" and "participation" within the same paragraph. n = 21 simply listed the two as part of the EcoHealth principles or mentioned the terms without further elaboration. Two articles outlined the synergy between "the EcoHealth and the One Health movement[s]", which is through their use of the main EcoHealth principles including systems thinking, disciplinary research and collaborative participation [33,79]. One article argued that EcoHealth should represent a globally inclusive community of researchers and practitioners covering real world issues such as climate change, biodiversity loss, land use change, emerging infectious diseases, global toxification, ecological health, and sustainability [105]. Spiegel *et al.* covered the term "sustain" together with participation [85] coming to the same conclusion as stated under Section 3.2.2.2. Yacoop *et al.* in their article *The EcoHealth System and the Community Engagement Movement in Foundations: A Case Study of Mutual Benefits from Grants Funded by the United Nations Foundation* concluded that donors and governments ignore the critical link between and alignment of control and responsibility for long-term sustainability when discussing civil society and participation [49].

3.2.2.5. "Sustain" and "Disciplinary"

n = 48 documents used the terms "sustain" and "disciplinary" within the same paragraph n = 211 times. The term "multidisciplinary" was used in n = 6 articles n = 10 times, "cross-disciplinary" n = 2 articles four times, "transdisciplinary" n = 31 articles 106 times and "interdisciplinary" n = 11 articles 26 times. As to the linkage between "sustain" and "transdisciplinary", only n = 7 articles mentioned the linkage more than once. Aside from the 2012 Charron article [34], the linkage appeared with sustainable livelihood [74], long term sustainable solutions [81], policies around sustainable futures in northern Australia [90], sustainability science [94,106], transdisciplinary education on sustainability for health [54] and sustainable dengue control [85].

3.2.2.6. "Sustain" and "Gender"

n = 21 documents covered "sustain" and "gender" within the same paragraph. Only one article mentioned it more than twice, covering the linkage between poverty, food security, food production sustainability, and gender equality as four determinants of health. The same paper stated that all of these determinants were covered by the Millennium Development Goals and that all coverage contained aspects of environmental sustainability, leading them to argue that they could capture the environmental determinants of health with these determinants [107].

4. Discussion

Words containing "sustain" were present in 47.9% of the n = 647 articles, while 40.18% of the articles contained the term "sustainable" or "sustainability". Excluding the three generic terms, only n = 12 sustainability related terms and phrases were mentioned in more than one percent of the n = 647 articles (Table 1). Of those n = 12 sustainability terms and phrases, n = 10 were engaged with conceptually. We found that n = 142 sustainability related terms and phrases were mentioned in less than one percent of the n = 647 articles and that n = 35 sustainability terms and phrases that were present in the Brundtland report or the 2015 outcome document *Transforming Our World: The 2030 Agenda for Sustainable Development* [3,31] were not present in the n = 647 articles (Table S1, Supplementary Materials). Our findings suggest that the n = 647 articles mention many sustainability related terms and phrases but do not engage with most of them extensively or in a conceptual way.

We found further that few articles engaged substantially with the linkage between sustainability concepts and the other five principles of EcoHealth, although many articles listed each of the principles of EcoHealth. This finding may be predictable given that the EcoHealth field has a diverse group of actors (see for example the exchange of views in [97,98]). Moreover, the field of EcoHealth is in constant flux, which can be observed in the recent interactions between EcoHealth and One Health [44,79]. Some actors find Ecohealth's engagement with sustainability, whether that be generally or with specific sustainability terms and concepts, practically or conceptually, to be more important than other actors in the field. In the next three sections we discuss our findings through the lens of (a) the vision of the 2014 [43] and 2016 [44] EcoHealth conference and (b) the paper "EcoHealth Research in Practice: Innovative Applications of an Ecosystem Approach to Health" which expands on sustainability as a principle for EcoHealth [34]. We also discuss our findings through the lens of the recent outcome document *Transforming Our World: The 2030 Agenda for Sustainable Development* [31] of the United Nations which is an opportunity to increase the visibility of the EcoHealth field by contributing to the 2030 Agenda for Sustainable Development discourses.

4.1. What Was Not Covered? The Issue of Linking Sustainability to the Other Five EcoHealth Principles

Few articles engaged substantially with the linkage between sustainability concepts and the other five principles of EcoHealth. Two articles highlighted the synergy between the EcoHealth and the One Health movements regarding some of the EcoHealth principles, which included systems thinking, transdisciplinary research, and collaborative participation [79], and that the outcomes of both One Health and EcoHealth research show a way toward more sustainable ecological, economic, and social development outcomes, including global health equity [50]. Given this sentiment, it seems to fit that the 2016 EcoHealth conference brings together both the One Health and EcoHealth community [44]. It will be interesting to explore the synergies that might arise between the two movements from the 2016 EcoHealth conference and whether it will generate new linkages between sustainability and the other five EcoHealth principles.

We posit that the 2030 Agenda for Sustainable Development, which is to be addressed within the next few years, might be an opportunity for the EcoHealth discourse to infuse their principles within the discourse based on the outcome document *Transforming Our World: The 2030 Agenda for Sustainable Development* [31] in two different ways. On the one hand the outcome document *Transforming Our*

World: The 2030 Agenda for Sustainable Development did not mention the terms "equity", "systems changes", "transdisciplinary", and "knowledge-to-action", suggesting that the ecosystem focus of the EcoHealth field, which is linked to these terms, may be beneficial for the 2030 Agenda for Sustainable Development discourse. On the other hand the outcome document *Transforming Our World: The 2030 Agenda for Sustainable Development* highlights the need for political participation, full participation in society, participation of developing countries, participation of local communities, and participation of stakeholders, which is in line with the EcoHealth field.

In the next section we discuss the three sustainability terms and phrases that were present in less than one percent of the n = 647 articles and why more coverage is warranted.

4.2. Which of the Sustainability Terms and Phrases Were Not Covered?

Many of the sustainability terms and phrases were not mentioned or were poorly mentioned. To comment further on just three of the sustainability terms and phrases that were not covered, bearing in mind (a) the 2014 EcoHealth conference [43], (b) the paper *"EcoHealth Research in Practice: Innovative Applications of an Ecosystem Approach to Health"*, which expands on sustainability as a principle for EcoHealth [34], and (c) the 2015 outcome document *"Transforming Our World: The 2030 Agenda for Sustainable Development"* [31] of the United Nations.

4.2.1. Sustainable Consumption

The term "sustainable consumption" was mentioned once (Table S1), stating that examples of sustainable consumption should be used as a basis for advisable harvests [108]. According to the *A/CONF.216/5–10-year Framework of Programmes on Sustainable Consumption and Production Patterns*, sustainable consumption "enhances the ability to meet the needs of future generations and conserves, protects and restores the health and integrity of the Earth's ecosystems" [109]. Therefore, sustainable consumption plays a role in the past, present and future trajectories of change within the Earth's ecosystems, which according to the 2014 EcoHealth conference, is one of three topics that EcoHealth seeks to address [43]. The outcome document *Transforming Our World: The 2030 Agenda for Sustainable Development* [31] mentioned sustainable consumption in various places. In one example, it talks about "implement[ing] the 10-year framework of programmes on sustainable consumption and production, all countries taking action, with developed countries taking the lead, taking into account the development and capabilities of developing countries" [31]. In addition, the outcome document asks for "support (for) developing countries, to strengthen their scientific and technological capacity to move towards more sustainable patterns of consumption and production" [31]. Although the outcome documents seem to focus mostly on sustainable consumption in relation to natural resources, there are many areas outside of natural resources where sustainable consumption plays a role, such as healthcare [109]. A recent article [110] showed that if one searches Google Scholar[TM] (Mountain View, CA, USA) for "sustainable consumption" that the target for sustainable consumption mentioned the most was natural resources (n = 109) [110]. However, the same article mentioned that other targets of sustainable consumption are also covered such as food (n = 108), environment (n = 72), water (n = 66), products for households and people (n = 48), energy (n = 46), economics/income (n = 15), shrimps, living sea resources and forests (n = 11), tourism and electronics/technology and employees (n = 10). Two targets were mentioned (n = 8); one target (n = 6) and one target (n = 5); six targets were mentioned (n = 4) and six (n = 3) [110]. There were 14 targets that were mentioned (n = 2) and 29 targets that were mentioned (n = 1) [110]. Many of the targets for sustainable consumption influence the field of EcoHealth's ability to fulfill their vision. According to the 647 articles we covered, it seems that EcoHealth as a field of practice and academic inquiry has an opportunity to investigate sustainable consumption beyond the limited focus on natural resources, thereby taking into account the triangle of human-animal-nature relationships and to contribute this knowledge to the 2030 Agenda for Sustainable Development discourses.

4.2.2. Sustainability Indicators

Another term that was poorly covered, with n = 4 hits within n = 3 articles, was "sustainability indicators" (Table S1). The n = 647 articles we covered mentioned that the purpose of sustainability indicators to assess sustainability levels [60], environmental conditions [53], and ecosystem health [111]. There was also mention of some applications of the indicators including the use of a DPSEE model by UN agencies [111]. According to the WHO indicators employed in sustainable development discourses are often used to transform raw data into synthesized information, which then enables decision-makers and stakeholders to interpret the data and reach a decision, typically in policy development [112]. The outcome document *Transforming Our world: The 2030 Agenda for Sustainable Development* [31] now gives concrete indicators and goals. We posit that this is a chance for the EcoHealth field to engage with the now agreed upon indicators and goals; to highlight how the EcoHealth field and the human-animal-environment relationship could engage with them as well as to monitor the progress of the indicators and goals that are important in fulfilling the vision of EcoHealth.

4.2.3. Social Sustainability

Finally, the phrase "social sustainability" appeared in n = 5 articles (Table S1). The coverage of "social sustainability" included the need for the evaluation of social sustainability, the need to use particular models such as the socioeconomic modeling approach [60], and the need for social sustainability alongside economic development in order to attain global health [37], and to enhance social sustainability and achieve sustainability enhancement [60]. Two articles linked social sustainability to the field of EcoHealth by stating that social sustainability underpins the field of EcoHealth and that the field of EcoHealth addresses issues that are related to social sustainability [37,113]. According to Vallance, social sustainability has three facets: "(a) 'development sustainability' addressing basic needs, the creation of social capital, justice and so on; (b) 'bridge sustainability' concerning changes in behavior so as to achieve bio-physical environmental goals and; (c) 'maintenance sustainability' referring to the preservation—or what can be sustained—of socio-cultural characteristics in the face of change, and the ways in which people actively embrace or resist those changes" [114]. According to Vallance's three-fold schema, social sustainability has aspects that can be linked to all six pillars of the EcoHealth field systems thinking, transdisciplinary research, participation, sustainability, gender, and social equity, and knowledge to action [34]. However, the 647 articles we covered only address social sustainability in relation to the fourth of the six pillars. The sustainability pillar maintains that social sustainability "is part of the change sought through EcoHealth research and action" [34], which suggests that more coverage and conceptual engagement with the phrase might be warranted.

The outcome document *Transforming Our World: The 2030 Agenda for Sustainable Development* which constitutes the new global sustainable development agenda makes it clear that the 17 Sustainable Development Goals and 169 targets are "integrated and indivisible and balance the three dimensions of sustainable development: the economic, social and environmental" [31]. The outcome document covers the linkage between the social, economic and environmental dimensions of sustainable development in numerous areas: it talks under prosperity that "economic, social and technological progress occurs in harmony with nature"; it talks about achieving "economic growth, social development, environmental protection and the eradication of poverty and hunger"; that "social and economic development depends on the sustainable management of our planet's natural resources"; to recognize "the link between sustainable development and other relevant ongoing processes in the economic, social and environmental fields"; that by 2030 "the resilience of the poor and those in vulnerable situations and reduce their exposure and vulnerability to climate-related extreme events and other economic, social and environmental shocks and disasters" should be built; that to "empower and promote the social, economic and political inclusion of all, irrespective of age, sex, disability, race, ethnicity, origin, religion or economic or other status" by 2030 and to "support positive economic, social and environmental links between urban, peri-urban and rural areas by strengthening national and regional development planning" [31].

We see the outcome document *Transforming Our World: The 2030 Agenda for Sustainable Development* as an opportunity for the field of EcoHealth to incorporate the balance between the economic, social and environmental aspects of sustainable development within the triangle of human-animal-environment relationships more visibly.

According to the outcome document *Transforming Our World: The 2030 Agenda for Sustainable Development* "each country has primary responsibility for its own economic and social development and that the role of national policies and development strategies cannot be overemphasized" [31]. This suggests that the role of national EcoHealth research and practice on the ground is important. The international EcoHealth conferences could be a platform to discuss ever-changing problems of such a global outcome, which is based on national actions. Indeed, the EcoHealth 2014 [43] and 2016 [44] conferences suggest a different focus in the discussion on EcoHealth.

The terms "sustainable consumption", "sustainability indicators", and "social sustainability" are just three sustainability terms that were inadequately mentioned in the 647 articles from our sample of the academic EcoHealth literature. It would be useful to engage with more sustainability terms and phrases in a conceptual way because the discourse around any given sustainability term or phrase could impact the ability of EcoHealth researchers and practitioners to fulfill their aim to make environmentally-stable and socially-appropriate changes [34,43].

In the next section we discuss three sustainability terms and phrases that were present in more than one percent of the n = 647 articles and why more coverage is warranted.

4.3. What Sustainability Terms and Phrases Were Covered but Could Have Had More Coverage?

There were n = 12 terms that were present in more than one percent of the n = 647 articles and n = 10 of these were dealt with conceptually. However, the coverage was still lacking as the term that was mentioned most—sustainable development—was still covered in less than 5% of the n = 647 articles. To discuss the three terms with the most hit counts further.

4.3.1. Sustainable Development

"Sustainable development" was used as a phrase n = 105 within n = 28 articles (Table 1). This coverage included the foundations and goals of sustainable development such as human health and biodiversity [41,50], and specific issues and applications of sustainable development, including pollution in China and the environmental action plan of the European Commission respectively [91,115]. The coverage of sustainable development within the 647 articles indicated that EcoHealth seeks to determine how (un)sustainable development impacts health and in turn, EcoHealth contributes to the goals outlined by the sustainable development discourse through their consideration of human health and the environment [37,56]. According to Charron, the term "sustainable development" takes into account the social and economic development needed to improve human lives as well as the irreversible ecosystem degradation this is causing [34]. This dilemma and the resulting desire to change the way in which people interact with the environment in order to achieve sustainable change in human health and well-being is the very basis of the field of EcoHealth and is demonstrated in the three main themes of the 2014 EcoHealth conference [43]. Given that the 2016 conference [44] aims to increase the collaboration between One Health and EcoHealth, it may be useful to investigate what this could mean and how to engage with the concept of sustainable development.

The outcome document *Transforming Our World: The 2030 Agenda for Sustainable Development* [31] gives guidance on the focus of sustainable development in the next 15 years. The document is an opportunity for the EcoHealth field to critically engage with that vision and explore what it means for its three constituencies (human, animal, and environment). Indeed animals are, for example, only mentioned once in the outcome document *Transforming Our World: The 2030 Agenda for Sustainable Development*, which could be challenged by the EcoHealth community.

4.3.2. Ecosystem Sustainability

The term "ecosystem sustainability" was mentioned n = 34 times in 20 articles (Table 1). The term was used to look at the aspects that benefit ecosystem sustainability, such as the recognition of the interdependence between long-term human existence and the health of ecosystems, as well as the aspects needed for ecosystem sustainability including scientists and the interconnections between them [63,67,69]. The coverage of ecosystem sustainability also highlighted many applications and programs including the Adaptive Methodology for Ecosystem Sustainability and Health (AMESH) [59], The Millennium Ecosystem Assessment [65], and The Network for Ecosystem Sustainability and Health [96], which utilize or work towards ecosystem sustainability. The coverage of the term "ecosystem sustainability" in the 647 articles indicates that the field of EcoHealth attempts to address issues pertaining to ecosystem sustainability, in particular the interrelationship between conservation medicine, human health and ecosystem sustainability [80,105]. However, given the importance of the concept for the fulfillment of the EcoHealth vision according to our guiding documents, to mention the term only 34 times in 20 articles seems to be insufficient. The coverage could have included agro-ecosystem sustainability in more depth, for example, given that according to Charron, farmer's health and agro-ecosystem sustainability is a major challenge that the field of EcoHealth seeks to tackle [34].

The outcome document *Transforming Our World: The 2030 Agenda for Sustainable Development* [31] mentions the term "ecosystem" n = 11 times; for example, the term ecosystem is used in Goal 15 as follows: "protect, restore and promote sustainable use of terrestrial ecosystems", "ensure sustainable food production systems and implement resilient agricultural practices that increase productivity and production, that help maintain ecosystems", "water-related ecosystems", "integrate ecosystem and biodiversity values into national and local planning, development processes, poverty reduction strategies and accounts", and "mobilize and significantly increase financial resources from all sources to conserve and sustainably use biodiversity and ecosystems" [31]. The outcome document *Transforming Our World: The 2030 Agenda for Sustainable Development* mentions various sustainability terms that are hardly visible in the n = 647 articles we covered and as such the outcome document is an opportunity for the EcoHealth community to look at whether or not their view and focus on the ecosystem is in sync with the ecosystem use within the outcome document.

4.3.3. Environmental Sustainability

The coverage of environmental sustainability highlighted the environmental sustainability framework or approach, which employs many goals, including poverty reduction and universal education [64,93]. In addition, the literature touched on competing priorities such as meeting the growing demands on resources and ecosystem services [76], as well as several application examples including Japan's desire for both industrial development and environmental sustainability [76], and the choices involving environmental sustainability made by water resource departments [116]. We see the coverage of environmental sustainability as an essential in order to elucidate the fundamental principles in the field of EcoHealth as outlined by Charron [34,37]. However only n = 6 articles engaged with environmental sustainability on a conceptual level, which we posit to be low given the importance of the term for the EcoHealth field.

The outcome document *Transforming Our World: The 2030 Agenda for Sustainable Development* [31] mentions the term "environmental" n = 19 times and the term "environment" n = 15 times, often as part of the three dimensions—social, economic, and environment of sustainable development. This is another opportunity for the EcoHealth field to compare their vision of environment (and environmental) with the vision of the outcome document.

5. Conclusions

Sustainability is seen as one principle that is used to inform the field of EcoHealth to make "ethical, positive and lasting changes (to environment-human interactions) which are environmentally sound and socially acceptable" [34]. Our findings suggest that within our sample of n = 647 articles, the academic EcoHealth literature has not yet engaged with many sustainability discourse terms. Furthermore, the sustainability terms that are mentioned are hardly interrogated in a conceptual way. This finding may be predictable given that the EcoHealth field has a diverse group of actors, many of whom do not necessarily prioritize EcoHealth's engagement with sustainability. As to future opportunities, Parkes mentioned the important influence of international conventions, declarations and assessments on the field of EcoHealth in 2012 [97]. The recent international outcome document *Transforming our world: the 2030 Agenda for Sustainable Development* [31] is an opportunity for the actors in the EcoHealth field that see sustainability as an important area of engagement to increase the theoretical and practical engagement with sustainability discourses and terms, and to increase the visibility of the EcoHealth field by contributing to the 2030 Agenda for Sustainable Development discourses.

Supplementary Materials: The following are available online at www.mdpi.com/2071-1050/8/3/202/s1, Table S1: Total hit count and number of articles for all sustainability terms and phrases we found in the 647 articles as well as sustainability terms that were present or not in the Brundtland report and the 2015 United Nations outcome document *Transforming Our World: The 2030 Agenda for Sustainable Development*.

Acknowledgments: We would like to thank the reviewers for their thoughtful and extensive comments.

Author Contributions: GW laid the conceptual groundwork for the article and was involved in the writing of the article as well as extensive editing of the article. AL conducted the research, wrote the original draft of the article and was involved in the editing of the article.

Conflicts of Interest: The authors declare no conflict of interest.

References

1. Godden, L. The Principle of Sustainability: Transforming Law and Governance. *Osgoode Hall Law J.* **2009**, *47*, 807.
2. Redclift, M. Sustainable development and global environmental change: Implications of a changing agenda. *Glob. Environ. Change* **1992**, *2*, 32–42. [CrossRef]
3. World Commission on Environment and Development (WCED). Our Common Future: Report of the World Commission on Environment and Development. Available online: http://www.worldinbalance.net/intagreements/1987-brundtland.php (accessed on 3 February 2016).
4. (2012) "Sustainability", Facilities. Available online: http://dx.doi.org/10.1108/f.2012.06930iaa.001 (accessed on 24 February 2016).
5. Barnes, P.M.; Hoerber, T.C. *Sustainable Development and Governance in Europe: The Evolution of the Discourse on Sustainability*; Routledge: London, UK, 2013; Volume 96.
6. Doyle, T. Sustainable development and Agenda 21: The secular bible of global free markets and pluralist democracy. *Third World Q.* **1998**, *19*, 771–786. [CrossRef]
7. Haque, M.S. The fate of sustainable development under neo-liberal regimes in developing countries. *Int. Polit. Sci. Rev.* **1999**, *20*, 197–218. [CrossRef]
8. Tisdell, C. Sustainable development: Differing perspectives of ecologists and economists, and relevance to LDCs. *World Dev.* **1988**, *16*, 373–384. [CrossRef]
9. Banerjee, S.B. Who sustains whose development? Sustainable development and the reinvention of nature. *Organ. Stud.* **2003**, *24*, 143–180. [CrossRef]
10. Daly, H.E. Toward some operational principles of sustainable development. *Ecol. Econ.* **1990**, *2*, 1–6. [CrossRef]
11. Daly, H.E. Sustainable Development: From Concept and Theory to Operational Principles. *Popul. Dev. Rev.* **1990**, *16*, 25–43. [CrossRef]
12. Daly, H.; Jacobs, M.; Skolimowski, H. Discussion of Beckerman's Critique of Sustainable Developemnt. *Environ. Values* **1995**, *4*, 49–70. [CrossRef]

13. Littig, B.; Griessler, E. Social Sustainability: A Catchword between Political Pragmatism and Social Theory. *Int. J. Sustain. Dev.* **2005**, *8*, 65–79. [CrossRef]

14. Esquer-Peralta, J.; Velazquez, L.; Munguia, N. Perceptions of core elements for sustainability management systems (SMS). *Manag. Decis.* **2008**, *46*, 1027–1038. [CrossRef]

15. Daly, H.E.; Cobb, J.B.; Cobb, C.W. For the Common Good: Redirecting the Economy Toward Community, the Environment, and A Sustainable Future. Beacon Press: Boston, MA, USA, 1994.

16. Barbanente, A.; Borri, D. Reviewing self-sustainability. *Plurimondi* **2000**, *4*, 5–19.

17. Hill, R.C.; Bowen, P.A. Sustainable construction: Principles and a framework for attainment. *Constr. Manag. Econ.* **1997**, *15*, 223–239. [CrossRef]

18. Mendler, S.; Odell, W. *The HOK Guidebook to Sustainable Design*; John Wiley & Sons: Hoboken, NJ, USA, 2000.

19. Ahern, J. Theories, methods and strategies for sustainable landscape planning. In *From Landscape Research to Landscape Planning: Aspects of Integration, Education and Application*; Springer: Dordrecht, The Netherlands, 2006; pp. 119–131.

20. Brown, B.J.; Hanson, M.E.; Liverman, D.M.; Merideth, R.W., Jr. Global sustainability: Toward definition. *Environ. Manag.* **1987**, *11*, 713–719. [CrossRef]

21. Roseland, M. *Toward Sustainable Communities: Solutions for Citizens and Their Governments*; New Society Publishers: Vancouver, BC, Canada, 2012; Volume 6.

22. Bell, S.; Morse, S. Sustainability Indicators: Measuring the Immeasurable? Earthscan: Abingdon, UK, 2008.

23. Clarke, J. A framework of approaches to sustainable tourism. *J. Sustain. Tour.* **1997**, *5*, 224–233. [CrossRef]

24. Anand, S.; Sen, A. Human development and economic sustainability. *World Dev.* **2000**, *28*, 2029–2049. [CrossRef]

25. Spaargaren, G. Sustainable consumption: A theoretical and environmental policy perspective. *Soc. Nat. Resour.* **2003**, *16*, 687–701. [CrossRef]

26. Lado, A.A.; Boyd, N.G.; Wright, P. A competency-based model of sustainable competitive advantage: Toward a conceptual integration. *J. Manag.* **1992**, *18*, 77–91. [CrossRef]

27. Kates, R.; Clark, W.C.; Hall, J.M.; Jaeger, C.; Lowe, I.; McCarthy, J.J.; Schellnhuber, H.J.; Bolin, B.; Dickson, N.M.; Faucheux, S.; *et al.* Sustainability science. *Science* **2001**, *292*, 641–642. [CrossRef] [PubMed]

28. Venkataraman, B. Education for sustainable development. *Environ. Sci. Policy Sustain. Dev.* **2009**, *51*, 8–10. [CrossRef]

29. Heiskanen, E.; Pantzar, M. Toward Sustainable Consumption: Two New Perspectives. *J. Consum. Policy* **1997**, *20*, 409–442. [CrossRef]

30. Brown, L.R. *Building A Sustainable Society*; W W Norton & Co Inc.: New York, NY, USA, 1981.

31. United Nation. Transforming Our World: The 2030 Agenda for Sustainable Development. Available online: https://sustainabledevelopment.un.org/post2015/transformingourworld (accessed on 3 February 2016).

32. Transforming Our World: The 2030 Agenda for Sustainable Development. Sustainable Development Knowledge Platform. Available online: https://sustainabledevelopment.un.org/ (accessed on 3 February 2016).

33. Leung, Z.; Middleton, D.; Morrison, K. One Health and EcoHealth in Ontario: A qualitative study exploring how holistic and integrative approaches are shaping public health practice in Ontario. *BMC Public Health* **2012**. [CrossRef] [PubMed]

34. Charron, D. Ecohealth Research in Practice. In *Ecohealth Research in Practice*; Charron, D.F., Ed.; Springer: New York, NY, USA, 2012; Volume 1, pp. 255–271.

35. Wilcox, B.A.; Aguirre, A.A.; Daszak, P.; Horwitz, P.; Martens, P.; Parkes, M.; Patz, J.A.; Waltner-Toews, D. EcoHealth: A transdisciplinary imperative for a sustainable future. *Ecohealth* **2004**, *1*, 3–5.

36. Butler, C.D.; Weinstein, P. Global ecology, global health, ecohealth. *Ecohealth* **2011**. [CrossRef] [PubMed]

37. Charron, D.F. Ecosystem Approaches to Health for a Global Sustainability Agenda. *Ecohealth* **2012**, *9*, 256–266. [CrossRef] [PubMed]

38. Wolbring, G. Ecohealth through an ability studies and disability studies lens. In *Ecological Health: Society, Ecology and Health*; Gislason, M.K., Ed.; Emerald: London, UK, 2013; Volume 15, pp. 91–107.

39. Unahalekhaka, A.; Pichpol, D.; Meeyam, T.; Chotinun, S.; Robert, G.; Robert, C. EcoHealth Manual. Available online: https://cgspace.cgiar.org/handle/10568/33566 (accessed on 22 February 2016).

40. Lebel, J. Ecohealth and the Developing World. *EcoHealth* **2004**, *1*, 325–326. [CrossRef]

41. Boischio, A.; Sánchez, A.; Orosz, Z.; Charron, D. Health and sustainable development: Challenges and opportunities of ecosystem approaches in the prevention and control of dengue and Chagas disease. *Cad. Saúde Públ.* **2009**, *25*, S149–S154. [CrossRef]

42. Kingsley, J.; Patrick, R.; Horwitz, P.; Parkes, M.; Jenkins, A.; Massy, C.; Henderson-Wilson, C.; Arabena, K. Exploring Ecosystems and Health by Shifting to a Regional Focus: Perspectives from the Oceania EcoHealth Chapter. *Int. J. Environ. Res. Public Health* **2015**, *12*, 12706–12722. [CrossRef] [PubMed]

43. Colloque EcoHealth, Themes. Available online: https://ecohealth2014.uqam.ca/en/a-propos/themes.html (accessed on 22 February 2016).

44. One Health EcoHealth 2016. Available online: http://oheh2016.org/ (accessed on 22 February 2016).

45. Vliegenthart, R.; van Zoonen, L. Power to the frame: Bringing sociology back to frame analysis. *Eur. J. Commun.* **2011**, *26*, 101–115. [CrossRef]

46. Entman, R.M. Framing: Towards clarification of a fractured paradigm. *McQuail's Read. Mass Commun. Theory* **1993**, *43*, 51–58. [CrossRef]

47. ATLAS-ti 7, ATLAS.ti Scientific Software Development GmbH, Berlin, Germany, 2013.

48. Rapport, D.J.; Singh, A. An ecohealth-based framework for state of environment reporting. *Ecol. Indic.* **2006**, *6*, 409–428. [CrossRef]

49. Yacoob, M.; Hetzler, B.; Langer, R. The Ecohealth System and the Community Engagement Movement in Foundations: A Case Study of Mutual Benefits from Grants Funded by the United Nations Foundation. Natural Resources Forum, Wiley Online Library: Hoboken, NJ, USA, 2004; pp. 133–143.

50. Langlois, E.V.; Campbell, K.; Prieur-Richard, A.-H.; Karesh, W.B.; Daszak, P. Towards a better integration of global health and biodiversity in the new sustainable development goals beyond Rio+ 20. *EcoHealth* **2012**, *9*, 381–385. [CrossRef] [PubMed]

51. Romanelli, C.; Corvalan, C.; Cooper, H.D.; Manga, L.; Maiero, M.; Campbell-Lendrum, D. From Manaus to Maputo: Toward a public health and biodiversity framework. *EcoHealth* **2014**, *11*, 292–299. [CrossRef] [PubMed]

52. Bunch, M.J.; Morrison, K.E.; Parkes, M.W.; Venema, H.D. Promoting health and well-being by managing for social–ecological resilience: The potential of integrating ecohealth and water resources management approaches. *Ecology Society* **2011**, *16*, 6.

53. Lam, S.; Leffley, A.; Cole, D.C. Applying an Ecohealth Perspective in a State of the Environment Report: Experiences of a Local Public Health Unit in Canada. *Int. J. Environ. Res. Public Health* **2014**, *12*, 16–31. [CrossRef] [PubMed]

54. Orozco, F.; Cole, D.C. Development of transdisciplinarity among students placed with a sustainability for health research project. *EcoHealth* **2008**, *5*, 491–503. [CrossRef] [PubMed]

55. Rao, S. Understanding Risk Culture and Developing a 'soft' Approach to Risk Assessment Methodologies. In *Environmental Security and Environmental Management: The Role of Risk Assessment*; Springer: Berlin/Heidelberg, Germany, 2006; pp. 201–210.

56. Custer, B.; Koné, B.; Kouassi, E.; Ontiri, E.; Watts, P.; Yi, Z.-F. News from the IAEH. *EcoHealth* **2014**, *11*, 286–289. [CrossRef] [PubMed]

57. Christens, B.; Speer, P.W. Review Essay: Tyranny/Transformation: Power and Paradox in Participatory Development. Available online: http://www.qualitative-research.net/index.php/fqs/article/view/91/190 (accessed on 22 February 2016).

58. Lipchin, C. A Future for the Dead Sea Basin: Options for a more sustainable water management. In *Environmental Security and Environmental Management: The Role of Risk Assessment*; Springer: Berlin/Heidelberg, Germany, 2006; pp. 79–91.

59. Berbés-Blázquez, M.; Oestreicher, J.S.; Mertens, F.; Saint-Charles, J. Ecohealth and resilience thinking: A dialog from experiences in research and practice. *Ecol. Soc.* **2014**, *19*, 24. [CrossRef]

60. Gilioli, G.; Baumgärtner, J. Adaptive ecosocial system sustainability enhancement in Sub-Saharan Africa. *EcoHealth* **2007**, *4*, 428–444. [CrossRef]

61. Harper, S.L.; Edge, V.L.; Cunsolo Willox, A. "Changing Climate, Changing Health, Changing Stories" Profile: Using an EcoHealth Approach to Explore Impacts of Climate Change on Inuit Health. *Ecohealth* **2012**, *9*, 89–101. [CrossRef] [PubMed]

62. Morrison, K.; Prieto, P.A.; Domínguez, A.C.; Waltner-Toews, D.; FitzGibbon, J. Ciguatera fish poisoning in La Habana, Cuba: A study of local social-ecological resilience. *EcoHealth* **2008**, *5*, 346–359. [CrossRef] [PubMed]

63. Harris, N.; Grootjans, J.; Wenham, K. Ecological aging: The settings approach in aged living and care accommodation. *EcoHealth* **2008**, *5*, 196–204. [CrossRef] [PubMed]

64. Mboera, L.E.; Mfinanga, S.G.; Karimuribo, E.D.; Rumisha, S.F.; Sindato, C. The changing landscape of public health in sub-Saharan Africa: Control and prevention of communicable diseases needs rethinking. *Onderstepoort J. Vet. Res.* **2014**, *81*, 1–6. [CrossRef] [PubMed]

65. Parkes, M. Personal Commentaries on "Ecosystems and Human Well-being: Health Synthesis—A Report of the Millennium Ecosystem Assessment". *EcoHealth* **2006**, *3*, 136–140. [CrossRef]

66. Stephens, C.; Parkes, M.W.; Chang, H. Indigenous perspectives on ecosystem sustainability and health. *EcoHealth* **2007**, *4*, 369–370. [CrossRef]

67. Nettleton, C.; Stephens, C.; Bristow, F.; Claro, S.; Hart, T.; McCausland, C.; Mijlof, I. Utz Wachil: Findings from an international study of indigenous perspectives on health and environment. *EcoHealth* **2007**, *4*, 461–471. [CrossRef]

68. Colwell, R.R.; Wilcox, B.A. Water, ecology, and health. *Ecohealth* **2010**, *7*, 151–152. [CrossRef] [PubMed]

69. Saint-Charles, J.; Surette, C.; Parkes, M.W.; Morrison, K.E. Connections for Health, Ecosystems and Society leading to Action and Change. *EcoHealth* **2014**, *11*, 279–280. [CrossRef] [PubMed]

70. Davies, K. Economic Costs of Childhood Diseases and Disabilities Attributable to Environmental Contaminants in Washington State, USA. *EcoHealth* **2006**, *3*, 86–94. [CrossRef]

71. Wernham, A. Inupiat health and proposed Alaskan oil development: Results of the first integrated Health Impact Assessment/Environmental Impact Statement for proposed oil development on Alaska's North Slope. *EcoHealth* **2007**, *4*, 500–513. [CrossRef]

72. Xu, J.; Sharma, R.; Fang, J.; Xu, Y. Critical linkages between land-use transition and human health in the Himalayan region. *Environ. Int.* **2008**, *34*, 239–247. [CrossRef] [PubMed]

73. Pesek, T.; Abramiuk, M.; Garagic, D.; Fini, N.; Meerman, J.; Cal, V. Sustaining plants and people: Traditional Q'eqchi'Maya botanical knowledge and interactive spatial modeling in prioritizing conservation of medicinal plants for culturally relative holistic health promotion. *EcoHealth* **2009**, *6*, 79–90. [CrossRef] [PubMed]

74. Connell, D.J. Sustainable livelihoods and ecosystem health: Exploring methodological relations as a source of synergy. *EcoHealth* **2010**, *7*, 351–360. [CrossRef] [PubMed]

75. Patrick, R.; Capetola, T.; Townsend, M.; Hanna, L. Incorporating sustainability into community-based healthcare practice. *EcoHealth* **2011**, *8*, 277–289. [CrossRef] [PubMed]

76. Asakura, T.; Mallee, H.; Tomokawa, S.; Moji, K.; Kobayashi, J. The ecosystem approach to health is a promising strategy in international development: Lessons from Japan and Laos. *Glob. Health* **2015**, *11*, 1–8. [CrossRef] [PubMed]

77. Crawshaw, L.; Fèvre, S.; Kaesombath, L.; Sivilai, B.; Boulom, S.; Southammavong, F. Lessons from an Integrated Community Health Education Initiative in Rural Laos. *World Dev.* **2014**, *64*, 487–502. [CrossRef]

78. Min, B.; Allen-Scott, L.; Buntain, B. Transdisciplinary research for complex One Health issues: A scoping review of key concepts. *Prev. Vet. Med.* **2013**, *112*, 222–229. [CrossRef] [PubMed]

79. Zinsstag, J. Convergence of ecohealth and one health. *EcoHealth* **2012**, *9*, 371–373. [CrossRef] [PubMed]

80. Patz, J. Launch of the International Association for Ecology and Health at Its First Biennial Conference: Message from the President Elect. *EcoHealth* **2007**, *4*, 6–9. [CrossRef]

81. Hall, D.C.; Le, Q.B. A Basic Strategy to Manage Global Health with Reference to Livestock Production in Asia. *Vet. Med. Int.* **2011**. [CrossRef] [PubMed]

82. Dolan, A.H.; Taylor, M.; Neis, B.; Ommer, R.; Eyles, J.; Schneider, D.; Montevecchi, B. Restructuring and health in Canadian coastal communities. *EcoHealth* **2005**, *2*, 195–208. [CrossRef]

83. Spiegel, S.J.; Veiga, M.M. Building capacity in small-scale mining communities: Health, ecosystem sustainability, and the Global Mercury Project. *EcoHealth* **2005**, *2*, 361–369. [CrossRef]

84. Fillion, M.; Philibert, A.; Mertens, F.; Lemire, M.; Passos, C.J.S.; Frenette, B.; Guimarães, J.R.D.; Mergler, D. Neurotoxic sequelae of mercury exposure: An intervention and follow-up study in the Brazilian Amazon. *Ecohealth* **2011**, *8*, 210–222. [CrossRef] [PubMed]

85. Spiegel, J.; Bennett, S.; Hattersley, L.; Hayden, M.H.; Kittayapong, P.; Nalim, S.; Wang, D.N.C.; Zielinski-Gutiérrez, E.; Gubler, D. Barriers and bridges to prevention and control of dengue: The need for a social-ecological approach. *EcoHealth* **2005**, *2*, 273–290. [CrossRef]

86. Forde, M.; Morrison, K.; Dewailly, E.; Badrie, N.; Robertson, L. Strengthening integrated research and capacity development within the Caribbean region. *BMC Int. Health and Hum. Rights* **2011**. [CrossRef] [PubMed]

87. Zaidi, M.K.; Ganoulis, J. Cooperative network for environmental risk analysis studies: The case of the Middle East region. In *Environmental Security and Environmental Management: The Role of Risk Assessment*; Springer: Berlin, Germany, 2006; pp. 315–320.

88. Stephen, C.; DiCicco, E.; Munk, B. British Columbia's Fish Health Regulatory Framework's Contribution to Sustainability Goals Related to Salmon Aquaculture. *Ecohealth* **2008**, *5*, 472–481. [CrossRef] [PubMed]

89. Hales, S.; Corvalan, C.A. Public Health Emergency on Planet Earth: Insights from the Millennium Ecosystem Assessment. *EcoHealth* **2006**, *3*, 130–135. [CrossRef]

90. Johnston, F.H.; Jacups, S.P.; Vickery, A.J.; Bowman, D.M. Ecohealth and Aboriginal testimony of the nexus between human health and place. *Ecohealth* **2007**, *4*, 489–499. [CrossRef]

91. Ali, R.; Zhao, H. Wuhan, China and Pittsburgh, USA: Urban environmental health past, present, and future. *EcoHealth* **2008**, *5*, 159–166. [CrossRef]

92. Añabieza, M.; Pajaro, M.; Reyes, G.; Tiburcio, F.; Watts, P. Philippine alliance of fisherfolk: Ecohealth practitioners for livelihood and food security. *EcoHealth* **2010**, *7*, 394–399. [CrossRef] [PubMed]

93. Brown, V.A. Principles for EcoHealth Action: Implications of the Health Synthesis Paper, the Millennium Ecosystem Assessment, and the Millennium Development Goals. Workshop Group, EcoHealth ONE, Madison, Wisconsin, October 2006. *EcoHealth* **2007**, *4*, 95–98. [CrossRef]

94. Rapport, D.J. Sustainability science: An ecohealth perspective. *Sustain. Sci.* **2007**, *2*, 77–84. [CrossRef]

95. Kittinger, J.N.; Coontz, K.M.; Yuan, Z.; Han, D.; Zhao, X.; Wilcox, B.A. Toward holistic evaluation and assessment: Linking ecosystems and human well-being for the Three Gorges Dam. *EcoHealth* **2009**, *6*, 601–613. [CrossRef] [PubMed]

96. Bunch, M.J.; Franklin, B.; Morley, D.; Kumaran, T.V.; Suresh, V.M. Research in turbulent environments: Slums in Chennai, India and the impact of the December 2004 Tsunami on an EcoHealth project. *EcoHealth* **2005**, *2*, 150–154. [CrossRef]

97. Parkes, M.W. Diversity, emergence, resilience: Guides for a new generation of ecohealth research and practice. *Ecohealth* **2012**, 1–3. [CrossRef] [PubMed]

98. Rapport, D.J. Response to EcoHealth editorial, Parkes MW (2011) Vol. 8, Issue 2. *EcoHealth* **2012**, *9*, 378–379. [PubMed]

99. Aoyama, M.; Hudson, M.J. No Better Medicine: Health in American Environmental Writing. *EcoHealth* **2014**, *11*, 461–463. [CrossRef] [PubMed]

100. Butler, C.D.; Friel, S. Time to regenerate: Ecosystems and health promotion. *PLoS Med.* **2006**. [CrossRef] [PubMed]

101. Bunch, M.J.; Parkes, M.; Zubrycki, K.; Venema, H.; Hallstrom, L.; Neudorffer, C.; Berbés-Blázquez, M.; Morrison, K. Watershed management and public health: An exploration of the intersection of two fields as reported in the literature from 2000 to 2010. *Environ. Manag.* **2014**, *54*, 240–254. [CrossRef] [PubMed]

102. Forster, P. Ten Years on: Generating Innovative Responses to Avian Influenza. *EcoHealth* **2014**, *11*, 15–21. [CrossRef] [PubMed]

103. Nguyen, V.; Nguyen-Viet, H.; Pham-Duc, P.; Stephen, C.; McEwen, S.A. Identifying the impediments and enablers of ecohealth for a case study on health and environmental sanitation in Hà Nam, Vietnam. *Infect. Dis. Poverty* **2014**, *3*, 36. [CrossRef] [PubMed]

104. Yee, S.H.; Bradley, P.; Fisher, W.S.; Perreault, S.D.; Quackenboss, J.; Johnson, E.D.; Bousquin, J.; Murphy, P.A. Integrating human health and environmental health into the DPSIR framework: A tool to identify research opportunities for sustainable and healthy communities. *EcoHealth* **2012**, *9*, 411–426. [CrossRef] [PubMed]

105. Aguirre, A.; Wilcox, B.A. EcoHealth: Envisioning and creating a truly global transdiscipline. *Ecohealth* **2008**, *5*, 238–239. [CrossRef] [PubMed]

106. Kaneshiro, K.Y.; Chinn, P.; Duin, K.N.; Hood, A.P.; Maly, K.; Wilcox, B.A. Hawaii's mountain-to-sea ecosystems: Social-ecological microcosms for sustainability science and practice. *EcoHealth* **2005**, *2*, 349–360. [CrossRef]

107. Burns, T.E.; Wade, J.; Stephen, C.; Toews, L. A Scoping Analysis of Peer-Reviewed Literature About Linkages Between Aquaculture and Determinants of Human Health. *EcoHealth* **2014**, *11*, 227–240. [CrossRef] [PubMed]
108. Fowler, C.W. Sustainability, health, and the human population. *EcoHealth* **2005**, *2*, 58–69. [CrossRef]
109. Mackay, R.; Wolbring, G. Sustainable consumption of healthcare: Linking sustainable consumption with sustainable healthcare and health consumer discourses. In Proceedings of the 3rd World Sustainability Forum, Basel, Switzerland, 1–30 November 2013.
110. Wolbring, G.; Mackay, R.; Rybchinski, T.; Noga, J. Disabled People and the Post-2015 Development Goal Agenda through a Disability Studies Lens. *Sustainability* **2013**, *5*, 4152–4182. [CrossRef]
111. Li, Y.-Y.; Dong, S.-K.; Wen, L.; Wang, X.-X.; Wu, Y. Three-dimensional framework of vigor, organization, and resilience (vor) for assessing rangeland health: A case study from the alpine meadow of the qinghai-tibetan plateau, China. *EcoHealth* **2013**, *10*, 423–433. [CrossRef] [PubMed]
112. Von Schirnding, Y. WHO World Health Organization Health in Sustainable Development Planning: The Role of Indicators. Available online: http://www.who.int/wssd/resources/indicators/en/ (accessed on 3 February 2016).
113. Saint-Charles, J.; Webb, J.; Sanchez, A.; Mallee, H.; de Joode, B.V.W.; Nguyen-Viet, H. Ecohealth as a field: Looking forward. *Ecohealth* **2014**, *11*, 300–307. [CrossRef] [PubMed]
114. Vallance, S.; Perkins, H.C.; Dixon, J.E. What is social sustainability? A clarification of concepts. *Geoforum* **2011**, *42*, 342–348. [CrossRef]
115. Koren, H.; Butler, C. The interconnection between the built environment ecology and health. In *Environmental Security and Environmental Management: The Role of Risk Assessment*; Springer: Berlin, Germany, 2006; pp. 109–125.
116. Fang, J.; Wu, X.; Xu, J.; Yang, X.; Song, X.; Wang, G.; Yan, M.; Yan, M.; Wang, D. Water management challenges in the context of agricultural intensification and endemic fluorosis: The case of Yuanmou County. *Ecohealth* **2011**, *8*, 444–455. [CrossRef] [PubMed]

![sustainability logo] *sustainability*

MDPI

Article

Sustaining without Changing: The Metabolic Rift of Certified Organic Farming

Julius Alexander McGee * and Camila Alvarez

Department of Sociology, University of Oregon, Eugene, OR 97402, USA; calvarez@uoregon.edu
* Correspondence: julimcgee9@gmail.com; Tel.: +1-916-873-7123

Academic Editor: Md Saidul Islam
Received: 21 October 2015; Accepted: 15 December 2015; Published: 27 January 2016

Abstract: Many proponents of organic farming claim that it is a sustainable alternative to conventional agriculture due to its reliance on natural agro-inputs, such as manure based fertilizers and organic pesticides. However, in this analysis we argue that although particular organic farming practices clearly benefit ecosystems and human consumers, the social context in which some organic farms develop, limit the potential environmental benefits of organic agriculture. Specifically, we argue that certified organic farming's increased reliance on agro-inputs, such as organic fertilizers and pesticides, reduces its ability to decrease global water pollution. We review recent research that demonstrates the environmental consequences of specific organic practices, as well as literature showing that global organic farming is increasing its reliance on agro-inputs, and contend that organic farming has its own metabolic rift with natural water systems similar to conventional agriculture. We use a fixed-effects panel regression model to explore how recent rises in certified organic farmland correlate to water pollution (measured as biochemical oxygen demand). Our findings indicate that increases in the proportion of organic farmland over time increases water pollution. We conclude that this may be a result of organic farms increasing their reliance on non-farm agro-inputs, such as fertilizers.

Keywords: organic farming; metabolic rift; conventionalization thesis

1. Introduction

Organic farming is often put forth as a sustainable alternative to conventional agriculture, claiming to rely on ecologically sustainable practices that are more in line with earth's natural ecology [1,2]. This has helped to increase the popularity of organic goods around the world, as sales on organic farms have risen five-fold over the past decade and a half [3]. The recent success of organic farming is also partially due to the rise in organic certification, a process whereby external entities, usually government organizations, create a unified definition of organic farming to regulate the practices used by farmers and help consumers identify organic goods [2,4,5]. While there are clear merits to having a cohesive definition of organic farming, some have argued that certification is being used to integrate the organic industry into to the agribusiness industry by regulating standards in a way that increases the economic viability of organic agriculture. Specifically, some researchers have suggested that organic certification leads to a "conventionalization" of the organic market, by watering down standards and increasing the use of inputs produced off farm, such as non-synthetic fertilizers and pesticides, to reduce the risk of direct farm investments [6–8]. If tilling methods and fertilizer management practices are being refashioned on organic farms to serve economic interests over ecological interests, then the ability of nations to reduce specific environmental hazards caused during agricultural production by shifting toward organic practices may be weakened. In particular, it has been noted that even though organic goods have clear environmental benefits in terms of biodiversity protection and human health [9–11], they can have similar, and in some instances higher levels of nitrate leaching as their conventional counterparts if certain practices (e.g., seasonal crop rotations and manure management) are not implemented properly [10–14].

To this end, we draw on an environmental sociological theory known as metabolic rift [15], to demonstrate how the conventionalization of the organic market may limit the ability of organic farming to address some of the ecological cost of agricultural production. Specifically, we contend that the conventionalization of organic farming reduces its ability to mend the metabolic rift between agriculture and natural water systems. We empirically test our assumption using a fixed-effects panel regression model to examine whether increases in the percentage of organic farmland cross-nationally from 2003–2007 reduced biochemical oxygen demand (BOD) in water, while controlling for population, percent urban population, and gross domestic product GDP (social components known to drive various environmental impacts including BOD).

2. Organic Farming and the Conventionalization Thesis

The rise of certified organic farming has been met with many criticism by social scientists. The most prevalent criticisms have been brought forth by scholars developing the conventionalization thesis, which hypothesizes that as certified organic farming grows, it begins to mimic conventional agricultural practices. The term conventionalization was first proposed by Buck *et al.* [6] to describe the changes occurring within organic agriculture in California. The authors utilized the concept to convey the transition of organic farming from an idealistically driven counter cultural movement, to a slight variant of conventional agriculture. Buck *et al.* [6] and Guthman [7], found that organic farming was increasingly becoming industrialized, relying on non-farm inputs, such as machinery, fertilizers, feed, agrochemicals, and resource substitutions, to stimulate production. This resulted in a bifurcation of the organic market, creating of two organic systems—one more in line with the original ideals of the movement that emphasized local small scale farming, direct consumer sales, and prohibited the use of non-farm inputs, and another economically driven market that helped to integrate organic agriculture into the agribusiness industry.

More recently, the conventionalization thesis has been expanded to focus on global organic practices. For example, Best [16] found that newer organic farms in Germany show signs of conventionalization, noting that newer organic farmers tended to use slightly larger farms and had more specialized operations. Additionally the author found that recent adopters did not share the same "pro-environmental" values as earlier farmers. Flaten *et al.* [17] similarly found that newer organic dairy farmers in Norway used more concentrates and had higher milk production yields, highlighting that while all organic farmers shared favorable views toward the environment, older farmers had much stronger views and placed more emphasis on soil fertility, fertilizers, and pollution. Läpple and Van Rensburg [18] in Ireland, also found that late adopters of organic farming expressed lower environmental values and were much more profit driven than early or medium adopters. In the Netherlands DeWit and Verhoog [19] found that conventional agro-food commodity chains were increasing and the use of non-farm inputs in organic farming. Specifically, the authors noted that conventional fertilizers were consistently being used in organic pig and poultry production.

These studies, although specific to particular locations, demonstrate a potential shift in organic farming practices globally. Furthermore, if these practices are becoming more prevalent globally, they may alter the ability of organic farms to reduce water pollution. Below we discuss the ecological implications of organic farming practices *versus* conventional farming specifically in regards to water pollution, to demonstrate the environmental impacts of organic agricultural practices.

3. Organic Agriculture and Water Pollution

Agriculture is one of the largest contributors to global water pollution. It increases the amount of organic contaminants found in natural water systems and produces chemical imbalances through the extensive use of pesticides and fertilizers [14]. Pesticide runoff is known to increase bioconcentration, which is the accumulation of chemicals on or in organisms, and biomagnification, where chemicals become more concentrated as they move up the food chain in ecosystems and may induce biodiversity loss [20]. While a lot of organic farms do use pesticides [2,4,5], organic pesticides have not been linked to water pollution, and there are currently no studies finding a clear relationship between organic pesticides and water pollution. Thus at this time, there is no reason to believe use of organic pesticides increases water pollution.

Organic fertilizers that contain nitrogen and phosphate on the other hand, can leach into soil and create algal blooms in surface water, causing overall oxygen levels in water to decline, which also can result in biodiversity loss in natural water systems [21]. This process often occurs when water drains through soil, taking with it the nitrates contained in the soil. Organic fertilizers, such as animal manures that contain nitrogen, have specifically been linked to nitrate leaching when nitrate is added to soil while drainage is occurring, when more nitrate is supplied than needed for a crop to grow, and when there is a lack of synchrony between nitrogen supply and crop uptake [9]. Shepard *et al.* [9] also notes that "if soils are left bare during fall or crops are poorly developed, there will not be an effective rooting system to utilise the soil N that is mineralised after harvest and this will be at risk of leaching over the winter" (p. 37).

Some studies that observe levels of nitrate leaching between organic and conventional farms argue that organic farms have lower levels of nitrate leaching due to overall lower inputs of nitrogen [9,22–24], however, the bulk of these studies relies on data from specific organic and conventional farms and were conducted prior to what recent research that is seen as the conventionalization period of organic practices. Furthermore, studies conducted during this same period noted that in some instances organic agriculture had similar or higher leaching rates than conventional farms. For instance, Kristensen *et al.* [25] showed that the average nitrate content in soils between conventional and organic farms that used manure-based fertilizers in fall was slightly higher in organic farms, and far higher in organic farms *versus* conventional farms that did not use manure-based-fertilizers. Condron *et al.* [26] found in simulations that nitrate losses were similar between conventional and organic farms during rotations in New Zealand. Stopes *et al.* [27], also found that during rotations nitrate leaching was similar for conventional and organic farms that used under 200 kilograms per hectare of fertilizer, but were greater for organic farms receiving more than 200 kilograms per hectare of fertilizer. More recent studies have also concluded that nitrate leaching is similar and in some instances slightly higher on organic farms [12,13]. For example, Tuomisto *et al.* [14] in a systematic study of research observing the environmental impacts between organic and conventional farms, concluded that nitrate leaching per unit of area was 31% lower on organic farms, but 49% higher per unit of product on organic farms.

Comprehensively, these studies demonstrate the degree to which water pollution derived from nitrate leaching is induced by conventional and organic farming. Furthermore, they reveal that in order for organic farms to have lower levels of nitrate leaching than conventional farms, they must use specific management practices, which include seasonally conscious crop rotations as well as careful and limited inputs of nitrate-based fertilizers. While organic farming is often promoted as an agricultural method more in line with Earth's natural ecology, the requisites for this are diverse and complex, and may be limited based on the social context in which organic farms are developed. For instance, the conventionalization thesis has revealed that over time organic farmers have become less concerned with the environment, less strict about farming practices, and more economically motivated [6,17,18]. These trends produce an organic agricultural system that is less cognizant of the practices necessary to reduce bio oxygen demand in water, due to decreasing concern about and application of methods necessary to combat nitrate leaching. Additionally, the processes of conventionalization work to increase the size of organic farms, and the concentration of inputs used on organic farms. Based on criticisms of proponents of the conventionalization thesis and the analyses of natural scientists regarding the practices necessary to reduce nitrate leaching, it is reasonable to believe that organic farming may not function as a counter-force to all forms of water pollution derived from agricultural production, but in fact perpetuate specific types of water contamination. Below we further develop this argument using the environmental sociological theory metabolic rift.

4. Organic Farming's Metabolic Rift

Metabolic rift was developed by John Bellamy Foster [15] to refer to Marx's expression of the "irreparable rift in the interdependent process of social metabolism" [28] (p. 949). The term is based on Marx's writings regarding metabolism and the development of soil chemistry and the use of fertilizer in agricultural production. Foster argues that Marx acknowledged the growing contradictions between capitalism and nature in his observation of Liebig's work and the British agricultural revolution. There, Marx proposes that capitalism is breaking the natural laws of sustainability in its use of fertilizers to restore nutrients to the soil that were lost during large scale agricultural production. Marx also accuses "large landed property" of "reducing the agricultural population to an ever decreasing minimum" and as a result, the concentration of populations in cities, leads to "a squandering of the vitality of the soil" (because all soil nutrients end up in city sewers rather than the land) [28] (p. 949). He further contends that "The way that the cultivation of particular crops depends on fluctuations in market prices and the constant change in cultivation with these prices—the entire spirit of capitalist production, which is oriented towards the most immediate monetary profits—stands in contradiction to agriculture, which has to concern itself with the whole gamut of permanent conditions of life required by the chain of successive generations" [28] (p. 754). In essence, as Foster [15] notes, Marx argues that the application of market values to agricultural production contradicts the ecological forces that sustain farm systems. This included the ever increasing size and scale of farms as well as enhanced reliance on non-farm inputs, such as nitrates, phosphates, and potassium derived from manure and guano that are added to soil to maintain and increase fertility.

While Marx's concern with the application of fertilizers was on soil sustainability rather than water pollution produced from nitrate leaching, the notion of metabolic rift has also been further developed to explore capitalism's inherent contradiction with sustainability. Clark and York [29] apply the term rifts and shifts to the process "whereby metabolic rifts are continually created and addressed (typically only after reaching crisis proportions) by shifting the type of rift generated" (p. 17). They argue that "To the myopic observer, capitalism may appear at any one moment to be addressing some environmental problems, since it does on occasion mitigate a crisis. However, a more far-sighted observer will recognize that new crises spring up where old ones are supposedly cut down" [29] (p. 17).

We expand on this argument, and contend that the socioeconomic conditions influencing organic agriculture mirror those influencing conventional agriculture, as a result, the environmental degradation developed by organic agriculture is similar to the environmental degradation of conventional agriculture. For instance, just as the metabolic rift observed by Marx was a result of the town-country divide, which was addressed by increasing the amount of non-farm inputs used in agriculture, we argue that conventional organic farming is a refashioning of this metabolic rift, relying on natural rather than synthetic inputs. This is to say that the production of industrial organic farming (the conventionalized cousin of the original organic movement) is simply a change in the technology used in agriculture's previous metabolic rift, shifting to the use of natural inputs (ironically the inputs observed in Marx's original analysis) instead of synthetic inputs. However, agriculture's metabolic rift was never about the inputs, but the structural processes necessary to maintain society's destructive relationship with nature. Thus in order to address industrial agriculture's rift with nature, nations must address the economic as well as technological context of agriculture. Before discussing how we model and test these assumptions, we briefly review previous research using metabolic rift theory and discuss how our research builds on this tradition.

Metabolic rift theory has been used by social scientists to contextualize the environmentally hazardous outcomes of various forms of social organization. For example, Mancus [30] examined the metabolic rift in global agriculture markets. He argues that structure of industrial agriculture, which is defined by the overuse and dependence of inorganic nitrogen fertilizer, has breached the social metabolism between society and the nitrogen cycle, creating massive environmental pollution in natural water ways and soil erosion. In a similar vein, Gunderson [31] applies metabolic rift theory to analyze large-scale livestock production, showing how the environmental impacts of industrial livestock production increase greenhouse gas emissions, and pollute natural water systems. Clausen and Clark [32] apply metabolic rift theory to marine systems, demonstrating how intensified production of aquaculture systems and overfishing practices pollute natural water systems and reduce aquatic biodiversity.

Others have expanded metabolic rift theory by focusing on the historical development of science and technology. For instance, Clark and York [33] focus on the historical development of science and technology to explain the metabolic rift between industrial civilization and the carbon cycle. Moore [34] provides a historical examination of environmental history using metabolic rift theory to explain the rise of global capitalism and the development of the world system.

In a fashion similar to these works, we apply metabolic rift theory to further explore the rift between modern social organizations and the natural environment. We expand the theory of metabolic rift by examining how it offers critical insights into mechanisms of sustainability, specifically, organic agriculture. Additionally, we adopt the conceptual framework of rifts and shifts to explain how organic farming is a result of shifting industrial agriculture's rift from synthetic agrochemicals to organic practices. We argue that the process of conventionalization, specifically, the vertical and horizontal integration of the organic market, mirrors the structure of the conventional agricultural industry by increasing organic farms' reliance on non-farm inputs. In turn, these inputs help to increase the economic viability of the organic market by increasing the financial gains of organic pesticide and fertilizer manufacturers [6]. This leads conventionalized organic farms to produce the same metabolic rift that Marx identified in his observations of the British agricultural revolution.

5. Hypotheses

Based on the theory discussed above we hypothesize that as the proportion of organic farming increases over time, it becomes more conventionalized, resulting in an expansion in industrial agriculture's rift to water ecosystems. To this end we ask if there is a positive correlation between organic farming and water pollution. The contrasting hypotheses we test are:

H1: Increases in the proportion of certified organic farmland is correlated positively with biochemical oxygen demand.

H2: Increases in the proportion of certified organic farmland is correlated negatively with biochemical oxygen demand.

We attribute hypothesis 1 to the conventionalization thesis and the theory of rifts and shifts, where the vast majority of certified organic farmland is increasing biochemical oxygen demand in water due to weak management practices and a shift in the technological methods used in farming. Hypothesis 2 assumes that certified organic farmland is in fact working as a counterforce to the environmental hazardous effects of agriculture and reducing water pollution such as biochemical oxygen demand.

6. Methods

To test our hypotheses we use a fixed-effects panel regression (for nations where sufficient data is available) including time dummies with robust standard errors adjusted for clustering by nation from 2002–2007. A fixed-effects panel model with time dummies controls for any unobserved, time-constant features particular to each nation, as well as events factors that change over time but that do not vary across nations, such as international commodity prices.

The logic of our modeling approach is based on the STIRPAT framework [35–43]. STIRPAT was first developed by Dietz and Rosa [44] as a reformulation of the popular IPAT equation to gauge how population (P), economic growth or affluence (A), and technology (T) affect the scale of environmental impacts (I). STIRPAT is a stochastic model that assumes environmental impacts are a multiplicative function of population, affluence, and technology, but does not assume that each factor has a proportional effect, STIRPAT thereby allows for hypothesis testing. In STIRPAT analyses each variable is converted to natural logarithmic form, since an additive model with logarithims is equivalent to a multiplicative model with variables in original units. STIRPAT is therefore an elasticity where beta coefficients represent a proportional rate in the dependent variable (here environmental impact) for every one-percent change in the independent variable corresponding to the beta coefficient [41,43]. The fixed-effects model specification is therefore:

$$\ln y_{it} = \beta_1 \ln(x_{it}) + \beta_2 \ln(x_{it}) \ldots \beta_k \ln(x_{itk}) + \mu_i + w_t + e_{it}$$

Here the subscript i represents each unit of analysis (nation) and the subscript t the time period, y_{it} is the dependent variable in original units for each nation at each point in time, x_{itk} represent the independent variables in original units for each nation at each point in time, β_k represents the elasticity coefficient for each independent variable, u_i is a nation specific disturbance term that is constant overtime (*i.e.*, the nation specific y-intercept), wt is a period specific disturbance term constant across nations, and e_{it} is the stochastic disturbance e term specific to each nation at each point in time. Our model is specified below:

Biochemical oxygen demand$_{it}$ = βpopulation$_{it}$ + βGDP per capita$_{it}$ + βpercent urban population$_{it}$ + βpercent organic hectares of total agricultural land$_{it}$ + μ_i + w_t + e_{it}

7. Dependent Variable

In this study, water pollution is the dependent variable and a proxy for environmental degradation. We measure water pollution via biochemical oxygen demand (BOD) (in thousands of kilograms per day) which is the amount of oxygen microorganisms in water needed to break down waste in natural water systems. Organic material in water comes from a variety of sources, such as plant, animal, and/or human waste and industrial activities. While the organic materials are in the water, metabolic processes of bacteria break down the waste over time [44]. During these process, a certain amount of dissolved oxygen is consumed. BOD measures the amount of oxygen consumed by microorganisms to decompose waste. Waters with high amounts of waste correspond to a high BOD because a large number of microorganisms are necessary to breakdown the waste. High BOD rates put other aquatic life at risk due to reduced oxygen availability. Nitrates and phosphates are important elements that contribute to the amount of BOD found in natural water systems [44]. BOD measurements are one of the most reliable pollution indicators because it is relatively inexpensive to measure. In addition, BOD measurements are traditional starters for industrial pollution control within nations and are widely used in across nations [25]. Our data for BOD comes from the World Bank's environmental indicators website [45]. The World Bank's data on BOD started as continuation of Hettige *et al.* [25] attempts to measure the amount of industrial pollutants found in natural water systems globally. To achieve this, the authors gather data on BOD levels in natural water systems from multiple nations, when/where data was available. The World Bank continued this aggregation through 2007.

8. Key Independent Variable

Our key independent variable in this analysis is proportion of organic farmland, which estimates the amount of the organic hectares divided by the total farming hectares. The data for organic agricultural land was obtained from Organic World Statistics [46]. Data on certified organic agriculture is obtained from the SOEL/FiBL/IFOAM survey. Certified organic farming refers to both the certified in conversion areas and the certified fully converted areas. A major drawback of this data is that definitions of organic may vary across countries and data are gathered using various methods (e.g., surveys, secondary data, experts, *etc.*) thus we interpret the results presented here cautiously.

9. Additional Independent Variables

GDP per capita is a control variable to account for a country's economic standing and was gathered from the World Bank [45]. The variable was measured in constant 2005 US dollars. GDP per capita is a standard control variable for most environmental impacts analyses. Environmental sociological theories of the treadmill of production and world-systems suggest economic development to be a major structural driver of environmental degradation [43]. Previous research on water pollution, ecological footprints, carbon dioxide emission, and energy consumption find GDP per capita to be a positive predictor [25,38–43,47] (Earlier models not shown here were estimated with a quadratic term for GDP per capita and urbanization, however neither was found significant in a two-tailed test).

Population and urbanization are additional control variables representing important national demographic factors and were collected via the World Bank. Previous research on nature/society have found population to be a significant factor [39–43,47]. Urbanization is included as a control variable to evaluate the level of a country's urbanization. Number of persons living in urban areas is estimated as the total persons living in urban areas divided by the total population. Additionally, we included urbanization as a control variable to serve as a proxy for the number of sewage systems and industrial processes that contribute to BOD [25]. Prior research has shown urbanization to be a significant predictor for environmental impacts. Table 1 includes a summary of descriptive statistics for all dependent and independent variables.

Table 1. Descriptive statistics of variables in raw form.

Variables	Mean	Standard Deviation	Minimum	Maximum
Biochemical oxygen demand	234,006.8	774,215.3	131.9	8,800,000
Proportion organic land	0.1	1.4	0.00003	14.5
Population	3.67×10^7	1.16×10^8	87,276	1.30×10^9
GDP per capita	11,297.4	12,804.6	118.1	74,220.4
Percent urban population	61.4	20.8	11.6	100

Note: N = 274.

10. Results

As noted above, the fixed-effects models presented below control for omitted factors that vary cross-nationally but are temporally invariant, such as geographic, climatic, and geological factors, as well as the effects of the historical legacy preceding the periods examined here (e.g., the era during which a nation began to industrialize agriculture). The models, therefore, control for temporally invariant characteristics unique to each nation. Additionally, the models control (via the time dummies) for cross-sectional invariant factors that change over time, such as international prices of resources. Thus, these models focus on change over time within nations, not on cross-sectional differences. All variables (except dummy variables) are in natural logarithmic form, which makes this an elasticity model.

The results from our analysis are reported in Table 2. We present R-squared within and the highest variance inflation factor (VIF) for each model. Within R-squared measures the variation of BOD within countries explained by the independent variables. In fixed-effect panel analyses, R-squared within is a better measurement than R-squared overall because fixed-effects disregards between unit variation [40]. The variance inflation factor measures the amount of multi-collinearity, note that none of our independent variables reached a VIF of 10 or higher. This means that our coefficients are not substantially affected by a collinear relationships [48].

Our results show support for H1, (although they do not confirm it) which provides evidence for our theoretical assumption that global conventionalization of organic farming is increasing, not reducing agriculture's metabolic rift with respect to water ecosystems. Specifically our model demonstrates that as a country's organic land increases there is a corresponding increase in BOD while holding constant population, urbanization, and GDP per capita, indicating that the rift of water pollution in the water cycle is enhanced through organic farming. It is important to note that our coefficient for proportion organic farmland is close to zero, meaning that organic farming may have a significant but negligible effect on BOD. Of course, importantly, the coefficient is not negative, clearly ruling out H2. While these results support our theoretical assumptions, they must be understood with caution as they do not assess the specific types of practices conducted on organic farms.

Table 2. Fixed-effects panel regression coefficients predicting Biochemical Oxygen Demand.

Independent Variables Logged	Coefficients (SE)
Population	1.308 *** (0.467)
Percent urban population	1.032 * (0.438)
GDP per capita squared	0.169 ** (0.054)
Proportion organic land	0.018 *** (0.003)
R-squared within	0.266
High VIF	1.003
N	277

* $p < 0.05$; ** $p < 0.01$; *** $p < 0.001$ (two-tailed tests).

Population, GDP per capita, and urban population were also found to be significant predictors on BOD, which is consistent with the findings of previous STIRPAT analyses [35–43]. Specifically we find that a one percent increase in GDP per capita corresponds with a .169 percent increase in BOD. We also find that a one percent increase in population results in a more than 1.3 percent increase in BOD, indicating that there an elastic relationship between BOD and population. Similarly, we find that a one unit increase in the percent of urban population corresponds to a one percent increase in BOD, meaning that not only is population a powerful contributor to BOD but specifically urban population. Previous research on BOD found similar results from control variables [47].

Our results support the findings of soil scientists who have found that specific organic management practices lead organic farms to have higher or similar levels of nitrate leaching as conventional farms [10,12,13,27]. Additionally our results support the findings of social scientists who argue that organic farming is becoming increasingly reliant on non-farm inputs such as organic fertilizers [6–8,16–19]. However, these results may also suggests that shifts toward organic farming are correlated with BOD but have not increased enough to counteract the amount nitrate leaching that occurs from conventional farming.

11. Discussion and Conclusions

Here we have reviewed literature that argues certified organic farms are becoming increasingly reliant on non-farm inputs, such as organic fertilizers [6–8,16–19], as well as literature demonstrating that some organic farming practices contribute to nitrate leaching [10,12,13,27]. We have also reviewed literature demonstrating how nitrate leaching contributes to water pollution and can increase the biochemical oxygen demand in natural water systems. Although shifting agricultural land toward organic land has the potential to reduce levels of BOD, due to specific organic management practices that limit the use of non-farm inputs, we have found that between 2002 and 2007 increasing the proportion of organic farmland has not reduced BOD. Specifically, we have measured the average rate per day of BOD in natural water systems within countries and have found that increasing the proportion of organic farmland increases BOD levels.

To better interpret this finding, we use the theory metabolic rift and argue that the conventionalization of organic farming reproduces industrial agriculture's rift with water ecosystems. Specifically, we contend that the increased use of non-farm inputs to maintain soil fertility on organic farms replicates conventional agriculture's metabolic rift, and as a result, the development of organic farming over time has only increased water pollution rather than reduce it. These results do not mean that shifting agricultural production toward organic practices will never reduce water pollution, however they do demonstrate a potential problem in current trends within the organic sector of the agricultural industry. Social science research conducted in different nations has found that over time new farmers participating in the organic industry are less cognizant of on-farm practices that maintain soil fertility and limit the necessity of ago-inputs [8,16,18]. This trend must be addressed if organic farming is going to be a sustainable alternative to conventional agriculture and limit water pollution. We believe future regulations aimed at reducing water pollution from agricultural production should address both the natural and social context in which agricultural systems progress in order to develop a more environmentally sustainable agricultural system.

Acknowledgments: Acknowledgments: The Authors would like to thank Richard York and the reviewers for their helpful comment.

Author Contributions: Author Contributions: Julius Alexander McGee conceptualized and designed the research. Julius Alexander McGee and Camila Alvarez wrote the paper and analyzed the data. Both authors have read and approved the final manuscript.

Conflicts of Interest: Conflicts of Interest: The authors declare no conflict of interest.

Appendix A

Table A1. Summary of countries and years.

Country	Year					
Albania	-	-	-	2005	2006	-
Argentina	2002	-	-	-	-	-
Austria	2002	2003	2004	2005	2006	-
Azerbaijan	-	2003	2004	2005	2006	2007
Belgium	-	-	2004	2005	2006	-
Bulgaria	2002	2003	2004	2005	2006	2007
Chile	2002	2003	2004	2005	2006	-
China	-	2003	2004	2005	2006	2007
Colombia	2002	2003	2004	2005	-	-
Croatia	2002	2003	2004	2005	2006	2007
Cyprus	2002	2003	2004	2005	2006	2007
Czech Republic	2002	2003	2004	2005	2006	-
Denmark	2002	2003	2004	2005	2006	-
Ecuador	2002	2003	2004	2005	-	-
Estonia	2002	-	2004	2005	2006	2007
Fiji	2002	2003	2004	-	-	-
Finland	2002	2003	2004	2005	2006	-
France	2002	2003	2004	2005	2006	-
Germany	2002	2003	2004	2005	2006	-
Greece	-	-	2004	2005	2006	-
Ghana	-	2003	-	-	-	-
Hungary	2002	2003	2004	2005	2006	-
Indonesia	-	2003	2004	2005	2006	-
Iran	2002	-	-	2005	-	-
Ireland	2002	2003	2004	2005	2006	-
Israel	2002	2003	-	-	-	-
Italy	2002	2003	2004	2005	2006	-
Japan	2002	2003	2004	2005	-	-
Jordan	-	-	-	2005	2006	2007
Kazakhstan	-	-	2004	2005	2006	2007
Kyrgyz Republic	-	-	-	-	-	2007
Lativa	2002	2003	2004	2005	2006	2007
Lithuania	2002	2003	2004	2005	2006	2007
Luxembourg	2002	2003	2004	2005	2006	-

Table A1. *Cont.*

Country	Year					
Macedonia, FYR	-	-	-	-	2006	2007
Madagascar	-	2003	2004	2005	2006	-
Malaysia	-	2003	-	-	2006	-
Malta	-	-	-	2005	-	-
Mauritius	2002	2003	2004	2005	2006	2007
Morocco	2002	2003	2004	2005	2006	2007
Netherlands	2002	2003	2004	2005	2006	-
New Zealand	2002	2003	2004	2005	2006	2007
Norway	2002	2003	2004	2005	2006	-
Pakistan	-	-	-	-	2006	-
Panama	-	-	2004	2005	-	-
Paraguay	2002	-	-	-	-	-
Philippines	-	2003		2005	-	-
Poland	2002	2003	2004	2005	2006	-
Portugal	2002	2003	2004	2005	2006	-
Romania	2002	2003	2004	2005	2006	2007
Russian Federation	2002	2003	2004	2005	2006	2007
Saudi Arabia	-	-	-	-	2006	-
Slovak Republic	-	-	-	-	2006	-
Slovenia	2002	2003	2004	2005	2006	2007
South Africa	2002	2003	2004	2005	2006	2007
South Korea	2002	2003	2004	-	2006	-
Spain	2002	2003	2004	2005	2006	-
Sri Lanka	-	-	-	-	2006	-
Sweden	2002	2003	2004	2005	2006	-
Syrian Arab Republic	2002	2003	2004	2005	2006	2007
Tanzania	-	2003	2004	2005	2006	2007
Thailand	2002	-	-	-	2006	-
Turkey	2002	2003	2004	2005	2006	-
Ukraine	-	2003	2004	2005	2006	2007
United Kingdom	2002	2003	2004	2005	-	-
United States	2002	-	2004	2005	2006	-
Vietnam	-	2003	2004	2005	2006	2007

References

1. FAO. What is Organic Agriculture? Available online: http://www.fao.org/organicag/oa-faq/oa-faq1/en/ (accessed on 23 April 2015).
2. USDA. United States Department of Agriculture. Available online: http://www.usda.gov (accessed on 10 April 2014).
3. FiBL (Research Institute of Organic Agriculture). Activity Report. Available online: http://www. fibl.org/fileadmin/documents/en/activityreport/FiBL_Taetigkeitsbericht2014_EN.pdf (accessed on 28 February 2015).
4. ECPA. Pesticides Used in Organic Farming. Available online: http://www.ecpa.eu/news-item/ agriculture-today/pesticides-used-organic-farming (accessed on 24 May 2015).
5. Soil Association. Organic Foods Taste Better Claims New Poll. Available online: http://www. foodnavigator.com/Market-Trends/Organic-foods-taste-better-claims-new-poll accessed May 25th 2015 (accessed on 1 July 2015).
6. Buck, D.; Getz, C.; Guthman, J. From Farm to Table: The Organic Vegetable Commodity Chain of Northern California. *Sociol. Rural.* **1997**, *37*, 3–20.
7. Guthman, J. *Agrarian Dreams: The Paradox of Organic Farming in California*; University of California Press: Berkeley, CA, USA, 2004.

8. Guthman, J. The Trouble with "Organic Lite": A Rejoinder to the "Conventionalization" Debate. *Sociol. Rural.* **2004**, *44*, 301–316. [CrossRef]

9. Shepherd, M.; Pearce, B.; Cormack, B.; Philipps, L.; Cuttle, S.; Bhogal, A.; Unwin, R. An assessment of the environmental impacts of organic farming. Available online: http://www.znrfak.ni.ac.rs/serbian/010-STUDIJE/OAS-3-2/PREDMETI/III%20GODINA/316-KOMUNALNI%20SISTEMI%20I%20ZIVOTNA%20SREDINA/SEMINARSKI%20RADOVI/2013%20OD%20141%20DO%20150%20%20%281%29.pdf (accessed on 1 July 2015).

10. Stolze, M.; Piorr, A.; Haring, A.; Dabbert, S. The Environmental Impacts of Organic Farming in Europe. Available online: http://orgprints.org/8400/1/Organic_Farming_in_Europe_Volume06_The_Environmental_Impacts_of_Organic_Farming_in_Europe.pdf (accessed on 15 June 2015).

11. Tuomisto, H.L.; Hodge, I.D.; Riordan, P.; Macdonald, D.W. Does Organic Farming Reduce Environmental Impacts?—A Meta-Analysis of European Research. *J. Environ. Manag.* **2013**, *112*, 309–320. [CrossRef] [PubMed]

12. Aronsson, H.; Torstensson, G.; Bergström, L. Leaching and Crop Uptake of N, P and K from Organic and Conventional Cropping Systems on a Clay Soil. *Soil Use Manag.* **2007**, *23*, 71–81. [CrossRef]

13. Syväsalo, E.; Regina, K.; Turtola, E.; Lemola, R.; Esala, M. Fluxes of Nitrous Oxide and Methane, and Nitrogen Leaching from Organically and Conventionally Cultivated Sandy Soil in Western Finland. *Agric. Ecosyst. Environ.* **2006**, *113*, 342–348. [CrossRef]

14. Torstensson, G.; Aronsson, H.; Bergström, L. Nutrient use efficiencies and leaching of organic and conventional cropping systems in Sweden. *Agron. J.* **2006**, *98*, 603–615. [CrossRef]

15. Foster, J.B. Marx's Theory of Metabolic Rift: Classical Foundations for Environmental Sociology. *Am. J. Sociol.* **1999**, *105*, 366–405. [CrossRef]

16. Best, H. Organic Agriculture and the Conventionalization Hypothesis: A Case Study from West Germany. *Agric. Hum. Values* **2008**, *25*, 95–106. [CrossRef]

17. Flaten, O.; Lien, G.; Ebbesvik, M.; Koesling, M.; Valle, P.S. Do the New Organic Producers Differ from the "Old Guard"? Empirical Results from Norwegian Dairy Farming. *Renew. Agric. Food Syst.* **2006**, *21*, 174–182. [CrossRef]

18. Läpple, D.; Rensburg, T.V. Adoption of Organic Farming: Are there Differences Between Early and Late Adoption? *Ecol. Econ.* **2011**, *70*, 1406–1414. [CrossRef]

19. De Wit, J.; Verhoog, H. Organic Values and the Conventionalization of Organic Agriculture. *NJAS-Wagening. J. Life Sci.* **2007**, *54*, 449–462. [CrossRef]

20. Ongley, E.D. Control of Water Pollution from Agriculture—RAO Irrigation and Drainage Paper 55. Food and Agriculture Organization of the United Nations Rome, 1996. Available online: http://www.fao.org/docrep/w2598e/w2598e04.htm (accessed on 1 June 2015).

21. EPA 2009 National Water Quality Report to Congress. Available online: http://water.epa.gov/lawsregs/guidance/cwa/305b/upload/2009_01_22_305b_2004report_2004_305Breport.pdf (accessed on 16 December 2015).

22. Edwards, C.A.; Lal, R.; Madden, P.; Miller, R.H.; House, G. *Research on Integrated Arable Farming and Organic Mixed Farming in the Netherlands. Sustainable Agricultural Systems;* Soil and Water Conservation Society: Ankeny, IA, USA, 1990; pp. 287–296.

23. Eltun, R. Comparisons of nitrogen leaching in ecological and conventional cropping systems. *Biol. Agric. Horticult.* **1995**, *11*, 103–114. [CrossRef]

24. Younie, D.; Watson, C.A. Soil Nitrate-N Levels in Organically and Intensively Managed Grassland Systems. *Asp. Appl. Biol.* **1992**, *30*, 235–238.

25. Hettige, H.; Mani, M.; Wheeler, D. Industrial Pollution in Economic Development: The Environmental Kuznets Curve Revisited. *J. Dev. Econ.* **2000**, *62*, 445–476. [CrossRef]

26. Condron, L.M.; Cameron, K.C.; Di, H.J.; Clough, T.J.; Forbes, E.A.; McLaren, R.G.; Silva, R.G. A comparison of soil and environmental quality under organic and conventional farming systems in New Zealand. *N. Z. J. Agric. Res.* **2000**, *43*, 443–466. [CrossRef]

27. Stopes, C.; Lord, E.I.; Philipps, L.; Woodward, L. Nitrate leaching from organic farms and conventional farms following best practice. *Soil Use Manag.* **2002**, *18*, 256–263. [CrossRef]

28. Marx, K. Capital, Vol. III. Available online: https://www.marxists.org/archive/marx/works/download/Marx_Capital_Vol_3.pdf (accessed on 21 December 2015).

29. Clark, B.; York, R. Rifts and Shifts: Getting to the Root of Environmental Crises. *Mon. Rev.* **2008**, *60*, 13–24. [CrossRef]

30. Mancus, P. Nitrogen Fertilizer Dependency and Its Contradictions: A Theoretical Exploration of Social-Ecological Metabolism. *Rural Sociol.* **2007**, *72*, 269–288. [CrossRef]

31. Gunderson, R. The Metabolic Rifts of Livestock Agribusiness. *Organ. Environ.* **2011**, *24*, 404–422. [CrossRef]

32. Clausen, R.; Clark, B. The Metabolic Rift and Marine Ecology: An Analysis of the Ocean Crisis Within Capitalist Production. *Organ. Environ.* **2005**, *18*, 422–444. [CrossRef]

33. Clark, B.; York, R. Carbon metabolism: Global capitalism, climate change, and the biospheric rift. *Theory Soc.* **2005**, *34*, 391–428. [CrossRef]

34. Moore, J.W. The Modern World System as Environmental History? Ecology and the Rise of Capitalism. *Theory Soc.* **2003**, *32*, 307–377. [CrossRef]

35. Cole, M.A.; Neumayer, E. Examining the Impact of Demographic Factors on Air Pollution. *Popul. Environ.* **2004**, *26*, 5–21. [CrossRef]

36. Cramer, J.C. Population Growth and Air Quality in California. *Demography* **1998**, *35*, 45–56. [CrossRef] [PubMed]

37. Rosa, E.A.; York, R.; Dietz, T. Tracking the Anthropogenic Drivers of Ecological Impacts. *Ambio* **2004**, *33*, 509–512. [CrossRef] [PubMed]

38. Shandra, J.M.; London, B.; Whooley, O.P.; Williamson, J.B. International Non-Governmental Organizations and Carbon Dioxide Emissions in the Developing World: A Quantitative, Cross-National Analysis. *Sociol. Inq.* **2004**, *74*, 520–545. [CrossRef]

39. Shi, A. The Impact of Population Pressure on Global Carbon Dioxide Emissions, 1975–1996: Evidence from Pooled Cross-Country Data. *Ecol. Econ.* **2003**, *44*, 29–42. [CrossRef]

40. York, R. De-Carbonization in Former Soviet Republics, 1992–2000: The Ecological Consequences of De-Modernization. *Soc. Probl.* **2008**, *55*, 370–390. [CrossRef]

41. York, R.; Rosa, E.A. Choking on Modernity: A Human Ecology of Air Pollution. *Soc. Probl.* **2012**, *59*, 282–300.

42. York, R.; Rosa, E.A.; Dietz, T. Footprints on the Earth: The Environmental Consequences of Modernity. *Am. Sociol. Rev.* **2003**, *68*, 279–300. [CrossRef]

43. York, R.; Rosa, E.A.; Dietz, T. A Rift in Modernity? Assessing the Anthropogenic Sources of Global Climate Change with the STIRPAT Model. *Int. J. Sociol. Soc. Policy* **2003**, *23*, 31–51. [CrossRef]

44. Penn, M.R.; Pauer, J.R.; Mihelcic, J.R. Biochemical Oxygen Demand. In *Environmental and Ecological Chemistry*; Sabljic, A., Ed.; EOLSS Publishers Company: Oxford, UK, 2003; Volume 2.

45. The World Bank. 2010 World Development Indicators. Available online: http://data.worldbank.org/sites/default/files/wdi-final.pdf (accessed on 1 March 2014).

46. Organic World Statistics. Organic World Statistics Data. Available online: http://www.organic-world.net/ (accessed on 30 December 2013).

47. Jorgenson, A.K. Does Foreign Investment Harm the Air We Breathe and the Water We Drink? *Organ. Environ.* **2007**, *20*, 137–156. [CrossRef]

48. Belsley, D.A.; Kuh, E.; Welsch, R.E. *Regression Diagnostics: Identifying Influential Data and Sources of Collinearity*; John Wiley & Sons: Hoboken, NJ, USA, 2005; Volume 571.

Article

Hybrid Arrangements as a Form of Ecological Modernization: The Case of the US Energy Efficiency Conservation Block Grants

Anya M. Galli and Dana R. Fisher *

Department of Sociology, University of Maryland, College Park, MD 20742, USA; galli@umd.edu
* Correspondence: drfisher@umd.edu; Tel.: +1-301-405-6469; Fax: +1-301-314-6892

Academic Editor: Md Saidul Islam
Received: 31 October 2015; Accepted: 12 January 2016; Published: 18 January 2016

Abstract: How are environmental policy goals implemented and sustained in the context of political stagnation surrounding national climate policies in the United States? In this paper, we discuss Ecological Modernization Theory as a tool for understanding the complexity of climate governance at the sub-national level. In particular, we explore the emergence of hybrid governance arrangements during the local implementation of federal energy efficiency programs in US cities. We analyze the formation and advancement of programs associated with one effort to establish a sub-national low carbon energy policy: the Energy Efficiency and Conservation Block Grant (EECBG) program administered by the US Department of Energy. Our findings highlight the diverse range of partnerships between state, private, and civil society actors that emerged through this program and point to some of the challenges associated with collaborative climate governance initiatives at the city level. Although some programs reflected ecologically modern outcomes, other cities were constrained in their ability to move beyond the status quo due to the demands of state bureaucracies and the challenges associated with inconsistent funding. We find that these programs cultivated hybrid arrangements in an effort to sustain the projects following the termination of federal grant funding. Overall, hybrid governance plays an important role in the implementation and long-term sustainability of climate-related policies.

Keywords: ecological modernization; environmental state; collaborative governance; hybrid arrangements

---

1. Introduction

What types of environmental governance arrangements are viable alternatives when top-down approaches are inadequate or unsuccessful in responding to environmental issues? Efforts to address environmental problems depend on the political and economic contexts in which they are developed and implemented. As recent United Nations climate negotiations have highlighted, nation-states vary in their ability to reach consensus on how to take action on climate issues as well as in their ability to enact effective policies. For example, despite its status as a world leader, the United States has struggled to implement substantive environmental policies in recent years. In particular, efforts to pass a national climate change policy through the US Congress have been unsuccessful. Since the Kyoto Protocol entered into legal force on 16 February 2005, the US Congress has repeatedly failed to enact proposed climate change policies [1].

At the same time, the Administrative branch of the US government has struggled to implement a climate policy that reduces carbon dioxide emissions. The Obama Administration has maintained a commitment to achieving significant greenhouse gas reductions (17 percent by 2020 and 32 percent by 2030), goals initially established in 2009 and reiterated again in the President's Climate Action

Plan in 2013 [2,3]. Because these regulations have faced staunch Republican opposition and have been the target of legal challenges led by energy industry interests, President Obama has relied on executive powers that can be exercised without the involvement of the US Congress [4,5]. For example, the White House has issued numerous executive orders addressing climate-related goals including state-level climate preparedness [6], climate-resilient international development [7], reduction of federal greenhouse gas emissions [8], and preparations for climate change impacts [9]. With the upcoming national election in 2016, it is unclear the degree to which the Obama Administration's efforts will be sustained. Only the passage of climate legislation through both houses of the US Congress will ensure that the Federal Government continues to take meaningful action to mitigate climate change.

In the absence of significant progress toward a national climate change policy in the United States, progress *has* taken place at the sub-national level [10,11]. Much of this progress has been achieved through collaborative partnerships between government agencies, civil society groups, and the private sector, or what some scholars call *hybrid arrangements* [12,13]. These hybrid arrangements are especially apparent in US cities that have implemented sustainability and carbon reduction initiatives [14–16]. In this study, we focus on one federal initiative that supports US cities in the implementation of low carbon policies: the Energy Efficiency and Conservation Block Grants (EECBGs) that were funded by the American Recovery and Reinvestment Act of 2009.

This paper explores how one type of environmental governance—hybrid arrangements at the sub-national level—can achieve environmental policy goals when state-led initiatives are unsuccessful. We apply the lens of Ecological Modernization Theory (EMT), which proposes that environmental crisis can be addressed through institutional reforms, technological innovation, and governance arrangements that link state, private sector, and civil society actors [17]. Previous research has shown that ecologically modern hybrid arrangements have emerged at the subnational level; this study builds on this work to consider the complexity of these arrangements, the consequences of collaboration between multiple state and non-state actors, and the effectiveness of such arrangements within the context of global climate governance regimes. We begin by reviewing debates within environmental sociology over the ability of the state to resolve environment crisis and present a brief overview of the broader literature on environmental governance. Then, we present an overview of EMT, focusing in particular on the notion of hybrid arrangements and their development at the city level. After presenting details about the data and methods used to assess how the low carbon energy policies initially funded by the EECBG program were implemented, we present findings from research on the city-level programs supported by the competitive grants component of this program. This paper concludes by discussing how these specific programs were developed and implemented through partnerships between federal energy efficiency programs, local agencies, organizations, and businesses. We explore the benefits and challenges associated with these partnerships and describe how they facilitated program sustainability after the federal funding ended.

1.1. Environmental Governance

The question of how best to respond to environmental crisis has been central to environmental sociology since the field emerged in the 1970s [18]. Whereas some scholars have argued that industrialization and economic growth are inherently harmful to the environment [19–23], others have proposed that continued economic development and modernization are prerequisites for environmental protection [24–27]. The question of whether the state can resolve environmental crisis has been a major focus of these debates. For example, the political economy perspective that state responses will always prioritize the economy over the environment stands in stark contrast to theories of the environmental state, which explore how environmental protection functions as an economically beneficial process and basic responsibility of industrialized nation-states [19,22,23,28–30].

In the context of 1990s debates over state failures in effectively coping with the challenges of modernity and industrialization, the bulk of the responsibility for environmental protection shifted toward private economic and civil society actors [31,32]. Although the environmental state has expanded its responses to environmental problems in recent decades [33], concerns about the efficacy of top-down environmental policy approaches remain highly relevant. Given what Fisher and colleagues identify as the "inability of national regulators to address successfully environmental problems in the decision-making process, and effectively enforce the decisions already made," alternative approaches to environmental governance will be crucial in moving forward with meaningful action on climate change, pollution control, and other urgent environmental issues [34] (p. 146).

It is rare for states to act alone in implementing and enforcing environmental policies. Instead, environmental governance is carried out through complex collaborations among state, market, and civil society actors [14,31,35–39]. Overall, scholars have documented a shift away from government toward governance of environmental issues in industrialized countries [40,41]. In contrast to government "command and control" over decision-making and policy implementation, governance refers to the complex, reciprocal array of arrangements between state, non-governmental, and individual actors that emerge through the definition and pursuit of collective political goals [11,14]. As Koontz and colleagues explain, "*government*, as a formal institution of the state, ceases to hold sole power through command and control mechanisms, thereby shifting to *governance*, a process that takes place through the collective action of a variety of participants, all of whom retain some control over decision making or implementation" [37] (p. 6, emphasis authors' own). In the context of this study, *environmental governance* refers to the "set of regulatory processes, mechanisms, and organizations through which political actors influence environmental actions and outcomes" [42] (p. 298).

Research on environmental governance spans a range of disciplines including ecology, economics, geography, political science, and sociology. At the broadest level, debates within this literature have centered around which actors are best suited to participate in environmental governance efforts. Advocates of community-based resource management, for example, contend that communities will be more sustainable and democratic in the management of their local environments than states or corporations [43–46]. From this perspective, communities have stronger interests in ensuring the quality of the resources upon which they depend, deeper knowledge of how best to manage those resources, and pre-established governance practices that are already understood by local actors [47–49]. However, critics of voluntary conservation measures have argued that the state and its policies play a crucial and protective role by maintaining environmental standards and limiting corporate access to natural resources [50,51]. The success of market-based approaches such as incentives, taxes, voluntary agreements, and certification programs has also been reliant on the ongoing presence of effective governmental leadership [52–55]. Further, scholars have questioned whether market-based approaches lead to equal benefits for all, or simply increased benefits from those who are already making a profit [56].

Governance approaches that establish partnerships and shared responsibility for environmental protection across state-market-society divisions, or what Lemos and Agrawal call "cogovernance" strategies, represent a middle ground between top-down (state-led), market-based (economic), and bottom-up (community-based) environmental governance [42]. By including a range of stakeholders from the start, these approaches facilitate the participation and cooperation of actors that may be excluded from more hierarchical arrangements [57–59]. One example is collaborative governance, which Ansell and Gash define as involving "one or more public agencies" working toward policy goals by collaborating with "non-state stakeholders in a collective decision-making process that is formal, consensus-oriented, and deliberative" [39] (p. 545). Moving beyond traditional public-private partnerships that focus predominantly on providing services to consumers, collaborative governance aims to set the agenda for policymaking and implementation. In particular, collaborative governance may emerge as a deliberate decision-making and management strategy in cases where consensus cannot be reached (what Ansell and Gash call "policy deadlock") or where policy makers foresee implementation as being potentially difficult (p. 553).

There are notable limitations to environmental governance. Some scholars argue that emergent governance arrangements are simply reorganizations of existing power distributions that do little to include new or underprivileged actors [60]. Scholars have also questioned whether democracy is limited when the complex bureaucracies produced under hybrid arrangements require stakeholders to possess a certain level of expertise in order to participate [61]. Further, the complexity of hybrid arrangements veil the fact that important actors are still excluded from the process [62]. It is important to consider whether governance arrangements fit the requirements of deliberative democracy, or whether elite actors are given more voice over environmental decision-making than actors who have little "veto power" [42]. Despite the challenges associated with forging partnerships across uneven levels of power, environmental governance strategies that link state, market, and civil society actors are becoming increasingly institutionalized, especially at the sub-national level.

One vein of literature that has been particularly critical in its analysis of environmental governance practices applies Foucault's concepts of "governmentality"—or the practices, ways of thinking, rationalities, and discourses through which subjects are governed—to highlight the social construction of environmental problems and the limitations of market-based governance practices [63,64]. This perspective considers the multiple and relational nature of power as it plays out between actors and arenas in environmental governance [65]. Scholars have also applied the concept of governmentality to discuss how neoliberal economic policies (regulatory approaches that promote privatization, deregulation, and free market practices) have influenced the ways in which we think about and govern the environment [66,67]. Here, market-based approaches to environmental governance are considered as the tools of what Oels calls "an advanced liberal government" [68]. From this perspective, hybrid governance arrangements are tied to the "hallmark" tendency of neoliberal policy to "move *outside* of the formal apparatus of the state [. . .] and achieve policy aims through the institutions of civil society" and the free market [69] (p. 504). Further, the multiplicity of governance approaches—e.g., the rising prevalence of hybrid arrangements—is considered to be a reflection of the adaptive nature of neoliberal policies, which tend to work around environmental crises rather than addressing them at their source [70,71].

1.2. Ecological Modernization Theory

This paper engages with concepts from Ecological Modernization Theory (EMT) to address questions about the efficacy and viability of hybrid environmental governance efforts at the sub-national level. EMT examines the transformations of social practices and institutions, or patterns of "ecological restructuring," that emerge from environmental concerns in industrialized countries when "the state can no longer be expected to design and prescribe the way society and economic interactions should be organized" [24,25,31]. In other words, EMT explores how economic growth and industrialization can be amenable to environmental protection and how solutions to environmental crises can evolve within, rather than outside of, the modern market economy [24,26]. This theory emerged in a Western European context and has been most applicable in cases within industrialized countries with established processes for environmental policymaking [17]. EMT has also expanded to consider global environmental governance and how new formations between science and technology, nation-states, and global markets can lead to environmental reform [25].

Although scholars have developed a strong theoretical framework for EMT, more empirical research is needed to resolve questions about the conditions under which ecologically modern outcomes are viable and successful [72]. Ecological modernization is often more reflexive in its theoretical form than it is in practice [26]. We assess whether the hybrid forms of climate partnerships emerging in US cities are examples of truly ecologically modern environmental governance, or whether they are simply another iteration of the "advanced liberal government" described by Oels and other critics of neoliberal environmental policies [68]. In other words, this project considers whether hybrid arrangements in US cities align with what Bäckstrand and Lövbrand identify as the "strong" version of ecological modernization, which transforms the dominant paradigms of social and institutional response to environmental problems, or whether they exemplify the "weak" version, which "does not involve any rethinking of societal institutions" and presents a false "win-win" storyline about the compatibility of economic growth and environmental protection [73] (p. 53).

Ecological Modernization Theory provides a framework for understanding environmental governance in the context of shifting boundaries between state, market, and civil society [27]. It describes how participatory governance practices emerged following the rise of the environmental state from the 1960s through the 1980s as a way of compensating for failures at the state level [31,34,74]. For example, Jänicke and Jörgens describe how environmental policy in Europe shifted away from the top-down approach of the 1960s and 70s as environmental organizations and industry groups began to interact and participate in policymaking [11]. This process of political modernization brought about a transition from a state-centered bureaucratic policy model to a more decentralized, consensus-oriented model of environmental governance [34,75].

Under political modernization, voluntary environmental protection measures tend to outnumber state-initiated actions as civil society and the private sector take on some of the responsibility for environmental regulation formerly shouldered by the state [25]. Van Tatenhove and Leroy contend that what they call the "societalization" of governance goes hand-in-hand with "marketization," or the delegation of responsibility for regulation away from the state to privatized agencies [35] (pp. 167–168). As state-market interactions shift, economic processes and actors take on increasing roles in environmental protection [76]. In Mol's words, market practices arise in which "economic processes of production and consumption are increasingly analyzed and judged, as well as designed and organized from both an economic *and* ecological point of view" [35] (p. 60, emphasis author's own). In this "ecologized economy," environmental protection is multidirectional, wherein the purchasing power of "citizen-consumers" combines with more top-down policies and economic tools to achieve environmental protection [24,77].

The central assumption of EMT—that ongoing processes of modernization and industrialization can solve, rather than exacerbate, environmental problems—stands in contrast to perspectives in environmental sociology that see economic growth as incompatible with environmental protection [19–23]. Thus, the most prevalent critiques of EMT have come from scholars who believe that the "sustainable capitalism" it promotes is not possible [78,79]. From a political ecology perspective, which focuses on the asymmetries of access to natural resources and exposure to environmental harm under capitalism, the market-based strategies associated with sustainable development and environmental protection efforts are inherently contradictory [80–83]. From this standpoint, technological innovations and "green" markets may alleviate specific aspects of environmental harm, but cannot resolve the underlying inequalities that produce environmental degradation at larger scales [57,84,85]. At the same time, however, institutions are a "necessary starting point" for understanding patterns of environmental inequality [86] (p. 268). In this context, EMT can provide a framework with which to consider contemporary governance arrangements that have the potential to redistribute, rather than exacerbate, power imbalances among economic interests, political actors, and communities.

As detailed by Leroy and Van Tatenhove, the institutionalization of "interference zones" between state, market, and civil society has created opportunities for new combinations of governance approaches and the emergence of unique policy arrangements [74]. Mol and Spaargaren have described these diverse forms of collaboration among social actors as "hybrid arrangements" [12]. The authors note that there has been "enmeshment and hybridization" between "formerly distinct entities" within the environmental state, pointing out that the "roles and responsibilities formerly reserved for the [state] are fulfilled by market actors and civil society groups and organizations, and vice versa" (p. 15). Thus, hybrid arrangements, which vary in terms of the actors and sectors involved, create new opportunities for innovative approaches to environmental governance [13]. Hybrid arrangements are highly contingent on the institutional, political and cultural contexts in which they emerge [33,34]. Rather than replacing more traditional approaches entirely, hybrid arrangements often function side-by-side with top-down environmental policy processes [87–89]. Although the role of the state may diminish as regulatory efforts are undertaken by other sectors, collaborative initiatives provide opportunities for the state to participate in environmental governance in a variety of capacities [32]. In other words, the role of the state may shift without disappearing entirely: for example, the state may function as a moderator and facilitator between different interests, rather than acting on those interests directly [31]. Because the lines of accountability within hybrid arrangements can be diffuse, state authority can also help to anchor environmental policies and provide incentives for effective implementation. For example, state-initiated regulation policies continue to play a role in providing resources, setting imperatives for regulation, incentivizing sustainable innovation, and assisting in the regulation process [90]. In their study of the impact of integrated pollution control in linking state and market actors in regulatory action in England and Wales, for example, Murphy and Gouldson find that these efforts were successful when there was collaboration between state regulators and companies [90]. State backing of environmental policies can provide much-needed accountability, or what some have called a "stick behind the door" in the event of noncompliance or policy failure [11]. In Mol's words, the state provides a "credible threat of regulation" that "may help ensure full commitment of all participants" in the governance and decision-making processes [33] (p. 345).

1.3. City-Level Hybrid Arrangements and Climate Governance

Hybrid arrangements have developed at multiple levels within the environmental state, indicating that civil society and private sector actors are aware of, and acting in response to, the limitations of top-down environmental governance [11]. Hybrid partnerships are especially central to climate governance, which is characterized by a multiplicity of actors with overlapping forms of authority across a range of political and social arenas [91]. Much of the literature on transnational climate governance has focused on proliferation of partnerships and collaboration between state, market, and civil society actors [65,92,93]. Transnational climate governance networks are characterized by their complexity and variety: they may be made up of purely public actors (governments, government sub-units, legislators, *etc.*), purely private actors (non-state entities and organizations), or, most commonly, a combination of the two [91]. In Bäckstrand's words, these partnerships "signify a shift to 'new' modes of governance, which build on non-hierarchical steering and are characterized by decentralized, voluntary, market-oriented interaction between public and private actors" [93] (pp. 74–75). At the same time, scholars have explored the increasing agency of non-state actors and private authority in climate governance at multiple levels [94,95].

The impacts of sub-national climate initiatives are smaller and more incremental than more sweeping transnational approaches or national policies, but may also be more successful. This paper focuses on city-level climate partnerships, which exemplify the hybrid arrangements that are at the core of ecologically modern responses to global climate change. For example, in light of the delayed response by national governments to global environmental issues such as climate change, many cities have implemented their own environmental protection programs, which Rabe calls "races to the top" [96,97]. In many cases, these programs are conceived as, or created as opportunities for, hybrid governance configurations [98].

Cities serve an important role as non-state actors in the transnational response to global climate change by connecting with local stakeholders, integrating climate change into pre-existing policies, and experimenting with innovative programs aimed at cost-effective greenhouse gas reduction and energy efficiency [15,95]. In other words, cities are nodes within transnational climate governance networks where collaborative initiatives—for example greenhouse gas reduction or energy conservation efforts—are conceived and implemented [99,100]. City-level environmental protection programs provide researchers with an opportunity to understand more fully how hybrid arrangements are formed and implemented. For example, Betsill and Bulkeley document how the Cities for Climate Protection program, enacted locally in cities across the world, includes a variety of state and non-state actors in its efforts to lower greenhouse gas emissions [14]. More recently, Bulkeley and Schroeder focus on the examples of London and Los Angeles, finding "new forms of public and private authority" in the urban governance of climate change [16] (p. 762). Looking at the case of environmental stewardship organizations in New York City, Fisher and Svendsen find a diversity of hybrid arrangements in practice [13].

When viewed through the lens of EMT, city-level sustainability initiatives demonstrate the diverse range of ways in which political, economic, and civil society actors collaborate to make environmental protection both economically and politically feasible. However, there is a need for more empirical assessments of hybrid arrangements (and ecologically modern outcomes more broadly), both in terms of the conditions under which they emerge and in terms of the challenges they face [13,72]. This paper uses the case of one particular environmental initiative in the United States to understand how a policy that was initiated by sub-national policy actors and financially supported by the federal government created opportunities for ecologically modern hybrid arrangements to emerge. We assess the role of hybrid arrangements in achieving initial policy goals as well as in sustaining these efforts after federal support had ended, and discuss the benefits and challenges of the specific structure of this program. We conclude our paper by discussing the implications of our findings for EMT and broader understandings of contemporary environmental governance.

2. Case

The EECBG program was funded through the American Recovery and Reinvestment Act, which was passed at the beginning of President Obama's first term in office in 2009 (for more details, see Fisher's study of the early stages of this program [101]). The program's website at the Department of Energy describes its goals: "It is intended to assist U.S. cities, counties, states, territories, and Indian tribes to develop, promote, implement, and manage energy efficiency and conservation projects and programs [. . .] through formula and competitive grants, the Program empowers local communities to make strategic investments to meet the nation's long-term goals for energy independence and leadership on climate change" [102]. Overall, the grants distributed $3.2 billion over two years (all grants were awarded by the end of 2011). The majority of the grants were distributed through a formula: 24% to states, 58% to cities/municipalities, and 1.7% to Indian tribes. The remaining 14% were allocated through competitive grants. Building off of previous research on these grants [101], this paper focuses specifically on the competitive grants awarded to cities in the United States to understand how the EECBG grants were implemented and sustained at the city level.

The competitive grant portion of the EECBG program, first called the "Retrofit Ramp-Up" program and later renamed "The Better Buildings Neighborhood Program" (BBNP), was announced after the formula grants were implemented in April 2010. Grants awarded under the BBNP were intended "fundamentally and permanently [to] transform energy markets in a way that makes energy efficiency and renewable energy the options of first choice" [103]. In summer, 2010, the Department of Energy (DOE) awarded $482 million in EECBG monies to 34 grantees across the state, county, and city levels [104]. An additional $26 million in grants funded by the State Energy Program were added later in 2010, with awards to seven states, which brought the total number of grant recipients under the BBNP program to 40. Efforts undertaken with BBNP funds included short-term market-based approaches such as incentives for installation of energy efficient technology, as well as longer-term initiatives such as training for contractors and "green" jobs creation.

In particular, the BBNP program supported projects that had explicit plans to "sustain themselves beyond the grant monies and the grant period." Initially, additional funding was likely to come from revenue generated from a carbon-trading program that was expected by many to pass through the Congress quickly after President Obama came into office. Although the bill passed through the House of Representatives in 2009, the companion bill never made it through the Senate [105]. Policymakers did not intend for the grants to be extended beyond the initial funding period stipulated by the DOE. Therefore, although some grantees organized campaigns to request additional federal funding, additional funds were not appropriated to extend EECBG-related efforts [106]. Federal funding for the program expired in 2014 [107]. Continuation of these programs after federal support ended was contingent on individual programs' ability to leverage and sustain the partnerships they had developed with private economic interests and local groups.

3. Data and Methods

This paper presents data from the city-level EECBG programs and the 40 BBNP program grants before discussing 17 cities that were awarded grants as part of the BBNP program in detail. We incorporate data from multiple sources. First, we use data from a 2012 survey conducted by the United States Conference of Mayors (UCSM) titled "Clean Energy Solutions for America's Cities" [108]. The UCSM is a national nonpartisan organization comprised of cities with populations over 30,000. Out of the 1200 cities associated with the UCSM, 1060 have signed the Conference Mayor's Climate Protection Agreement, and the organization is dedicated to "leadership on energy and climate protection" [109]. As has been noted in previous research, the USCM played an integral role in securing federal funding for the EECBG program [101]. A private research company conducted the survey of all members of the US Conference of Mayors online during spring 2011 [108]. Data from this survey provide a picture of the city-level impact of the EECBG program as a whole.

Second, we collected data in 2011 and 2012 from the websites for all 40 recipients of BBNP grants, as well as the program website at the DOE. We coded this content to assess program goals and to investigate the types of partnerships that emerged from the program. Website content was supplemented with publically available data from DOE and other sources, which are cited throughout this paper.

Third and finally, we conducted in-depth interviews with the directors of the city-level BBNP-funded programs in fall 2012. Contact information for respondents was obtained from listings on the DOE website. We contacted all city-level grant recipients via email to request that they participate in interviews. Twelve cities (or 71% of the cities participating in the program) responded to these requests. This paper also incorporates information from multiple conversations with DOE staff and public officials with knowledge of the BBNP and EECBG programs. We conducted the majority of interviews via phone, as programs were located across the country. Interviews followed an open-ended, semi-structured format and asked questions about the history and progress of the BBNP program, the role of the Federal government, and the future of BBNP programs following the expiration of the EECBG funds [110]. Interviews lasted between 15 min and 1.5 h. We recorded interviews digitally and took extensive notes and memos during all conversations. To analyze the qualitative data collected for this project we used an open coding technique, which allowed themes to emerge from the data. Interviewees participated with the understanding that their words and comments would not be directly attributed. As a result, we cite only their general affiliations throughout the remainder of this paper.

4. Results

In the pages that follow, we present our findings from this study. First, we discuss the findings from the UCSM survey of cities to understand the impact of the overall EECBG program at the city level, which includes both formula and competitive grants. Then, we analyze data from the DOE and BBNP grantee websites to understand the types of partnerships emerging from the competitive grant portion of the EECBG program. Finally, we discuss the results of our interviews with city-level recipients of BBNP grants to provide a more in-depth picture of how the grant programs were implemented and how cities planned to sustain their programs after the expiration of the EECBG grants.

4.1. EECBG Impacts at the City Level

Consistent with findings of previous studies of sub-national efforts to address climate change [13–15,99,100], we find that the cities represented in the UCSM Survey were committed to energy efficiency and conservation. For example, 75 percent reported planning to increase their use of clean energy technologies and 25 percent had set targets for the use of future renewable energy.

However, cities also identified "financial constraints" as the most significant barrier to improving and continuing their energy efficiency and conservation efforts. In the context of budgetary constraints, high up-front costs, and uncertainty about the economic returns of clean energy technologies were frequently noted.

In terms of how cities spent EECBG funds, the overwhelming majority of cities (83%) reported that they were using the money to implement new energy technologies. There was a wide range of technologies deployed under the EECBG grants. The most common technologies included: installing LED and energy-efficient lighting (73% of cities), implementing new building technologies (40%), and installing photovoltaic (solar energy) systems (31%). In all, city representatives saw the EECBG program as being incredibly important to their efforts toward energy efficiency and conservation. In response to a question about whether the "initial EECBG funds [were] important to city's efforts," the overwhelming majority (85%) of respondents agreed that the funds had been important and only 5 percent disagreed.

Cities mentioned the EECBG program (including both the formula grants and the competitive BBNP grants) as the most widely used funding source for efforts to expand clean energy and energy efficiency programs. When asked about their next steps, most of the city representatives (87%) reported that additional EECBG funding was needed for the continued deployment of clean energy technologies. Looking toward the expiration of the EEBCG grants in 2013, about half of respondents (51%) reported that they would seek future funding from the federal government, and 42 percent stated that they would look for money from their state governments.

4.2. Program Partnerships

A review of the program partnerships reported by BBNP funding recipients, shows that collaborations among state, private, and civil society actors were central to the city-level implementation of energy efficiency initiatives under the EECBG program. Beyond working with the federal government to implement the grants, which were seen as the most important form of funding (71%), the majority of cities responding to the USCM survey reported partnering with members of the private sector (59%). When we look specifically at the 40 competitive grants awarded under the BBNP program, we are able to understand the partnerships more clearly. As has been previously mentioned, the BBNP program supported projects at many levels: seventeen of the grants were city-level (42.5%), thirteen were state-level (32.5%), nine were county-level (22.5%), and one was regional. The regional grant involved sub-grantees across ten states in the Southeastern US [111]. Figure 1 presents this distribution.

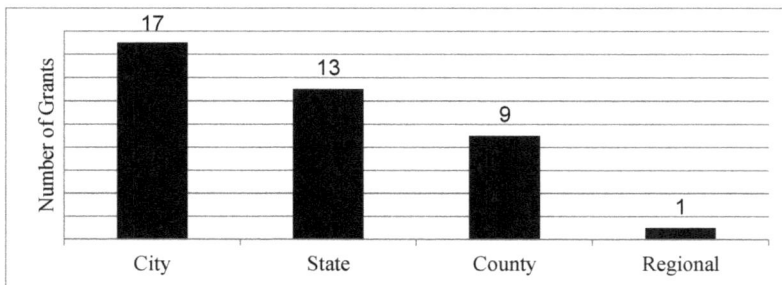

Figure 1. The "Better Buildings Neighborhood Program" (BBNP) Grants Allocation by Level.

All of the BBNP grants were implemented through partnerships. In fact, the relevant program websites at the Department of Energy (DOE) and individual programs' sites listed 337 separate partners across seven categories. Overall, the vast majority of these partnerships were with market actors rather than civil society collaborators. The most common program partners were local businesses and business alliances (such as green technology businesses, chambers of commerce, and sustainable development groups), which constitute over 26% of the total partnerships listed. Energy companies—including both public and private utilities—were the second-most common category, representing 22% of the partnerships. Non-profits, including environmental groups and community organizations and alliances, were the third-most-common partnership (20%). Financial institutions (banks, credit-unions, and other lenders) constituted about 16% of the total partners listed. BBNP programs also partnered with local contractors to assist homeowners with energy retrofits and with colleges and universities to provide "green jobs" training (just over 6% and 4% of partnerships, respectively). About 5% of partnerships fell into other categories, including sports teams and consulting agencies, which are depicted as "other" in Figure 2.

Given the prevalence of partnerships with energy companies and other private sector actors, it is important to consider the impact of industry interests on the implementation of the BBNP grant programs. Consistent with previous research that has found that climate-related programs and policies are less likely to be supported by states that extract coal we look at how BBNP partnerships are related to the natural resource endowment of the states in which they are situated [101,112]. In 2010, half of the states in the US extracted some coal [113]. Of the twenty-five states that extract coal, ten of them are considered "major coal producing states," extracting more than 25,000 short tons of coal in 2010. Although the percentage of grant recipients is somewhat consistent across coal and non-coal extracting states (43% *versus* 57% accordingly), there are differences when we look specifically at the types of partnerships forged by grant-recipients in major coal states *versus* non coal-extracting states. ("Non-coal extracting states" are being operationalized here as states that extract less than 25,000 short tons of coal a year. Such a low level of extraction does not contribute significantly to the states' overall economies.) Table 1 presents these partnerships comparing major coal extracting to non-coal extracting states and the percentages of each type of partnership across coal and non-coal states. Table 1 also indicates the total percentage of each type of partnership for all states. As can be seen, partnerships with energy companies were relatively equally distributed in coal extracting and non-coal extracting states. Non-coal extracting states were more likely to partner with businesses, educational institutions, financial institutions, local contractors, and other groups. Coal extracting states were more likely to partner with non-profit organizations.

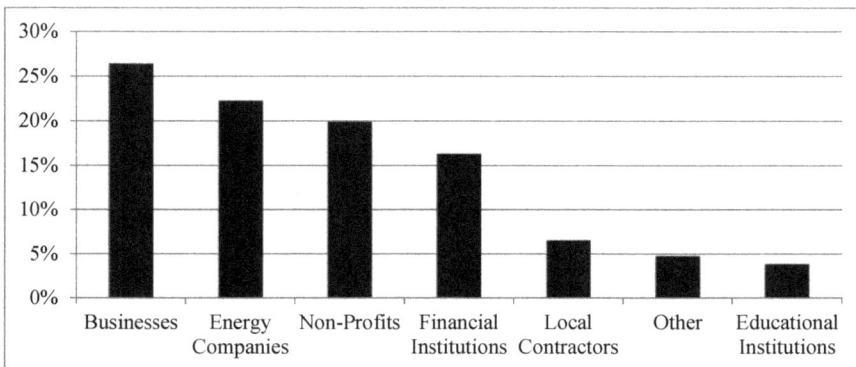

Figure 2. BBNP Partnerships by Type.

Table 1. BBNP Partnerships.

Partnership Type	Major Coal Extracting States	Non-Coal Extracting States	Total N (Percentage of Total)
Energy Companies	19 (28%)	56 (26%)	75 (27%)
Financial Institutions	11 (16%)	44 (21%)	55 (20%)
Non-Profits	19 (28%)	48 (23%)	67 (24%)
Local Contractors	4 (6%)	18 (8%)	22 (8%)
Businesses	11 (16%)	21 (10%)	32 (11%)
Educational Institutions	2 (3%)	12 (6%)	14 (5%)
Other	3 (4%)	13 (6%)	16 (6%)
Total	69	212	281

4.3. Emerging Hybrid Arrangements in Sub-national Climate Governance

Through an analysis of program websites, we are able to see partnerships forming across social actors involved in the Better Buildings Neighborhood Program (BBNP). These hybrid arrangements become even clearer in the results of our open-ended semi-structured interviews, which were conducted with representatives of the programs in each city that received a BBNP grant. These data provide clear examples of how government agencies are collaborating with the private sector and civil society to implement what the US Department of Energy (DOE) calls "innovative ways to engage, inform, and motivate Americans to increase energy efficiency" through this program [114]. Specifically, local governments worked with civic groups and businesses in a manner consistent with the work on ecologically modern hybrid arrangements, which argues for the need "to rethink the role of the state, market, and civil society actors in environmental governance efforts" [89] (p. 15).

In some cities, the competitive BBNP grants were the first energy efficiency initiatives ever implemented. Several respondents from cities new to the energy efficiency arena asserted that the BBNP funding allowed them to establish programs that would have otherwise been impossible to get off the ground. For these cities, extension of the BBNP was seen as the only opportunity to address energy efficiency in the coming years. For other cities, the grants provided supplementary funding for well-established climate protection or environmental sustainability plans. Cities with ongoing energy efficiency programs tended to partner with utilities, city government, and local nonprofits from the outset, whereas less-experienced cities had fewer opportunities to create partnerships early in the grant period. Because cities with pre-existing green development and sustainability policies had already done the hard work of forging partnerships with private sector actors (utility companies, chambers of commerce, banks, *etc.*), they entered into the BBNP grant with knowledge about which actors were already on board and what kinds of partnerships were most beneficial for the specific circumstances of their cities. For example, one city in the Midwest had recently gone through the lengthy process of adopting a climate protection plan. As the program director explained, partnerships that were established through the climate protection plan proved to be beneficial in expanding and sustaining projects under the BBNP grant. In this case, federal grant money helped to carry over existing programs and partnerships during an "economically challenging time."

In contrast, partnerships in cities with less experience "evolved over time" as programs sought to find the most effective ways of implementing their goals. For some cities, newly-forged partnerships fell through, while for others, awareness of the needs and demands within their communities did not fully develop until halfway through the grant period. Two program directors, both from cities lacking formal energy efficiency policies prior to the BBNP grant, described having to overhaul their programs halfway through the grant period because of low adoption. They reported that this process consumed much of the time they had dedicated for cultivating partnerships. Another representative explained that his office had to "push local utilities to be more sustainable" through education about the economic benefits of energy efficiency before they could discuss setting up rebate programs.

Because of the grants' short timeline of only three years, nearly every respondent described having to "hit the ground running" and "learn on the fly" once funds arrived. Regardless of whether grants funded pre-existing programs or entirely new projects, the funding timeline often constrained opportunities for partnerships to develop fully. In the words of a program manager of a city in the Southwest, cities struggled to "get everyone on board" in the limited time available, a challenge that was especially daunting given what another director described as "the inherent conflicts between private and public sectors." Additionally, the pressure to report positive outcomes meant that programs were implemented before staff could work out the basic details of how partners would collaborate. As one BBNP director described, "this whole portfolio had the feeling of sailing the ship out of the port while you were still building the ship . . . it was frankly too early for most of the grantees because building a state-wide or region-wide energy efficiency program is a lot of work and it takes years to put the footing in place." Further, as another program director noted, it was a challenge to "get everyone to work together under one brand" when partnerships bridged numerous established organizations and companies. These challenges were especially significant for programs established as entirely new energy efficiency initiatives, as there was even more to accomplish within an already short allotment of time.

The Federal Government directed the implementation of BBNP programs from a distance. Rather than maintaining tight oversight of day-to-day operations in each city, DOE provided broad guidance about how to develop, sustain, and maintain the compliance of BBNP programs. This fact did not mean, however, that cities were free to develop their programs without constraints. In addition to the timeline challenges described above, respondents identified paperwork and reporting requirements as significant barriers to efficient program implementation. Overall, respondents noted that there was a need to streamline the federal compliance process. As one director from a city with a long history of energy efficiency programs said, the "administration side was pretty bad and sucked up a lot of resources . . . [there was a] need for innovation and speed at the same time as intensive reporting." She added that compliance-related guidance from DOE was often "unintelligible" and that it "took forever to resolve questions about what regulations applied to the loan program," creating a situation in which "everyone was afraid to go awry [and spent] lot of very fruitless energy . . . [trying to] accomplish what DOE wanted and still do what we were trying to do."

As detailed earlier, respondents to the USCM survey indicated that they hoped for additional federal funding to continue their projects. A year after the survey was conducted, interview respondents were much less optimistic. Some city representatives said that they were holding out hope depending on the outcome of the 2012 Presidential election, but the overall expectations were consistent with the position presented by a DOE representative: the support from the federal government would not continue after the initial grants expired [55]. As one BBNP program director explained, grantees "are operating under the understanding that there are no more dollars from the Federal government." Another program director described the future of the BBNP program in his city as "one big question mark." Most grantees were equally pessimistic about funds from city governments, most of which were in the process of dealing with financial crises by cutting, rather than expanding their budgets. Instead, city representatives specifically discussed expanding existing partnerships and initiating new partnerships with local organizations, businesses, and utilities when they discussed their plans for the future.

Overall, forging long-term partnerships to implement goals and sustain future funding was in the best interest of BBNP grantees, who sought to continue providing services beyond the lifespan of the grant program. Sustaining these programs was also good for DOE and the Obama Administration, which benefited from having supported projects that were successful, both within and beyond the grant period. Thus, the Federal Government played an active role in cultivating relationships among the BBNP and utilities, non-profits, and local associations. In fact, DOE itself coached grant recipients as they initiated and enhanced partnerships with non-profits, utilities, contractors, and other local groups during the start-up phase of their projects. Halfway through the grant period, DOE held a workshop for grantees encouraging them to develop partnerships with utilities, integration with local non-profits, and business partnerships as methods of sustaining their operations. A representative from DOE noted during an interview that forging partnerships, especially with utilities, provided an opportunity to extend the life of the projects. In addition, the representative pointed out that some cities' programs included spin-off nonprofit organizations that were looking for other revenue. This type of a hybrid arrangement, wherein one partnership leads to the creation of a new program in order to address a specific issue (in this case, the need for additional funding for energy efficiency programs), is consistent with what Fisher and Svendsen call a "nested governing arrangement" [13].

In terms of the types of partnerships that were expected to succeed in the post-grant period, most programs reported looking to utility companies and the private sector. For example, one city representative explained that their program was planning to split into a partnership program with the local municipal utility and a nonprofit program building off of an organization established through a statewide formula grant. Another city representative described their program as a "venture capital" investment focused on attracting private investors to contribute to lending programs. Similar to several other cases, the hope in this city was that the existing program would mature into an established non-profit bank, or "green lender." Alternatively, some cities reported turning toward increasing rebates and incentives in order to encourage uptake of energy efficiency upgrades among single-family homeowners. Representatives noted that partnerships with utilities were necessary to subsidize further rebates, which had been some of the most successful citywide initiatives. Overall, arrangements such as those explored by BBNP grantees offer opportunities for sustained funding and support in light of uncertainty about future grants and limitations on federal funding. In other words, these government-initiated programs intentionally capitalized on hybrid arrangements to sustain their projects.

5. Discussion and Conclusions

Hybrid governance arrangements provide opportunities for environmental policymaking and implementation in cases where top-down approaches have failed due to policy deadlock. Although there are numerous challenges to collaboration between state and non-state actors across multiple levels of authority, hybrid arrangements such as those observed in this study provide an opportunity for meaningful climate governance at the sub-national level. In the face of uncertainty about future action on environmental policies at the federal level in the United States, sustainable development efforts are increasingly implemented via collaborative governance processes [34–39]. These partnerships among government, market, and civil society actors are creating innovative organizational and civic forms that blur the lines between public and private [13]. Our findings contribute to the broader literature on environmental governance, the majority of which has focused on transnational governance networks, by highlighting an important case of hybrid arrangements at the city/sub-national level. We find that city-level grant recipients in the Energy Efficiency and Conservation Block Grant program cultivated collaborative partnerships as they sought to implement and sustain their programs. These conclusions are consistent with the expectations of Ecological Modernization Theory, which proposes that environmental protection can be achieved via hybrid arrangements that bridge state, market, and society actors [33,34,38,77].

At the same time, we find that the process of developing partnerships under the EECBG program was neither uncomplicated nor without its challenges, especially when it came to working with the state. Although some cities were able to establish programs that fit the expectations of a "strong" interpretation of EMT (increasing public participation, spurring technological innovation, transforming governance approaches) [68,73], other cities were constrained in their ability to move beyond the status quo due to the demands of state bureaucracies and the challenges associated with inconsistent funding. Overall, we find that some hybrid arrangements reflected ecologically modern outcomes, while others took less reflexive, albeit hybrid, forms [66,67,70].

Hybrid arrangements at the city level take different forms based on the specific constellation of actors, political and economic contexts, and pre-existing partnerships [14,15,99,100]. These unique partnerships among state, private, and civil society actors were catalyzed by the availability of federal funding. Consistent with the findings of previous research on hybrid arrangements, we find that city-level energy efficiency projects that were funded by the EECBGs initiated "new forms of public and private authority" in the urban governance of climate change [13,16]. These programs exemplify what Mol describes as the ecologically modern "ecologized economy:" participation in energy efficiency efforts was incentivized through funding and partnerships leveraged across government agencies, local businesses, and energy companies [24,77]. However, the overall approach of these programs went beyond market-based strategies to engage with local communities while also receiving guidance from the state.

Our findings also highlight the ongoing role of the state in emergent hybrid arrangements at the sub-national level. The local energy efficiency programs we studied through the EECBG program used federal funds as a method of "getting programs up and running" or expanding their existing capacities by creating partnerships with businesses, financial institutions, and other local groups. Consistent with previous research, we find that continued state involvement provided accountability and support during policy implementation [31,32,78,89]. At the same time, we find that some challenges did arise as cities developed collaborative initiatives under the constraints of a federal grant program. Interview data show that the development of partnerships within these programs was constrained by the tight timeline and strict reporting guidelines associated with federal grant funding.

Our research also finds that, although these city-level environmental initiatives benefitted from funding established through federal policies and received guidance and program assistance from federal agencies, they did not rely on the state for long-term support or funding. The inconsistency of federal support for environmental sustainability efforts created an imperative for programs established with EECBG funding to establish other means of sustaining their operations. Organizers of local BBNP initiatives did not expect federal support to be consistent or reliable beyond the short scope of the grant period. To fill this gap, hybrid arrangements were cultivated as a way of sustaining local initiatives once federal funding expired in 2014 [107]. In fact, representatives from the federal government themselves reported that they encouraged these arrangements to make implementation possible and provide opportunities for future monetary support.

This paper has explored a case in which hybrid arrangements were effective in implementing energy efficiency programs at the local level despite broader political and economic contexts that pose significant challenges to environmental policy goals. The sustainability of the EECBG programs in US cities was dependent on the specific partnerships and forms of collaboration that developed over the course of the grant period. Given that climate policies are currently unlikely to advance at the federal level in the United States, sub-national energy efficiency efforts such as the programs described in this paper are especially important. Because hybrid arrangements support these efforts, more information is needed about the specific types of collaboration that are most successful over time. Our findings should also be tested in other political contexts to understand the degree to which the US is, indeed, exceptional [115]. Such specificity will also enable us to understand the ways that Ecological Modernization Theory fits policy implementation of sustainability initiatives in greater detail.

Acknowledgments: This project was funded through grant number 199880, Research Council of Norway.

Author Contributions: Dana R. Fisher conceived of this research as part of grant with the Center for International Climate and Environmental Research—Oslo. Dana R. Fisher designed the study. Anya M. Galli collected the program-level data and conducted phone interviews. Anya M. Galli analyzed the data with input from Dana R. Fisher. The authors worked together to write this paper.

Conflicts of Interest: The authors declare no conflict of interest. The founding sponsors had no role in the design of the study; in the collection, analyses, or interpretation of data; in the writing of the manuscript, and in the decision to publish the results.

References

1. Center for Climate and Energy Solutions. United States Congress. Available online: http://www.c2es.org/federal/congress (accessed on 27 October 2015).

2. The White House. Climate Action Plan. Available online: http://www.whitehouse.gov/sites/default/files/image/president27sclimateactionplan.pdf (accessed on 30 September 2013).

3. The White House. Climate Change and President Obama's Action Plan. Available online: https://www.whitehouse.gov/climate-change (accessed on 27 October 2015).

4. Davenport, C. Numerous States Prepare Lawsuits Against Obama's Climate Policy. New York Times, 22 October 2015, A22.

5. Tubman, M. *President Obama's Climate Action Plan: Two Years Later*; Center for Climate and Energy Solutions: Washington, DC, USA, 2015.

6. The White House. *Highlighting Federal Actions Addressing the Recommendations of the State, Local, and Tribal Leaders Task Force on Climate Preparedness and Resilience*; The White House: Washington, DC, USA, 2015.

7. The White House. *Executive Order 13677—Climate-Resilient International Development*; The White House: Washington, DC, USA, 2014; Volume 79.

8. The White House. *Executive Order 13693—Planning for Federal Sustainability in the Next Decade*; The White House: Washington, DC, USA, 2015; Volume 80.

9. The White House. *The White House. Executive Order 13653—Preparing the United States for the Impacts of Climate Change*; The White House: Washington, DC, USA, 2013; Volume 78.

10. Mayntz, R. Governing Failures and the Problem of Governability: Some Comments on a Theoretical Paradigm. In *Modern Governance*; Kooiman, J., Ed.; Sage: London, UK, 1993; pp. 9–20.

11. Jänicke, M.; Jörgens, H. New Approaches to Environmental Governance. In *The Ecological Modernization Reader*; Mol, A.P.J., Sonnenfeld, D.A., Spaargaren, G., Eds.; Routledge: London, UK, 2009; pp. 157–187.

12. Mol, A.P.J.; Spaargaren, G. Towards a Sociology of Environmental Flows. A New Agenda for Twenty-first Century Environmental Sociology. In *Governing Environmental Flows: Global Challenges for Social Theory*; Spaargaren, G., Mol, A.P.J., Buttel, F.H., Eds.; MIT Press: Cambridge, MA, USA, 2006; pp. 39–84.

13. Fisher, D.R.; Svendsen, E.S. Hybrid Arrangements within the Environmental State. In *Routledge International Handbook of Social and Environmental Change*; Lockie, S., Sonnenfeld, D.A., Fisher, D.R., Eds.; Routledge: New York, NY, USA, 2013.

14. Betsill, M.M.; Bulkeley, H. Cities and the Multilevel Governance of Global Climate Change. *Glob. Gov.* **2006**, *12*, 141–159.

15. Corfee-Morlot, J.; Cochran, I.; Teasdale, P.-J. Cities and Climate Change: Harnessing the Potential for Local Action. In Proceedings of the OECD Conference on Competitive Cities and Climate Change, Milan, Italy, 9–10 October 2008; pp. 78–104.

16. Bulkeley, H.; Schroeder, H. Beyond state/non-state divides: Global cities and the governing of climate change. *Eur. J. Int. Relat.* **2012**, *18*, 743–766. [CrossRef]

17. Spaargaren, G.; Mol, A.P.J. Sociology, Environment, and Modernity: Ecological Modernization as a Theory of Social Change. *Soc. Nat. Resour.* **1992**, *5*, 323–344. [CrossRef]

18. Dunlap, R.E.; Catton, W.R. Struggling with Human Exemptionaltsm: The Rise, Decline and Revitalization of Environmental Sociology. *Am. Sociol.* **1994**, *5*, 243–273. [CrossRef]

19. Catton, W.R. *Overshoot: The Ecological Basis of Revolutionary Change*; University of Illinois Press: Chicago, IL, USA, 1980.

20. Foster, J.B. The Absolute Law of Environmental Degradation Under Capitalism. *Capital. Nat. Soc.* **1992**, *3*, 77–82. [CrossRef]

21. O'Connor, J. On the Two Contradictions of Capitalism. *Capital. Nat. Social.* **1992**, *2*, 107–109. [CrossRef]

22. Schnaiberg, A. *The Environment: From Surplus to Scarcity*; Oxford University Press: New York, NY, USA, 1980.

23. Schnaiberg, A.; Gould, K.A. *Environment and Society: The Enduring Conflict*; St. Martin'ss Press: New York, NY, USA, 1994.

24. Mol, A.P.J.; Spaargaren, G.; Sonnenfeld, D.A. Ecological Modernization: Three Decades of Policy, Practice, and Theoretical Reflection. In *The Ecological Modernization Reader*; Routledge: London, UK, 2009; pp. 3–16.

25. Mol, A.P.J. *Globalization and Environmental Reform*; The MIT Press: Cambridge, MA, USA, 2001.

26. Hajer, M.A. *The Politics of Environmental Discourse: Ecological Modernization and the Policy Process*; Oxford University Press: New York, NY, USA, 1995.

27. Mol, A.P.J.; Janicke, M. The Origins and Theoretical Foundations of Ecological Modernization Theory. In *The Ecological Modernization Reader*; Mol, A.P.J., Sonnenfeld, D.A., Spaargaren, G., Eds.; Routledge: London, UK, 2009; pp. 17–27.

28. Frank, D.J.; Hironaka, A.; Schofer, E. The Nation-State and the Natural Environment over the Twentieth Century. *Am. Sociol. Rev.* **2000**, *65*, 96–116. [CrossRef]

29. Giddens, A. *The Third Way*; Polity Press: Cambridge, MA, USA, 1998.

30. Giddens, A. *The Consequences of Modernity*; Polity Press: Malden, MA, USA, 1990.

31. Mol, A.P.J.; Buttel, F.H. The Environmental State Under Pressure: An Introduction. *Res. Soc. Probl. Public Policy* **2002**, *10*, 1–11.

32. Jänicke, M. On Ecological and Political Modernization (Über ökologische und politishe). *Mod. Z. Umweltpolit. Umweltr.* **1993**, *2*, 159–175.

33. Mol, A.P.J. Joint Environmental Policymaking in Europe: Between Deregulation and Political Modernization. *Soc. Nat. Resour.* **2003**, *16*, 335–348. [CrossRef]

34. Fisher, D.R.; Fritsch, O.; Andersen, M.S. Transformations in Environmental Governance and Participation. In *The Ecological Modernization Reader*; Mol, A.P.J., Sonnenfeld, D.A., Spaargaren, G., Eds.; Routledge: London, UK, 2009; pp. 143–155.

35. Van Tatenhove, J.P.M.; Leroy, P. Environment and Participation in a Context of Political Modernization. *Environ. Values* **2003**, *12*, 155–174. [CrossRef]

36. Mol, A.P.J.; Spaargaren, G.; Sonnenfeld, D.A. *The Ecological Modernisation Reader*; Routledge: London, UK, 2009.

37. Koontz, T.M.; Steelman, T.A.; Carmin, J.; Korfmacher, K.S.; Moseley, C.; Thomas, C.W. *Collaborative Environmental Management: What Roles for Government?*; Resources for the Future: Washington, DC, USA, 2004.

38. Sirianni, C. The Civic Mission of a Federal Agency in the Age of Networked Governance: U.S. Environmental Protection Agency. *Am. Behav. Sci.* **2009**, *52*, 933–952. [CrossRef]

39. Ansell, C.; Gash, A. Collaborative Governance in Theory and Practice. *J. Public Adm. Res. Theory* **2007**, *18*, 543–571. [CrossRef]

40. Boyte, H.C. Reframing Democracy: Governance, Civic Agency, and Politics. *Public Adm. Rev.* **2005**, *65*, 536–546. [CrossRef]

41. Gross, M., Heinrichs, H., Eds.; *Environmental Sociology: European Perspectives and Interdisciplinary Challenges*; Springer: New York, NY, USA, 2010.

42. Lemos, M.C.; Agrawal, A. Environmental Governance. *Annu. Rev. Environ. Resour.* **2006**, *31*, 297–325. [CrossRef]

43. Ostrom, E. *Governing the Commons: The Evolution of Institutions for Collective Action*; Cambridge University Press: New York, NY, USA, 1990.

44. Peluso, N.L. *Rich Forests, Poor People: Resource Control and Resistance in Java*; Univeristy of California Press: Berkeley, CA, USA, 1992.

45. Brosius, P.; Cannon, T.; Davis, I.; Wisner, B. *At Risk*; Routledge: London, UK, 1994.

46. Ostrom, E., Dietz, T., Dolsak, N., Stern, P., Stonich, S.E., Eds.; *The Drama of the Commons*; National Academies Press: Washington, DC, USA, 2000.

47. Agrawal, A. The Regulatory Community: Decentralization and the Environment in the Van Panchayats (Forest Councils) of Kumaon. *Mt. Res. Dev.* **2001**, *21*, 208–211. [CrossRef]

48. Agrawal, A. *Environmentality: Technologies of Government and the Making of Subjects*; Duke University Press: Durham, NC, USA, 2005.

49. Neumann, R. *Imposing Wilderness*; Univeristy of California Press: Berkeley, CA, USA, 1999.

50. McCarthy, J. States of nature: Theorizing the state in environmental governance. *Rev. Int. Polit. Econ.* **2007**, *14*, 176–194. [CrossRef]

51. Watts, M. Antinomies of Community: Some Thoughts on Geography, Resources and Empire. *Trans. Inst. Br. Geogr.* **2004**, *29*, 195–216. [CrossRef]

52. Cashore, B. Legitimacy and the privatizationof environmental governance: How nonstate market driven (NSMD) governance systems gain rule-making authority. *Governance* **2002**, *15*, 503–529. [CrossRef]

53. Tews, K.; Busch, P.O.; Jorgens, H. The diffusion of new environmental policy instruments. *Eur. J. Polit. Res.* **2003**, *42*, 569–600. [CrossRef]
54. MacKendrick, N.M. The role of the state in voluntary environmental reform: A case study of public land. *Policy Sci.* **2005**, *3*, 21–44. [CrossRef]
55. Durant, R.F.; Chun, Y.P.; Kim, B.; Lee, S. Toward a new governance paradigm for environmental and natural resource management in the 21st century? *Adm. Soc.* **2004**, *35*, 643–682. [CrossRef]
56. Liverman, D. Who governs, at what scale, and at what price? Geography, environmental governance, and the commodification of nature. *Ann. Assoc. Am. Geogr.* **2004**, *94*, 734–738.
57. Evans, P. Government action, social capital and development: Reviewing the evidence on synergy. *World Dev.* **1996**, *24*, 1119–1132. [CrossRef]
58. Clark, W. Environmental Globalization. In *Governance in a Globalizing World*; Held, I.D., McGrew, A., Eds.; Brookings Institute: Washington, DC, USA, 2000; pp. 86–108.
59. Haas, P. Addressing the global governance deficit. *Glob. Environ. Polit.* **2004**, *4*, 1–15. [CrossRef]
60. Ford, L.H. Challenging global environmental governance: Social movement agency and global civil society. *Glob. Environ. Polit.* **2003**, *3*, 120–134. [CrossRef]
61. Ribot, J.C.; Peluso, N.L. A theory of access. *Rural Sociol.* **2003**, *68*, 153–181. [CrossRef]
62. Papadopoulos, Y. Cooperative forms of governance: Problems of democratic accountability in complex environments. *Eur. J. Polit. Res.* **2003**, *42*, 473–501. [CrossRef]
63. Luke, T.W. On environmentality: Geo-power and eco-knowledge in the discourses of contemporary environmentalism. *Cult. Crit.* **1995**, *31*, 57–81. [CrossRef]
64. Luke, T.W. Neither sustainable nor development: Reconsidering sustainability in development. *Sustain. Dev.* **2005**, *13*, 228–238. [CrossRef]
65. Okereke, C.; Bulkeley, H.; Schroeder, H. Conceptualizing climate governance beyond the international regime. *Glob. Environ. Polit.* **2009**, *9*, 58–78. [CrossRef]
66. Fletcher, R. Neoliberal environmentality: Toward a postructuralist political ecology of the conservation debate. *Conserv. Soc.* **2010**, *8*, 171–181. [CrossRef]
67. Lockie, S. Neoliberal regimes of environmental governance: Climate change, biodiversity and agriculture in Australia. In *The International Handbook of Environmental Sociology*, 2nd ed.; Reclift, M.R., Woodgate, G., Eds.; Edward Elgar: Northamton, MA, USA, 2010.
68. Oels, A. Rendering climate change governable: From biopower to advanced liberal government. *J. Environ. Policy Plan.* **2005**, *7*, 185–207. [CrossRef]
69. Robertson, M. Discovering price in all the wrong places: The work of commodity definition and price in neoliberal environmental policy. *Antipode* **2007**, *39*, 500–526. [CrossRef]
70. Lockie, S.; Higgins, V. Roll-out neoliberalism and hybrid practices of regulation in Australian agri-environmental governance. *J. Rural Stud.* **2007**, *23*, 1–11. [CrossRef]
71. Higgins, V.; Lockie, S. Re-discovering the social: Neo-liberalism and hybrid practices of governing in rural natural resource management. *J. Rural Stud.* **2002**, *18*, 419–428. [CrossRef]
72. Fisher, D.R.; Freudenburg, W.R. Insights and applications ecological modernization and its critics: Assessing the past and looking toward the future. *Soc. Nat. Resour.* **2001**, *14*, 701–709. [CrossRef]
73. Bäckstrand, K.; Lövbrand, E. Planting trees to mitigate climate change: Contested discourses of ecological modernization, green governmentality and civic environmentalism. *Glob. Environ. Polit.* **2006**, *6*, 50–74. [CrossRef]
74. Leroy, P.; van Tatenhove, J.P.M. Political Modernization and Environmental Politics. In *Environment and Global Modernity*; Spaargaren, G., Mol, A.P.J., Buttel, F.H., Eds.; SAGE Publications Inc.: Thousand Oaks, CA, 2000; pp. 187–208.
75. Sonnenfeld, D.; Mol, A. Globalization and the Transformation of Environmental Governance: An Introduction. *Am. Behav. Sci.* **2002**, *45*, 1318–1339. [CrossRef]
76. Huber, J. *Die verlorene Unchuld der Okologie, Neue Technologien und Superindustrielle Entwicklung*; Fisher Verlag: Frankfurt am Main, Germany, 1982.
77. Mol, A.P.J. Ecological Modernization and the Global Economy. *Glob. Environ. Polit.* **2002**, *2*, 92–115. [CrossRef]
78. O'Connor, J. *Natural Causes*; Guilford: New York, NY, USA, 1999.

79. Pellow, D. Environmental Inequality Formation: Toward a Theory. *Am. Behav. Sci.* **2000**, *43*, 581–601. [CrossRef]

80. Martinez-Alier, J. Political ecology, distributional conflicts, and economic incommensurability. *New Left Rev.* **1995**, *9*, 70–88.

81. Jorgenson, A.; Clark, B. Are the Economy and the Environment Decoupling? A Comparative International Study, 1960–2005. *Am. J. Sociol.* **2012**, *118*, 1–44. [CrossRef]

82. Watts, M. *Silent Violence*; Univeristy of California Press: Berkeley, CA, USA, 1983.

83. Redclift, M. Sustainable Development (1987–2005): An Oxymoron Comes of Age. *Sustain. Dev.* **2005**, *13*, 212–227. [CrossRef]

84. Ribot, J. Theorizing access: Forest profits along senegal's charcoal commodity chain. *Dev. Chang.* **1998**, *29*, 307–341. [CrossRef]

85. Blaikie, P.; Brookfield, H. *Land Degradation and Society*; Routledge: London, UK, 1987.

86. Watts, M.J. Contested communities, malignant markets, and gilded governance: Justice, resource extraction, and conservation in the tropics. In *People, Plants, and Justice: The Politics of Nature Conservation*; Columbia University Press: New York, NY, USA, 2000; pp. 21–51.

87. Oosterveer, P. Environmental Governance of Global Food Flows: The Case of Labeling Strategies. In *Governing Environmental Flows: Global Changes to Social Theory*; Spaargaren, G., Mol, A.P.J., Buttel, F.H., Eds.; MIT Press: Cambridge, MA, USA, 2006; pp. 267–302.

88. Mol, A.P.J.; Spaargaren, G.; Sonnenfeld, D.A. Ecological Modernization Theory: Taking Stock, Moving Forward. In *Routledge International Handbook of Social and Environmental Change*; Lockie, S., Sonnenfeld, D.A., Fisher, D.R., Eds.; Routledge: New York, NY, USA, 2014; pp. 15–30.

89. Spaargaren, G.; Mol, A.P.J.; Bruyninckx, H. Introduction: Governing Environmental Flows in Global Modernity. In *Governing Environmental Flows: Global Challenges to Social Theory*; Spaargaren, G., Mol, A.P.J., Buttel, F.H., Eds.; MIT Press: Cambridge, MA, USA, 2006.

90. Murphy, J.; Gouldson, A. Environmental policy and industrial innovation: Integrating environment and economy through ecological modernisation. *Geoforum* **2000**, *31*, 33–44. [CrossRef]

91. Andonova, L.B.; Betsill, M.M.; Bulkeley, H. Transnational climate governance. *Glob. Environ. Polit.* **2009**, *9*, 52–73. [CrossRef]

92. Pattiberg, P. Public-private partnerships in global climate governance. *WIREs Clim. Chang.* **2010**, *1*, 279–287. [CrossRef]

93. Bäckstrand, K. Accountability of networked climate governance: The rise of transnational climate partnerships. *Glob. Environ. Polit.* **2008**, *8*, 74–102. [CrossRef]

94. Jagers, S.C.; Stripple, J. Climate Governance beyond the state. *Glob. Gov.* **2003**, *9*, 385–399.

95. Pattiberg, P.; Stripple, J. Beyond the public and private divide: Remapping transnational climate governance in the 21st century. *Int. Environ. Agreem.* **2008**, *8*, 367–388. [CrossRef]

96. Rabe, B. Racing to the top, the bottom, or the middle of the pack? The evolving state government role in environmental protection. In *Environmental Policy: New Directions for the 21st Century*; Vig, N., Craft, M., Eds.; CQ Press: Washington, DC, USA, 2013; pp. 30–53.

97. Rabe, B.; Borick, C. Conventional Politics for Unconventional Drilling? Lessons from Pennsylvania's Early Move into Fracking Policy Development. *Rev. Policy Res.* **2013**, *30*, 321–340. [CrossRef]

98. Bulkeley, H. Reconfiguring Environmental Governance: Towards a Politics of Scales and Networks. *Polit. Geogr.* **2005**, *24*, 875–902. [CrossRef]

99. Bulkeley, H.; Betsill, M. Rethinking Sustainable Cities: Multilevel Governance and the "Urban" Politics of Climate Change. *Environ. Polit.* **2005**, *14*, 42–63. [CrossRef]

100. Betsill, M.; Bulkeley, H. Looking Back and Thinking Ahead: A Decade of Cities and Climate Change Research. *Local Environ.* **2007**, *12*, 447–456. [CrossRef]

101. Fisher, D.R. Understanding the Relationship Between Sub-National and National Climate Change Politics in the United States: Toward a Theory of Boomerang Federalism. *Environ. Plan. C Gov. Policy* **2013**, *31*, 769–784. [CrossRef]

102. Office of Energy Efficiency & Renewable Energy. Conservation Block Grants. Available online: http://www1.eere.energy.gov/wip/eecbg.html (accessed on 10 October 2012).

103. United States Departmet of Energy. Energy Efficiency Conservation Block Grant Allocation. Available online: http://www1.eere.energy.gov/wip/eecbg_grants.html (accessed on 10 September 2012).

104. United States Departmet of Energy. Better Buildings Neighborhoods History. Available online: http:// www1.eere.energy.gov/buildings/betterbuildings/neighborhoods/history.html (accessed on 25 September 2012).

105. Center for Climate and Energy Solutions. 111th U.S. Congress. Available online: http://www.c2es.org/ federal/congress/111 (accessed on 31 September 2013).

106. Energy Block Grants. Available online: http://www.energyblockgrants.org/ (accessed on 4 September 2013).

107. United States Departmet of Energy. Energy Efficiency and Conservation Block Grant Financing Programs After Grant Retirement. Available online: http://energy.gov/eere/wipo/articles/energy-efficiency-and-conservation-block-grant-financing-programs-after-grant (accessed on 28 October 2015).

108. The United States Conference of Mayors. *Clean Energy Solutions for America's Cities*; The United States Conference of Mayors: Washington, DC, USA, 2012.

109. United States Conference of Mayors. Conference of Mayors Climate Protection Agreement: List of Participating Mayors. Available online: http://www.usmayors.org/climateprotection/list.asp (accessed on 28 October 2015).

110. Lofland, J.; Lofland, L.H. *Analyzing Social Settings: A Guide to Qualitative Observation and Analysis*; Wadsworth: Belmont, CA, USA, 1995.

111. United States Departmet of Energy. Better Buildings Neighborhood Partners. Available online: http: //www1.eere.energy.gov/buildings/betterbuildings/neighborhoods/partners.html (accessed on 2 October 2012).

112. Fisher, D.R. Bringing the material back in: Understanding the U.S. position on climate change. *Sociol. Forum* **2006**, *21*, 467–494. [CrossRef]

113. United States Energy Information Administration. *Annual Coal Report 2010 (DOE/EIA-0584)*; United States Energy Information Administration: Washington, DC, USA, 2011.

114. United States Departmet of Energy. Better Buildings Grant Recipients. Available online: http://www1.eere. energy.gov/buildings/betterbuildings/grant_recipients.html (accessed on 15 June 2011).

115. Voss, K. *The Making of American Exceptionalism: The Knights of Labor and Class Formation in the Nineteenth Century*; Cornell University Press: Ithaca, NY, USA, 1993.

MDPI

Article

Certification of Markets, Markets of Certificates: Tracing Sustainability in Global Agro-Food Value Chains

Arthur P. J. Mol and Peter Oosterveer *

Environmental Policy Group, Wageningen University, P.O. Box 8130, Wageningen 6700 EW, The Netherlands; arthur.mol@wur.nl

* Author to whom correspondence should be addressed; peter.oosterveer@wur.nl; Tel.: +31-317-427494.

Academic Editor: Md Saidul Islam
Received: 30 July 2015; Accepted: 2 September 2015; Published: 8 September 2015

Abstract: There is a blossoming of voluntary certification initiatives for sustainable agro-food products and production processes. With these certification initiatives come traceability in supply chains, to guarantee the sustainability of the products consumed. No systematic analysis exists of traceability systems for sustainability in agro-food supply chains. Hence, the purpose of this article is to analyze the prevalence of four different traceability systems to guarantee sustainability; to identify the factors that determine the kind of traceability systems applied in particular supply chains; and to assess what the emergence of economic and market logics in traceability mean for sustainability. Two conclusions are drawn. Globalizing markets for sustainable agro-food products induces the emergence of book-and-claim traceability systems, but the other three systems (identity preservation, segregation and mass balance) will continue to exist as different factors drive traceability requirements in different supply chains. Secondly, traceability itself is becoming a market driven by economic and market logics, and this may have consequences for sustainability in agro-food supply chains in the future.

Keywords: voluntary certification initiatives; agro-food supply chains; traceability; sustainability; marketization

---

1. Introduction

Over the past decades, increasing globalization in agro-food trade has been paralleled with a growing importance attached to sustainability of products and of the circumstances under which product have been produced. This is especially mounting in global value chains and networks [1,2] that trade products to the wealthier markets in the North, where among others ethical, environmental, health, animal welfare, and (child) labor values of products and production processes play a growing role in driving consumer demand. However, also in newly emerging and transitional economies such as Brazil and China a growing (upper) middle class is starting to show an interest in ethical and sustainability aspects of products and production circumstances, although there is still a world to win here (see on aquaculture fish: [3]).

Since the mid-1990s and following this growing demand for sustainability in transnational value chains, a rapid increase in the design and implementation of all kinds of (mostly voluntary) public, private and hybrid standards and certification schemes can be witnessed, to ensure that sustainability of products and production circumstances are communicated towards customers and consumers downstream the global value chains. Hence, we see the blossoming of sustainability labels and certification systems in global value chains of food (fish, coffee, tea, cocoa, vegetables, *etc.*), (bio) fuel/energy (electricity, liquid biofuels) and agro-industrial commodities (timber, cotton, textiles).

Most of these labelled and certified products started as niche markets and included only a limited number of producer and consumer countries (as was initially the case for fair trade coffee, and for organic products). However, over the years these certified products developed into quite substantial markets with truly global reach [4]. Key in these certification schemes is that sustainability claims put on final consumer products can be traced back through the global chain of custody to initial (primary) production circumstances. This requires transparency of the value chain, traceability of products [5] and verification of sustainability claims, especially when product attributes themselves do not allow distinguishing between sustainably and not-sustainably produced products (credence goods). Different tracking and tracing systems have been developed and applied in agro-food value chains to relate sustainability claims made on final products (for instance through a label or product information) back to the initial agricultural production circumstances.

This article has three goals. We (i) analyze the prevalence of the various systems to trace sustainable products through value chains; (ii) identify the factors that determine which kind of traceability system is applied for different agro-food commodities; and (iii) assess the consequences of emerging trade in sustainability certificates (rather than sustainable products) in value chains. For this we have reviewed global traceability systems currently applied for key traded agro-food commodities, analyzed relevant scientific literature and compared agro-food traceability systems with those in other markets. After introducing four systems for tracing sustainably produced products, the prevalence of the various systems in different commodity markets is analyzed. Subsequently, we analyze the factors behind applying different traceability systems and focus especially on the role of book and claim systems (as they create a market of certificates). The final section draws conclusions.

2. Tracking and Tracing Sustainability in Value Chains

2.1. Sustainability Governance through Information

Especially in the field of environmental governance, the notion of regulation or governance through information has emerged over the last two decades [6–8]. Informational regulation or governance refers to the idea that information (and informational processes, technologies, institutions and resources linked to it) is fundamentally restructuring processes, institutions and practices of governance, making these governance processes essentially different from conventional modes of governance. Where conventional governance highly relies on authoritative resources, belief in information control, and state power, in informational governance information is becoming a crucial (re)source with transformative powers in specified practices, although nobody is in control of information. Information processes (e.g., on collection, monitoring, disclosure, dissemination, framing, verification) now start to become acts of governance with transformative power, instead of just enabling processes for formulating and implementing authoritative state policies. This counts especially in contexts where environmental/sustainability governance transcends the nation-state and becomes international/global. Hence, state agencies, international organizations, companies, utilities, NGOs, retailers, consumers and the like govern—and are being governed—through the production, use, release, framing, accessibility, demand, and verification of information.

In directing (transnational) agro-food chains towards sustainability the collection, processing, verification, disclosing and disseminating of information, by value chain actors and stakeholder related to value chains (NGOs, policy-makers, *etc.*), are crucial acts of governance that transform social practices of production and consumption. This becomes evident when one thinks of, for instance, guaranteeing sources of origin, tracking and tracing of animal diseases, eco-labelling and certification [9], corporate social responsibility reporting and auditing, separation of GMO and non-GMO products, media controversies on the sustainability of food products. Through these informational processes and practices actors profile and advance sustainability of agro-food products. With information moving center stage in the growing market for sustainable agro-food products, new practices, actors and power relations emerge in sustainable value chains. Sustainability information becomes of value in agro-food chains independent from products themselves and may even become a commodity in a separate market. Tracing sustainability of products is a new practice that is rapidly moving to the center of sustainability governance of global agro-food chains and this comes along with new actor roles and changing power relations.

2.2. Value Chain Traceability Typology

While there is now a growing literature on traceability in agro-food supply chains (see [10] for a useful review), most studies concentrate on food safety and food quality (where food product characteristics are essential). Very few studies address traceability systems for sustainability (where production process characteristics and cultivation of crops/raising of animals are key). In addition, in the rapidly growing literature on sustainable or certified agro-food chains [4,9,11–14], only very limited attention has been paid to the kinds of systems for tracking and tracing sustainability through such global value chains.

Following earlier work of transparency in agro-food value chains [15] traceability systems can serve different purposes. Four ideal-types of traceability can be distinguished (Table 1), where information on the quality of products and production processes is traced for different target groups. The first (also in time) type of traceability in value chains relates to logistics, total quality management of chains and products and verification of product specifications. As such it is a restricted form of traceability for a limited number of economic actors in value chains, primarily motivated by economic interests, and focused on tracing product quality. This so-called management-traceability has its origins in management sciences and logistics [16] and is not directed at, nor does it involve, citizen-consumers or public authorities. A second type of traceability in value chains relates to legal and policy requirements of public authorities on especially food safety and product qualities, such as in the EU tracking and tracing policies. Here public bodies and authorities demand tracking and documenting of information along the value chain, to be disclosed to regulators and inspectors when asked for. Mad cow disease, bird flu, swine fever and other highly contagious animal diseases have rapidly diffused state requirements for this kind of tracking and tracing around the globe [17]. A third and wider form of traceability in value chains relates to quality and sustainability of production processes and products as articulated in public or private labeling and certification. Here traceability is meant to track and verify information along the value chain for consumers and public and private certification bodies and is strongly associated with the consumerist turn and what some call an alternative food economy [18]. Organic, green, sustainable, fair trade and all kind of other sustainability product and production process claims are articulated in standards, disclosed in labels and information systems, guaranteed through certification, and aimed at price premiums and niche market competitiveness. These claims on the consumer product need to be verified and trusted through traceability systems. This can be labeled consumer traceability. Finally, global agro-food value chains are subject of and involved in public scrutiny of their sustainability performance. Information on the sustainability of production processes and product characteristics needs to be traceable to safeguard reputational capital of chain actors and to gain a competitive advantage in the public domain. The Carbon Disclosure Project, the activities of Transparency International, television programs on tracing production and product quality claims, but also claims of certification and labeling bodies and the subsequent certification of certification systems through ISEAL, are examples of public tracking and tracing.

Sustainability tracking and tracing, which involves both product and production process information, is more prevalent in consumer and public traceability types and less in management and regulatory ones. Hence, we will especially focus on the consumer and public traceability. The scarce literature points out that consumer and public traceability forms can differ, among others, in their organizational structure; the involvement of different public and/or private actors; the system of information collection, reporting and flow along the value chain; and the rules and procedures of conformity/verification assessment, and subsequent certification.

Table 1. Four ideal types of traceability in value chains and networks.

Ideal Type Value Chain Traceability	Tracing Information from	Tracing Information for	Focus on Product/Process
Management traceability	Upstream * economic producers in chains	Downstream * economic customers in chains	Product quality
Regulatory traceability	Economic actors in chains	Regulatory and inspection bodies	Product quality
Consumer traceability	Economic actors in chains	Consumers and certification bodies	Product & processes quality and sustainability
Public traceability	Economic actors in chains, certification bodies	Public (citizen-consumers, NGOs, media)	Product & processes quality and sustainability

* Upstream refers to chain actors higher up in the value chain such as primary producers and raw material processors. Downstream refers to chain actors lower in the value chain such as final processors, customers and retailers.

2.3. Power and Markets

With the emerging importance of certified sustainability as a preferential product quality, a potential for premium prices, an access requirement for some markets and a core element of reputational capital, traceability systems have become more consequential, more complex and more costly. Tracking, tracing, verifying and certifying information on the sustainability of agro-food production processes and products for consumers and the wider public involve new practices, power relations and power brokers. New practices of separating (sustainably and non-sustainably produced) product flows, of monitoring, registering and reporting, of verification, of handling out labels and certificates, and of trading in sustainability certificates have emerged in agro-food chains. With these new practices, new actors and power brokers have emerged along the traced value chain, such as verification and certification agencies, trading platforms, registries, and traders in certificates. In addition, new powers relations have been formed, with a power shift towards actors at the consumer end of value chains, but also towards actors outside the value chain such as tracing and certifying agencies.

One of the major changes in practices, actors and power relations in traced sustainability markets comes when sustainability certificates emerge. As we see, for instance, with electricity, carbon and fisheries [19–23], sustainability certification can lead to a market for certificates, relatively decoupled from the material sustainability of primary production and products. With sustainability certificates in agro-food, tracing sustainable products is no longer tracing the product through the value chain to its origin, but the establishment of a new market with new rules, new resources and new actors. As often with new markets, there are market winners, losers and advocates, and major debates on the direct and long-term sustainability performance of markets in sustainability certificates are taking place. One of these debates concerns the use of market-based tools as instruments for environmental policy which is criticized by some scholars as they fear the appropriation by dominant market actors [24] and that their structural limitations will prevent the radical transformations in current supply systems they consider necessary [25].

3. Models of Tracing Sustainability

A large variety of certification systems have been developed to guarantee downstream consumers that products are produced upstream in a sustainable manner. In the literature and in the practice of different global commodity chains these certification systems prove to make use of one or more of four models of sustainability tracing [26–29]. Figure 1 clarifies the differences.

The first model, *identity preserved* or *track and trace*, ensures that the certified product delivered to the end user (customer or consumer) is uniquely identifiable and can be related to the identity of producer and resource base. Initially, it was especially developed and applied to distinguish GM from nonGM products [30], but this has widened more recently to tracing sustainably produced products.

In order to be able to preserve the identity of the certified resource, half-product and final product they have to be kept physically isolated and separated from non-certified equivalents at each stage of the value chain, as well as from certified equivalents from another resource base. Only then full traceability can be organized from the origin of the product to the point of delivery to the end user; and the end user can be assured of the origin and identity of that specific product. In consumer-facing certification systems barcodes or other identification systems, often supported by recent developments in ICT, allow the consumer to trace the origin of the product to the producer [12]. The logistics, monitoring, reporting and verification required for identity preserved systems result in high costs along the value chain, to be compensated through price premiums, unique market access or larger market share. Due to the high costs, identity preserved systems are only applied when considered necessary, either because these are explicitly prescribed or when there is a clear market demand from consumers for such far-reaching certification systems. In most commodity value chains other traceability systems often prevail due to lower costs, reduced complexity, lower data availability requirements, and business preferences [30].

The second model, segregation, is more than incidentally lumped together with identity preserved systems [27], but they are different. In addition, in a segregated system of sustainability certification it is assured to the end-user that a certified product consists of natural resources and production processes (storage, transportation, processing, trading, packaging, selling) that fulfil all the requirements of the certification scheme. At every step certified produce is kept separate from non-certified produce. However, the final certified product cannot be uniquely identified and related to a single identifiable producer and resource base. Transporters, traders and processors mix produce from different certified producers and resource bases, which makes tracking and tracing the final product to a single initial producer/site no longer possible. However, costs are lower due to economies of scale and increased competition.

In the mass balance model the traded volume of certified sustainable produce is administratively monitored throughout the entire value chain to ensure that the volume of certified products downstream equals the volume of certified resource base upstream that very same value chain [31]. The mass balance system allows, however, for the mixing of certified and non-certified produce at any stage of the value chain after the certified produce has been registered and left the farm gate. For the end-user, there exists no longer a one-to-one physical or chemical tie or relation between the consumed certified end product and the certified resource base at the primary producer. Certified end products most likely also consist of non-certified resource base. However, at each stage there is a reconciliation between the quantity of certified material bought and the quantity of certified material sold, verified by a certification agency. As no separate storage, transport or production processes are needed for certified products and less verification, monitoring and control is required, costs are lower compared to the first two systems.

Finally, the book and claim model moves away from any physical/material link between the certified resource or primary produced crop and the final certified product. Operators under this model register the sustainable resource/produce upstream which is booked in a central registry at a trading platform, and for which the operator receives a tradable certificate. The producer then sells his certificates on the (global) market to interested companies through a credit trading platform. For each unit of certified sustainable product that is sold to customers/consumers, final manufacturers need to buy certificates from this platform. The price of a certificate depends on supply and demand and may therefore vary widely over the years as the experience of GreenPalm has shown (the price dropped from 81.58 USD in December 2014 to 35 USD in September 2015 [32]. The major advantage of book and claim systems is that no segregation, monitoring and registering is needed of sustainable produce (after leaving the farm gate). Any final producer who wants to sell certified sustainable products, or any actor that wants to support the production of sustainable primary resource base/products, can do so via buying certificates on the credit trading platform. This reduces costs and complexities in organizing the chain and allows for easy trade of larger volumes of sustainable products. A well-functioning farm-gate and end-user monitoring and registration system, a market of certificates, and a central registry are crucial preconditions to let this system function. The system is more vulnerable for fraud, especially when geographies and numbers of buyers and sellers expand; for consumers contesting the sustainability of products; and for seizing price premiums by selling certificates.

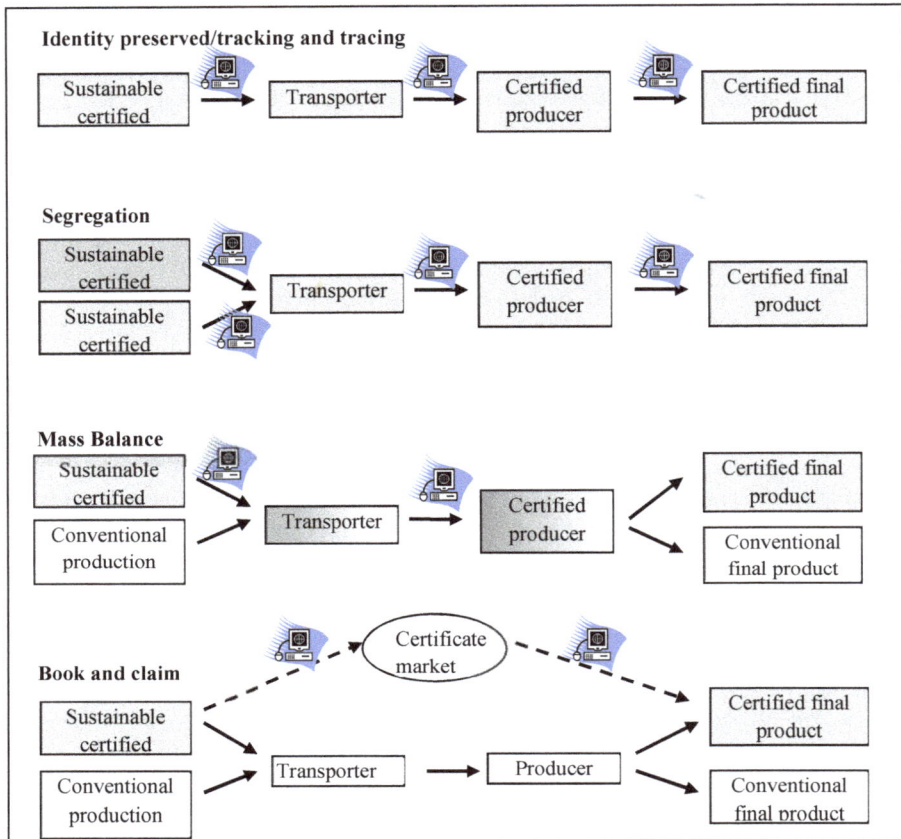

Figure 1. Four sustainability tracing models (adopted from [28,29]).

4. Explaining Prevalence of Traceability Models in Different Markets

The early voluntary certification initiatives usually cover various commodities (IFOAM, Fair Trade, UTZ), while the recent ones are more often focused on one specific commodity. While increasingly commodities are dominated by one certification initiative, such as the 4C Association in coffee, ProTerra in soy, BCI (Better Cotton Initiative) in cotton, Bonsucro (2013) in cane sugar and RSPO (Roundtable on Sustainable Palm Oil) in palm oil, there is not often a complete monopoly [4]. However, even if one certification initiative is developed for one product category, it often applies different traceability systems to fulfil traceability requirements/preferences of different market segments. Table 2 illustrates the diversity in certification schemes in agro-food provision, with the used/allowed traceability systems (see Appendix A). With regard to traceability, they all allow for segregation, most for identity preserved, several for mass balance and only a few for book and claim systems. In addition, the spreading in terms of market share over the allowed traceability systems is not equal, with usually small shares of the marketed certified products having identity preserved traceability. What are the factors explaining preference for identity preserved and mass balance for most certification systems, in most markets for the major share, while only for some certification systems book and claim systems are an option and few marketed products have identity preserved traceability?

Table 2. Product categories, voluntary labels and traceability models (see Appendix A).

Product	Label/Certification Organization/System	Traceability Model Allowed [1]				Year of Introduction
		Identity Preserved	Segregation	Mass Balance	Book and Claim	
Palm oil	RSPO [2]	X	x	x	X	2004
Soy	RTRS [3]		x	X	X	2006
	ProTerra	x	X			2012
Sugar	Fair Trade	x	X	x		1997
	Bonsucro		x	X	X	2006
Cotton	Fair Trade		X			1997
	Better Cotton Initiative [13]	x	X	X		2005
Marine fish	MSC [4]		X			1997
	This Fish	X				2010
Aquaculture fish	ASC [5]		X			2011
Timber	FSC [6]	x	X	x		1993
	PEFC [7]	x	x	X		1999
Biofuels EU market	15 different schemes	x	x	X		2009
(non)GMO crops	EU [8]		X			1997/2004
Biofuels	RSB [9]	x	x	X		2007
Agricultural products	IFOAM [10]	x	X			1972
	Rainforest Alliance	x	X	X		1987
	Organic label US and EU		X			1990/1991
Tea	Fair Trade	x	X	x		1997
	UTZ	X	X			2002
	Ethical Tea Partnership		X			2009
Cocoa	Fair Trade	x	X	x		1997
	UTZ	x	X	X		2002
Coffee	Fair Trade	x	X			1997(1988) [12]
	UTZ	x	X			2002
	4C association [11]	x	X	x		2006
Meat	GRSB	X	X			(2016) [14]

[1]. A capital and bold **X** means used for the major share of the market; small x means less often used; [2]. RSPO: Roundtable on Sustainable Palm Oil; [3]. RTRS: Round Table Responsible Soy; [4]. MSC: Marine Stewardship Council; [5]. ASC: Aquaculture Stewardship Council; [6]. FSC: Forest Stewardship Council; [7]. PEFC: Programme for the Endorsement of Forest Certification; [8]. EU: European Union; [9]. RSB: Roundtable on Sustainable Biomaterials; [10]. IFOAM: International Federation of Organic Agriculture Movements; [11]. 4C Association uses mass balance but the license/certificate must be passed on with the coffee up to the final buyer; [12]. Fair Trade originates from the Dutch Max Havelaar certification scheme for coffee, which started in 1988; [13]. BCI uses a combination of segregation (up until the ginner) mass balance (after the ginner). [14]. GRSB has developed a standards which McDonalds intends to implement in 2016 [33].

4.1. Historical Sequence?

The first traceability systems in sustainability certification of supply chains in the early 1990s resembled an identity preserved or a segregation system, where products could be traced back to sustainable production of the resource base. In general, one would expect that with the further globalization of value chains and networks, the mainstreaming of sustainability in larger markets, and the inclusion of more product categories in sustainability certification, sustainability traceability would increasingly develop from identity preserved models to book and claim systems. The latter type of system is especially apt for large volumes, lowers traceability costs, makes sustainably produced products more competitive with conventional products, and is more concerned with global sustainability and less with identity formation of smaller groups of (dark) green consumers. In that sense the more recent emergence of mass balance and book and claim systems in certification initiatives makes sense.

At the same time, others argue that a reversed trend would make more sense [34]. Initially, mass balance and book and claim systems allow for encouraging producers to produce more sustainably and reward them for it, without necessarily involving additional costs for consumers and other value chain actors. Only when a sustainably produced product is recognized and valued on the market, companies can obtain a somewhat higher price from selling a clearly identifiable product from a single certified resource base. This enables identity preserved traceability systems.

In looking at the distinct products and certification initiatives, it becomes clear that there exists no easy relation between the allowed and prevailing traceability system on the one hand and the time lapse since the start of the certification initiative on the other. Although book and claim systems are emerging more recently for some products and supply chains and then take a significant market share (Appendix A), this is not an evolutionary development. We cannot conclude that once certification matures in a specified market, book and claim systems massively replace identity preserved, segregation and mass balance systems. Nor can we easily conclude that book and claim systems form a starting point for traceability, to be taken over by segregation and identity preserved systems once the market matures and price premiums are possible. Obviously there are (also) other factors involved in determining the prevalence of a traceability system for a sustainably produced agro-food products in a specific market.

4.2. Determining Factors

When comparing the different products, certification initiatives and traceability systems presented in Table 1, at least five factors play a major role in the allowed application and the prevalence (in market share of certified products) of the different traceability systems.

In those markets where products are consumed that are recognizable for individual consumers and where consumer identity through consuming labelled products plays a major role, identity preserved or segregation are more likely to prevail. Regarding final consumer products, such as coffee, vegetables, fish, wood and sugar, identity preserved or segregation is preferred above book and claim and even mass balance. When consumers cannot easily identify sustainability properties of products and cannot distinguish themselves through buying and consuming certified products, mass balance and book and claim systems are more likely to emerge, such as in the case of sustainable palm oil and biofuels.

Second, in markets/products where clear inherent product quality differences between sustainably produced and non-sustainably produced products exist (or are perceived to exist), identity preserved or segregation are likely to prevail. This is often the case with respect to organic vegetables, fruits and meat, and non-GM food products. Product markets where (perceived) product quality differences are absent, and sustainability claims are only related to production processes, are more likely to apply mass balance and book and claim systems, as in the case of liquid biofuels. This differentiation enhances when transport routes of product flows cannot be easily separated, for instance when sustainable and non-sustainable products have to use the same transport infrastructure. Electricity transported through the grid is a typical example [21], as would be any future traceability system for sustainable biogas transported through piped gas systems [35]. Segregation or identity preserved is then only possible for decentralized local systems, with direct connections between producers and consumers of products.

Third, when the lead firm in a global value chain is (perceived to be) quite vulnerable for sustainability questions and accusations from the public, consumers and consumer/environmental non-governmental organizations, one can expect identity preserved and segregation systems to prevail over mass balance and book and claim systems. A clear example is certified capture fish (MSC or This Fish) in value chains where major retailers are lead firms and demand fully segregated chains or even identity preserved [36]. Unilever announced in 2012 that it had set itself the target to buy all of its palm oil from traceable sources by 2015 to 2020, instead of buying it via the book and claim system of GreenPalm. The executive director of New Britain Palm Oil Limited NBPOL claimed with respect to GreenPalm book and claim certificates: "We feel that this is not widely understood and we do not think it is what consumers want in their products (…). Additionally, the entire system including all the associated claims is unaudited and therefore open to abuse. We feel the concept is flawed and potentially misleading" [37]. Identity preserved and segregation systems are superior in guaranteeing individual global value chain companies sustainable resource bases within their chain of custody. Book-and-claim systems are not able to fully guarantee sustainable production of the actual products sold by the lead firm.

Fourth, if the main players around a global value chain are institutional actors (processing companies, traders, major environmental and consumer NGOs, states), which are only to a limited extent directly dependent on consumer legitimacy or citizen membership, one can expect book and claim systems to prevail. Institutional actors focus more strongly on higher level aggregated sustainability effects and less on the sustainability of individual, identifiable products. The systems' perspective of mass balance and book-and-claim systems, with their focus on "aggregated" sustainability, higher levels of efficiency, lower complexities, lower transaction costs (and thus better competitiveness vis-à-vis conventional products), are then often prevailing.

Finally, more extended supply chains, in terms of geographical reach, size of markets, number of actors in global production networks, and 'social distance' between initial producer and final consumer, 'prefer' book and claim and mass balance systems. While in shorter supply chains, with closer social proximity between producer and consumer and smaller markets, identity preserved and segregation systems of traceability are more likely to prevail.

Figure 2 puts together these five different dimensions that jointly influence what type of traceability system is allowed in certification and prevailing in the certified market of distinct commodities, giving examples for RSPO certified palm oil for food products, aquaculture fish labelled through ASC, and USDA organic food products. The larger the surface of the 5-edged figure, the more likely it is that book and claim systems emerge; the smaller the surface the more likely segregation and identity preserved systems dominate.

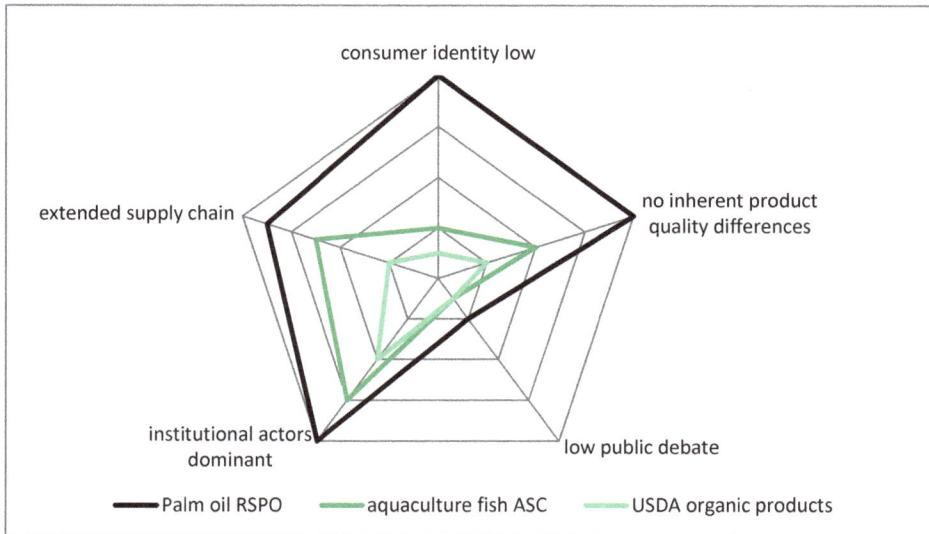

Figure 2. Determinants of traceability systems.

5. New Markets of Traceability

Mass balance or book-and-claim systems seem most promising if one wants to make cost-efficient contributions at sizeable scale to sustainability of agro-food products. When widely introduced, these systems (i) lower the costs of traceability because they require no separate systems of storage, transportation and processing; (ii) are less complex (and thus less costly) in implementation, monitoring, auditing and certification for all intermediate value chain actors, and (iii) make sustainably produced products really competitive with conventional ones. Only in this way sustainably produced products can seize significant market shares beyond niche markets, as also a large share of the middle class consumers, emerging economies, and major institutional actors such as mainstream retailers and lead firms are seduced to articulate demand for sustainably produced (semi-)products. It can be noted that in those global supply chains where multiple certification systems co-exist at the same time (e.g., palm oil, soy, sugar cane), book and claim systems handle the largest market share of certified products, compared to the other systems.

Book and claim systems differ from the other models because here certificates are developed as a new symbolic token that codifies sustainability, provides it with a monetary value and allows it to be traded over long geographical stretches independently from material (product) flows. In that sense book and claim systems reflect global ecological modernization, where "ecology" becomes articulated, forms a separate global "ecological" flow, and becomes "economized" [22,38]. From a global system perspective book and claim is an ecologically and economically rational design for greening global supply networks, as it incentivizes certified production of fresh produce, reduces transaction costs for value chain actors in tracing, and thus makes certified products competitive with non-certified alternatives. In its operationalization, however, these book and claim systems run against a number of challenges.

First, book and claim systems bring in a different set of actors in sustainable global value chains/networks, each with their own role and with their own interests and rationalities. Private brokers of green certificates such as GreenPalm in the sustainable palm oil value chain, private consultancies that set up trading platforms and systems such as Book & Claim [40], and various financial institutions that also function in carbon markets all have an interest in maximizing trade in these certificates. Trading certificates is business. Financial institutions trading certificates have a clear interest in price settings and manipulation, introducing all kinds of new financial products related to these certificates [39]. This makes the sustainability of the certificates more and more competing with their market value, introducing stronger economic logics and rationalities in these sustainability markets. As a consequence traceability becomes a market in itself; traceability is traded, new companies emerge that make a profit out of trading traceability, or from setting up systems and companies that become traceability brokers. An illustration for this trend is the advertisement by the consultancy firm Book & Claim offering assistance in setting up book and claim systems and trading platforms in any industry [40]. The relation with on the ground sustainability of primary production then becomes increasingly indirect or "footloose". Consequently, a stronger element of futures trading and speculation is brought into the sustainability commodity market, which allows middlemen and traders to take a larger proportion of the price and increases the risks of price volatility. Both consequences are considered to be problematic for the poor, producers as well as consumers [41,42].

Second, it is widely conceived that book-and-claim systems are more vulnerable to fraud than identity preserved and segregation systems, with mass balance systems in-between [29]. As the administration of sustainable primary production and the final certified products that are sold are decoupled, more vulnerabilities emerge in terms of illegal introduction of non-sustainable products, creation of certificates, fraud in monitoring and registration, *etc.* Experiences with the carbon credit systems show that this is not just a potential drawback, but that multiple forms of fraud do occur in trading systems of certificates [43,44]. Identity preserved and segregation systems are not immune to such fraud either (as several cases of organic product fraud have shown [45,46]), but such vulnerabilities are considered a lower risk as verification/certification takes place at different stages of the supply chain.

A third challenge is related to the division of responsibilities in some book and claim systems. Especially where buying and selling of sustainability certificates is detached from the organization that sets, monitors and verifies the production standards, responsibilities become blurred [47]. This is especially relevant in cases of fraud or illegality. Is the RSPO responsible for the green certificates handed out and traded by GreenPalm, the private organization that performs the book and claim system? Most actors involved in this book and claim system would probably consider this is not to be the case; but the RSPO will be the first to bear the consequences when certificates are not backed by sustainable oil palm plantations.

Fourth, book and claim systems rarely operate in markets without alternative traceability systems; hence, they have to compete with them. In this competition, efficiency and costs are important but responsibility claims, assurance and trust as well. Some consumers are not impressed by the cost-efficiency of book and claim systems and prefer the greater transparency towards initial producers and resources offered by identity preserved and segregation systems. Such consumer preferences are increasingly articulated by large (and thus powerful) retailers, who sense consumer preferences and know what reputational damage can do.

Finally, book and claim systems have a lower level of environmental effectiveness through the equivalent of the "hot air" mechanism that prevailed in the flexible mechanisms in greenhouse gas emission reduction [19]. All production that fulfils sustainability criteria will be used in a book and claim traceability system, while in segregation systems and identity preserved systems volumes of sustainably produced primary commodities will exist that are not certified, due to the costs and the management complexities involved [2,18]. The latter situation results in higher volumes of sustainably produced primary commodities than certified in a market.

As book and claim systems are relatively new in agro-food markets we have paid special attention to the challenges these traceability systems face. However, articulating these challenges should not be interpreted as predicting a dark future for or discrediting book and claim systems. For one, the other traceability models each have their own strengths and weaknesses. Second, several of these challenges will not be decisive in decision-making processes on which traceability model to apply in certification of products for specific markets. What constitute challenges or even disadvantages for some actors in certified global production networks are sources of (economic) value, market share, low transaction costs or aggregated environmental gains for others.

6. Conclusions

The growing demand for sustainably produced agro-food products in an increasingly global market has resulted in developing distinct systems for certifying sustainability claims, which fit in a wider tendency of governance through information. Traceability forms a key element in these—mostly voluntary—sustainability certification initiatives. The four models of traceability (mass balance, segregation, identity preserved (or tracking and tracing) and book and claim) differ in how sustainability certification of a final product is related to the sustainability qualities of production circumstances and products at different stages of the value chain. In this paper, we focused on explaining the prevalence of different traceability models for different commodities and markets. There proves to be no simple historical or evolutionary development in the prevalence of traceability models in specific commodities and/or markets. In contrast to a straightforward evolutionary development in traceability models, five factors were identified that are co-determining the kind of traceability model applied and prevailing in a specific commodity-market combination. Analyzing these factors provides better predictive power of likely traceability systems than an evolutionary view.

Particularly interesting is the book and claim traceability model because it is rather new in agricultural and food markets and allows for decoupling the sustainably produced material flow from the flow of sustainability claims, making it particularly apt for global value chains. Products are traded separately from sustainability certificates. We identified competing claims among academic scholars, sustainability NGOs and value chain practitioners on the desirability and future outlook of book and claim systems, in "competition" with the three other traceability models. Some consider book and claim to be the most appropriate model in the context of globalizing markets and the proliferation of sustainability preferences among increasing segments of consumers due to its high efficiency, low costs, large aggregate sustainability gains and adequate fit with global value chains. Others believe and/or strongly prefer that the book and claim models will only play a temporary and transitional role in traceability system and will disappear over time. According to them, book and claim systems do not create a market for the sustainably produced products themselves but only for sustainability certificates, making their sustainability claims too vulnerable for fraud and consumer/public mistrust, endangering the entire landscape of sustainably certified products and markets. Our analysis showed that the prevalence of any traceability model remains strongly related to the architecture of the supply chain/network serving a specified commodity market: where is the power located in the chain/network, are lead firms to be found upstream or downstream, can the certified product play a role in consumer identity formation, are collective actors outside the value chain interfering strongly (regulatory bodies, NGOs), and how much premium is to be gained through different traceability modes?

Deciding on the most appropriate traceability system is not a straightforward process on the basis of (economic and environmental) costs and benefits, but relates as much to fundamental consumer identities, ideologies and power relations in chains. While technical-scientific claims can be decided on the basis of 'right' or 'wrong', debates involving fundamental ideologies and power inequalities are never resolved or closed easily. Hence, the debate on traceability system is likely to continue for some time. Despite the criticism we expect book and claim models to remain part and parcel of the options for sustainability certification, be it mainly for commodities that cannot easily (thus: at relatively low costs) be kept physically separate throughout a globalized supply chain for a particular market (e.g., palm oil, soybeans, biofuels).

Acknowledgments: The authors kindly acknowledge the input from two anonymous reviewers that has contributed to this article.

Author Contributions: This paper has been the result of collaborative efforts from both authors. Each of them has participated equally in designing the paper, performing the research and writing the paper. Both authors have read and approved the final manuscript.

Conflicts of Interest: The authors declare no conflict of interest.

Appendix A. Traceability Systems of Different Commodities/Markets

Fair Trade certification, including more than 15 product categories such as coffee, tea, bananas, vegetables and cotton, follows mostly a segregation chain of custody traceability design, although for special markets identity preserved is offered, and for cocoa, sugar cane and tea mass balance systems operate to ensure that Fair Trade producers receive their premium. Tracing back to the origin of production is secondary to production procedures and product characteristics.

UTZ certified allows both mass balance and segregation certification systems in cocoa trade, but only segregation in coffee and tea. The organization gives clear reasons why it continues to have a mass balance system besides a segregation system on cocoa, because "while volumes of certified cocoa are still limited but growing and the processing is quite complex, keeping all certified cocoa separated would imply high investments in the supply chain" and UTZ rather invests in "training of farmers and actual purchase of UTZ certified cocoa" [48]. UTZ expects that with the maturation of the market certification will shift more towards segregation systems. The Sustainable Agriculture Network/Rainforest Alliance has developed a kind of "in-between" category in chain of custody certification, which they label "controlled blending" (only for cocoa until now). Controlled blending differs from mass balance in that it monitors sustainable products up till the gate of the manufacturer, whereas mass balance only monitors sustainable produce until it leaves the farm gate [49]. This enables the system to control the proportion of certified cocoa in each final product. Rainforest Alliance allows only for sugar cane and palm oil a mass balance chain of custody certification, after written permission from the Rainforest Alliance. Segregation and identity preservation are allowed on all product categories [50].

The Stewardship Council systems (such as those of Marine Stewardship Council MSC, Forest Stewardship Council FSC, and Aquaculture Stewardship Council ASC) all work with a segregated system of chain of custody traceability (although FSC has also developed a mass balance system). In Stewardship Council systems certified natural resources are kept segregated throughout the supply chain from non-certified equivalents, up till the final consumer. Sometimes in these markets smaller alternative certification initiatives opt for an identity preserved system of traceability, such as This Fish in capture fisheries.

Under the Round Table for Sustainable Palm Oil (RSPO; Kuala Lumpur, Malaysia, established in 2004) identity preserved, segregated, mass balance and book and claim traceability systems (or modules as the RSPO calls them) operate at the same time for different palm oil markets. Each of the four systems comes with different allowable claims regarding sustainability on the final products, and with different regimes of traceability at the various stages of the value chain [28]. To facilitate the traceability of identity preserved, physically segregated, or mass balanced RSPO certified sustainable palm oil, a new RSPO eTrace system has been launched. The system is designed to improve transparency and efficiency. For facilitating book and claim modules the private company GreenPalm (Hull, United Kingdom, established in 2008) has set up a credit trading platform for the RSPO. With over 750 members, sales of GreenPalm certificates over the first quarter of 2014 ranged to 850,00MT, up 54% compared to the same period in 2013 [51]. Certificate trading via book and claim makes up almost 63% of RSPO Crude Sustainable Palm Oil (CSPO) trading. Although these four different "modules" of traceability and certification are indicated on the Chain of Custody certificate delivered by the certifier, the system (or module) used is not always communicated clearly to the final consumer. Together, annual available certified sustainable palm oil in 2012 makes up 10% of the global market (around 5 million tonnes), but only about 70% of this volume has actually been traded as such.

The Round Table on Responsible Soy RTRS to some extent resembles developments in the RSPO, although it has been established much later. With the first producer was certified only early 2011, in 2014 it had 31 producer members (with a production area of over 450,000 hectares) and 96 members from industry, trade and finance. Besides the segregated and mass balance modules, the RTRS has established a book and claim system of trading responsible soy. Under the RTRS, soy companies, but also other interested companies and organizations not having access to soy value chains, can purchase "responsible soy production credits" directly from soybean growers on the Soy Credit Trading Platform (under the authority of RTRS), with one credit equalling the responsible production of one metric ton of soybeans. Credits can be bought, sold and re-sold, but once validated they can no longer be re-sold [52]. This resembles similar systems as the credit trading platforms of sustainable palm oil of GreenPalm and for carbon credits of ISCC (International Sustainability and Carbon Certification). The different models (segregated, mass balance and book and claim) come with different claims on the products, and even with different logo's to be used [53]. In 2014 over 1.3 million metric tons of responsible soy was sold (including segregated, mass balance and traded credits), in a global market for soy of nearly 240 million tons (FAO Statistics). The recently established ProTerra certification initiative for soy only operates identity preserved and segregated traceability systems.

Better Sugarcane Initiative or Bonsucro (London, United Kingdom, established in 2006) operationalizes several certification systems for sugarcane supply chains [54]. Next to "physically segregated shipment" and mass balance, Bonsucro has a credit trading system where companies wanting to make a claim of sustainable sugar can purchase credits of responsibly produced sugar from certified mills [55]. This facility is only accessible for Bonsucro members (34 mills and 181 other members in 27 countries) and comes together with registration and verification systems at the mills. There is also ample possibility of buying and reselling certificates, making the system into a real market.

Prices are set between buyer and seller and Bonsucro charges a USD $1.3/ton fee. Bonsucro certifies only sugarcane and by early 2014 the organization has certified 3.32% of global sugarcane production (55 million tonnes) and 3.66% (which equals 870,000 hectares) of the total land area under sugarcane. Of the end products, Bonsucro has certified around 3.8 million tons of sugar and some 2.6 million m^3 of ethanol.

Biofuels for the European market need to be sustainable under the EU RED directive (2009/28/EC) in order to allow these biofuels to count in the compulsory percentages of renewable fuel mixing in transport fuel [56]. To date, 19 different certification schemes have been allowed, using a mass balance traceability system (allowing also for segregation and identity preserved) [57]. The allowed systems include RSPO, RTRS, Roundtable for Sustainable Biofuels RSB and Bonsucro, which have developed special mass balance systems for RED-certification, (besides other segregation, identity preserved and/or book and claim traceability systems for other markets) [58]. Since 2008 the English RTFO (Renewable Transport Fuel Obligation) contained a book and claim traceability system for biofuels, but this was discontinued when RTFO had to be harmonized with the EU RED in 2011/12. Staaij and colleagues [29] note the complications of such a large and diversified mass balance system for biofuel traceability, as voluntary certification initiatives vary and EU countries differ in implementing and recording mass balance traceability. For instance, so-called second generation biofuels based on waste and residues are preferred in EU policy and can be counted double in the accounting sheets; but it differs per certification initiative what is seen as waste and residues and what not [59].

A recent initiative (started in 2011) is the Global Roundtable for Sustainable Beef (GRSB), aiming at achieving continuous improvement in the sustainability of beef production systems and value chains around the globe [60]. The GRSB is a multi-stakeholder initiative with representatives from producers and consumers around the world, such as Cargill, McDonalds, the Brazilian Roundtable on Sustainable Livestock, Solidaridad and WWF. The GRSB intends to formulate principles and criteria for global sustainable beef with the help of its members and other stakeholders while considering the indicators and the means of verification to be context-specific and not feasible as elements of a global standard. These important issues are left to local, national and regional groups. GRSB has not yet touched upon issues of traceability of sustainable beef, but it is unlikely that it will introduce mass balance or book and claim systems. One of GRSB's members, McDonalds, has announced that it will begin to purchase sustainable beef verified against these principles and criteria by 2016, after having developed specific targets, to ensure that sustainable beef is verifiable and transparent, making the need for a traceability system pressing [33].

References

1. Hatanaka, M.; Bain, C.; Busch, L. Third-party certification in the global agrifood system. *Food Policy* **2005**, *30*, 354–369. [CrossRef]
2. Veldstra, M.D.; Alexander, C.; Marshall, M.I. To certify or not to certify? Separating the organic production and certification decisions. *Food Policy* **2014**, *49*, 429–436. [CrossRef]
3. Bush, S.R.; Belton, B.; Hall, D.; Vandergeest, P.; Murray, F.J.; Ponte, S.; Oosterveer, P.; Islam, M.S.; Mol, A.P.J.; Hatanaka, M.; *et al.* Certify sustainable aquaculture? *Science* **2013**, *341*, 1067–1068. [CrossRef] [PubMed]
4. Potts, J.; Lynch, M.; Wilkings, A.; Huppe, G.; Cunningham, M.; Voora, V. *The State of Sustainability Initiatives Reviews 2014-Standards and the Green Economy*; International Institute for Sustainable Development: Manitoba, MB, Canada, 2014.
5. Opara, L.U. Traceability in agriculture and food supply chain: A review of basic concepts, technological implications, and future prospects. *Food Agric. Environ.* **2003**, *1*, 101–106.
6. Esty, D. Environmental Protection in the Information Age. *SSRN Electron. J.* 2003. [CrossRef]
7. Kleindorfer, P.R.; Orts, E.W. Informational regulation of environmental risks. *R. Anal.* **1999**, *18*, 155–170. [CrossRef]
8. Mol, A.P.J. Environmental Governance in the Information Age: The Emergence of Informational Governance. *Environ. Plan. C* **2006**, *24*, 497–514.

9. Boström, M.; Klintman, M. *Eco-Standards, Product Labelling and Green Consumerism*; Palgrave McMillan: London, UK, 2008.

10. Ringsberg, H. Perspectives of Food Traceability: A Systematic Literature Review. *Supply Chain Manag.* **2014**, *19*, 558–576. [CrossRef]

11. Bush, S.R.; Oosterveer, P.; Bailey, M.; Mol, A.P.J. Governing sustainable value chains: Review and future outlook. *J. Clean. Prod.* **2015**. [CrossRef]

12. Miller, A.M.M. *Governance Innovation Networks for Sustainable Tuna*; Wageningen University Press: Wageningen, The Netherlands, 2014.

13. Smyth, S.; Phillips, P.W.B. Product Differentiation Alternatives: Identity Preservation, Segregation, and Traceability. *AgBioForum* **2002**, *5*, 30–42.

14. Stetter, A.; Zangl, B. *Certifying Natural Resources—A Comparative Study on Global Standards and Certification Schemes for Sustainability, Part II—Empirical Assessment of Case Studies*; Deutsche Rohstoffagentur: Berlin, Germany, 2012.

15. Mol, A.P.J. The Role of Transparency in Governing China's Food Quality: A review. *Food Control* **2014**, *43*, 49–56. [CrossRef]

16. Schaltegger, S.; Burritt, R. Measuring and managing sustainability performance of supply chains, Review and sustainability supply chain management framework. *Supply Chain Manag. Int. J.* **2014**, *19*, 232–241. [CrossRef]

17. Kjaernes, U.; Harvey, M.; Warde, A. *Trust in Food: A Comparative and Institutional Analysis*; Palgrave MacMillan: London, UK, 2007.

18. Glin, L.; Mol, A.P.J.; Oosterveer, P. Conventionalization of the organic sesame network from Burkina Faso: Shrinking into mainstream. *Agric. Hum. Values* **2013**, *30*, 539–554. [CrossRef]

19. Mol, A.P.J. Carbon flows, financial markets and the challenge of global environmental governance. *Environ. Dev.* **2012**, *1*, 10–24. [CrossRef]

20. Newell, P.; Paterson, M. *Climate Capitalism-Global Warming and the Transformation of the Global Economy*; CUP: Cambridge, UK, 2010.

21. Raadal, H.L.; Dotzauer, E.; Hanssen, O.J.; Kildal, H.P. The interaction between Electricity Disclosure and Tradable Green Certificates. *Energy Policy* **2012**, *42*, 419–428. [CrossRef]

22. Spaargaren, G.; Mol, A.P.J. Carbon Flows, Carbon Markets and Low-Carbon Lifestyles. *Environ. Polit.* **2013**, *22*, 174–193.

23. Van Riel, M.C.; Bush, S.R.; van Zwieten, P.A.M.; Mol, A.P.J. Understanding fisheries credit systems: Potentials and pitfalls of managing catch efficiency. *J. Environ. Policy Plan.* **2015**, *16*, 453–470.

24. Fuchs, D.; Kalfagianni, A.; Havinga, T. Actors in private food governance: the legitimacy of retail standards and multistakeholder initiatives with civil society participation. *Agric. Hum. Values* **2011**, *28*, 353–367. [CrossRef]

25. Castree, N. Neoliberalising nature: Processes, effects, and evaluations. *Environ. Plan. A* **2008**, *40*, 153–173. [CrossRef]

26. Commission of the European Communities. *Annex to the Impact Assessment. Document Accompanying the Package of Implementation Measures for the EU's Objectives on Climate Change and Renewable Energy for 2020*; SEC: Brussels, Belgium, 2008.

27. Pacini, H.; Silveira, S.; da Silvo Filho, A.C. The European Biofuels Policy: From where and where to? *Eur. Energy J.* **2013**, *3*, 17–36.

28. RSPO. *RSPO Supply Chain Certification Standard-Final Document*; RSPO: Kuala Lumpur, Malaysia, 2012.

29. Staaij, J.; van den Bos, A.; Toop, G.; Alberici, S.; Yildiz, I. *Analysis of the Operation of the Mass Balance System and Alternatives*; Ecofys: Utrecht, The Netherlands, 2012.

30. Bullock, D.S.; Desquilbet, M. The economics of non-GMO segregation and identity preservation. *Food Policy* **2002**, *27*, 81–99. [CrossRef]

31. Manning, L.; Soon, J.M. Developing systems to control food adulteration. *Food Policy* **2014**, *49*, 23–32. [CrossRef]

32. Greenpalm. Market Volume and Price Charts. 2015. Available online: http://greenpalm.org/the-market/market-overview/market-volume-and-price-charts (accessed on 1 September 2015).

33. McDonalds. Our Journey to Sustainable Beef. 2015. Available online: http://www.aboutmcdonalds.com/mcd/sustainability/signature_programs/beef-sustainability.html (accessed on 3 September 2015).

34. *Traceability and Market claim Working Group and Steering Committee Traceability Delivered. A Strategic Recommendation for Credible and Cost-Efficient Supply Chain Traceability and Labelling Systems in the Soy Supply Chain*; RTRS: Buenos Aires, Argentina, 2010.

35. Mol, A.P.J. Bounded biofuels? Sustainability of Global Biogas Developments. *Sociol. Rural.* **2014**, *54*, 1–20. [CrossRef]

36. Fiorillo, J. Are the World's Retailers and Restaurants Delivering on their Sustainable Seafood Promises? Available online: http://seafoodinternationaldigital.com/are-the-worlds-retailers-and-restaurants-delivering-on-their-sustainable-seafood-promises (accessed on 3 September 2015).

37. Byrne, J. New CSPO deal means palm oil certs no longer needed, says NBPOL. 2011. Available online: http://www.confectionerynews.com/Commodities/New-CSPO-deal-means-palm-oil-certs-no-longer-needed-says-NBPOL (accessed on 15 April 2014).

38. Mol, A.P.J. Ecological modernisation and institutional reflexivity. Environmental reform in the late modern age. *Environ. Polit.* **1996**, *5*, 302–323. [CrossRef]

39. Richardson, B. Making a Market for Sustainability: The Commodification of Certified Palm Oil. *New Polit. Econ.* **2015**, *20*, 545–568. [CrossRef]

40. Book & Claim. Trade Certificates Support the Environment. 2014. Available online: http://www.bookandclaim.co.uk (accessed on 19 May 2014).

41. Beall, E. *Smallholders in Global Bioenergy Value Chains and Certification-Evidence from Three Case Studies, Environment and Natural Resources Working Paper No.50*; FAO: Rome, Italy, 2012.

42. Dallinger, J. Oil palm development in Thailand: Economic, social and environmental considerations. In *Oil Palm Expansion in South East Asia: Trends and Implications for Local Communities and Indigenous Peoples*; Colchester, M., Chao, S., Eds.; Forest Peoples Programme/Perkumpulan Sawit Watc: Bogor, Indonesia, 2011; pp. 24–51.

43. INTERPOL. *Guide to Carbon Trading Crime*; Interpol Environmental Crime Program: Lyon, France, 2013.

44. Nellemann, C. *Green Carbon, Black Trade: Illegal Logging, Tax Fraud and Laundering in the World's Tropical Forests*; GRID: Arendal, Norway, 2012.

45. Neuendorff, J.; Fischer, U. Maintaining organic integrity: Tackling fraud in organics. In *Quality Management in Food Chains*; Theuvsen, L., Spiller, A., Peupert, M., Jahn, G., Eds.; Wageningen Academic Publishers: Wageningen, The Netherlands, 2007; pp. 209–217.

46. Shears, P. Food fraud—A current issue but an old problem. *Br. Food J.* **2010**, *112*, 198–213. [CrossRef]

47. Partzsch, L. The legitimacy of biofuel certification. *Agric. Hum. Values* **2011**, *28*, 413–425. [CrossRef]

48. UTZ. Cocoa. 2014. Available online: https://www.utzcertified.org/ (accessed on 14 October 2014).

49. Rainforest Alliance. Rainforest Alliance Controlled Blending. An Insider's Look at Cocoa Certification. 2014. Available online: http://www.rainforest-alliance.org/agriculture/faq-controlled-blending (accessed on 14 October 2014).

50. Sustainable Agriculture Network (SAN) Secretariat. *List of Permitted Mass Balance Products*; SAN: San José, Costa Rica, 2012.

51. Greenpalm. Record demand in 2014 for RSPO certified palm oil and palm kernel oil. 2014. Available online: http://www.greenpalm.org/en/blog-press/blog/record-demand-in-2014-for-rspo-certified-palm-oil-and-palm-kernel-oil (accessed on 14 October 2014).

52. RTRS. Scope of the Overall Supply Chain. 2015. Available online: http://www.responsiblesoy.org/en/certification/tipos-de-certificacion/cadena-de-custodia/ (accessed on 3 September 2015).

53. RTRS. *RTRS Use of the Logo & Claims Policy*, 4th ed.; RTRS: Buenos Aires, Argentina, 2014.

54. Bonsucro. *A Guide to Bonsucro*; Bonsucro: London, UK, 2013.

55. Bonsucro. Bonsucro Credit Trading System. 2014. Available online: http://www.bonsucro.com/credit_trading_system/ (accessed on 14 October 2014).

56. European Commission. Voluntary schemes. 2015. Available online: http://ec.europa.eu/energy/en/topics/renewable-energy/biofuels/voluntary-schemes (accessed on 15 April 2015).

57. Scarlat, N.; Dallemand, J.-F. Recent developments of biofuels/bioenergy sustainability certification: A global overview. *Energy Policy* **2011**, *39*, 1630–1646. [CrossRef]

58. Meyer, S.; Schmidhuber, J.; Barreiro-Hurlé, J. *Global Biofuel Trade. How Uncoordinated Biofuel Policy Fuels Resource Use and GHG Emissons*; ICTSD: Geneva, Switzerland, 2013.

59. Laurent, B. The politics of European agencements: Constructing a market of sustainable biofuels. *Environ. Polit.* **2015**, *24*, 138–155. [CrossRef]

60. Global Roundtable for Sustainable Beef. *Draft Principles & Criteria for Global Sustainable Beef*; GRSB: Geesteren, The Netherlands, 2014.

sustainability

MDPI

Article

Capitalism with a Human Face: Debates on Contemporary Globalization and Sustainability

Chua Yuhan and Md Saidul Islam *

Division of Sociology, Nanyang Technological University Singapore, 14 Nanyang Drive, Singapore 637332, Singapore; YCHUA014@e.ntu.edu.sg
* Correspondence: msaidul@ntu.edu.sg

Abstract: This paper aims to explore the possibilities of a humane form of capitalism—one that allows for the pursuit of environmental and social justice, without significant curtailment of economic growth. The nature of capitalism has been explored through a discourse analytical framework, with three main discourses being identified—the Neoliberal perspective, the Neo-Marxist perspective and the Sustainable Development discourse. This paper argues that the Sustainable Development discourse that frames the economy of certain countries has been significantly influenced by the Ecological Modernisation Theories, and such conflation of discourses has problematized the practice of capitalism in certain regions. Such a proposition has been explored through the use of the Singapore case study, and various socio-economic and environmental policies and indices have been examined, in order to determine if this specific brand of sustainable capitalism could be humane in nature. Detaching from the localized evaluation of sustainability in Singapore, the concept of a sustainable society within the context of contemporary globalization is also explored briefly, with it revealing greater insights into the true nature of capitalism.

Keywords: capitalism with human face; sustainable capitalism; discourse analysis; Singapore

---

1. Introduction

Capitalism has been the dominant economic system for most nation states in the past few centuries. Emphasizing the concept of the free market, capitalism posits that the most socially efficient division of resources, should in theory, come about naturally from the self-allocative mechanism of the market [1]. This should result in a fair form of economic growth, bringing about the most socially beneficial outcome for members of the society [1]. However, since its inception, critics have pointed out the innate contradictions present within the system, and capitalism has been attributed as the source of the social inequality, injustice, and environmental degradation observed in society [2,3]. Such discourses have arguably influenced the manner in which capitalism is perceived, and in turn shaped the various economic policies and systems in various countries.

These two opposing perspectives could loosely be categorized under the Neoliberal discourse, which largely support free market capitalism, and the Neo-Marxist discourse, which largely disagree with such an economic system. Between the two discourses, lies the Sustainable Development Discourse, which arguably takes a more nuanced stance in the argument. Such a discourse recognized the many problems present in the current capitalistic system, but puts forth the notion that capitalism is flexible enough to be pushed towards an economic model that allows for the co-development of the social, economic and environmental spheres, a form of humane capitalism that is of interest in this paper [4]. The sustainable development discourse has arguably gained relative dominance in the shaping of policies within various countries, most notably in the Scandinavian regions such as Sweden, as well as various smaller nation states such as Singapore, and has in turn shaped the nature of capitalism within these places.

This paper aims to explore the possibilities of a humane form of capitalism, and to determine if the sustainable development discourse has sufficiently altered the economic systems of various purported sustainable states to a form of capitalism with a human face. This objective has been achieved through the use of a discourse analytical framework, applied on to the specific case study of Singapore—a country that has a history of active pursuit towards sustainable goals. The examination of the Singapore case study could allow for greater insights into the nature of capitalism, allowing us to determine if capitalism could be truly humane in nature.

The first section of the paper will put forth the aims of the paper, as well as highlight the limitations present in the study. The second section will briefly review the various literature present in the study of capitalism, as well as highlight some rationales for the study. The third section will highlight some key definitions and concepts, giving a brief explanation on the concept of capitalism, as well as what constitutes capitalism with human face—a term that is of main focus in the paper. The fourth section would discuss the methodologies used in the study, and explain how and why the study has been carried out in a certain manner. The fifth section will discuss the three main discourses of capitalism, identified as the Neoliberal discourse, the Neo-Marxist discourse, and the Sustainable Development discourse. Various sub discourses will also be discussed, with emphasis placed on the Environmental Modernisation Theories (EMT)—a framework that has arguably influenced the form of sustainable capitalism currently present in the Singapore economy. Such a claim will be explored in details in the following section of the paper. This sixth section will examine the Singapore case study, wherein the dominant discourse influencing the Singapore's capitalism, would first be identified, before discussing if such a form of capitalism could provide for the humane type of economic system that is of interest in this paper, especially when the context of globalization is taken into account. The last section of the paper will then give a brief conclusion, and summarize the various points posited in the earlier sections.

1.1. Research Aim

(1) To explore the contemporary discourses surrounding the concept of sustainability, and through it provide insights into the nature of capitalism
(2) To apply an environmental sociological lens on the study of Singapore's brand of sustainable capitalism, and provide an evaluation of its efficacy and its nature, taking into account its context in a globalizing world.

1.2. Limitations

Before delving into the main section of paper, it is necessary to put forth some of the limitations of the paper. Due to the limited scope of the paper, wherein only one single case study was examined, the study could only provide an exploratory view of one specific form of capitalism. Various models of sustainable capitalism exist in different regions and through different times. Findings from this study therefore cannot be used to generalize and provide a conclusive statement on the overall nature of Capitalism. Nevertheless, such a study can still be useful in providing some insights into specific forms of sustainable capitalism, and perhaps suggest some hints of its true character. The employment of the case study has provided some empirical evidence to this aspect, with findings highlighting the complexities of sustainable capitalistic system.

The study has also placed more focus on the internal dynamics of Singapore's economy, with less focus placed on its relations with other world economies. As such even if Singapore's form of capitalism is demonstrated to be humane in nature, such a conclusion could only be posited within the confines of the country. When placed within the context of a larger global economy and society, Singapore's economic system could still have certain exploitative aspects. Due to the need to keep brevity and focus for this paper, this claim has only been briefly discussed, and future studies could perhaps expound more on this perspective.

2. Literature Review

Much literature have discussed and examined the nature of capitalism, since its conception. Three main arguments can be identified, loosely categorized under the Classical and Neoliberal perspectives, the Marxists and Neo-Marxists criticism on capitalism, and the Sustainable Development views [1–5]. These discourses vary in their outlook on capitalism, arguably forming a theoretical debate on their perceived effects of capitalism on society and the environment. These discourses have arguably shaped and influenced policy decisions and economic practices in various societies, and across different epochs. Various studies have examined these various discourses in details, and applied them to case studies of variants of capitalism practiced in different regions, with findings providing different interpretations of the nature and character of capitalism [6–8].

However, few studies have recognized that the various discourses could have co-varying effects on social and economic policies in different economies. In this sense, capitalism in praxis, could exhibit evidence of exploitation and sustainability at the same time. This paper thus places more focus on this issue, and aims to examine how a conflation of discourses, could exist in certain cases, such as the Singapore case study, providing more insights into the nature of capitalism, and to allow some indication on whether capitalism could indeed have a human face.

3. Conceptual Threads

3.1. Capitalism

In order to delve into the nature of capitalism, it is necessary to first conceptualize and put forth clearly an exact definition of capitalism. Various definitions of capitalism has been put forth in previous literature, but such definitions have been relatively unclear in identifying the exact nature of capitalism. Attempts to shed light on this concept has been put forth by Sternberg [9], who offered an essentialist definition on this form of economic system—identifying capitalism by its three main features—that is the use of "private property, free-market pricing, and the absence of coercion".

Within this operational definition, only one form of capitalism exist. Other variants—such as state capitalism—are seen as deviants, attributed as the source of economic failures contributing to problems within society [9]. Problems traditionally associated with capitalism are deemed to be an effect of attempts to control the economic system [9], and capitalism in itself is placed in a positive light—as a panacea to problems located in the economy and in society.

While such a definition is indeed useful in delineating the scope of capitalism, providing a clear explanation of capitalism as an economic system, such a definition fails to recognize that capitalism is ultimately a system embedded within a larger social and political system [10]. In praxis, the essentialist definition of capitalism is ultimately purely theoretical, and capitalism is and will always be affected by social and political forces. Several assumptions that contribute to the proper workings of the system more often tend to fail, such as the assumption of the rational nature of humans in carrying out economic actions in the market. As such, this conception of capitalism is limited in its understanding of capitalism, providing only a theoretical examination of the concept, and thus is less relevant in the examination of capitalism within this paper, which seeks to explore into the nature of capitalism in praxis.

Recognizing that capitalism is ultimately an economic system embedded within a larger social and political system, this paper seeks to stray away from the essentialist definition of capitalism, instead taking the stance that various forms of capitalism can exist, in different societies and different timelines. An operational definition would thus have to take into account the variation in practices for capitalism in different societies, while at the same time delineating a clear boundary for which capitalism could be distinguished from other forms of economic systems, such as socialism.

While there are several variants of capitalism present both historically and through different societies, their main commonalities could be distilled into the systems that display the follow characteristics:

(1)   The use of the market as the main redistributive mechanism
(2)   Concept of private property

As such, capitalism would be defined, in this paper, as an economic system, where the main pricing mechanism is derived from the workings of the market, and the use of private property is upheld within the system.

The first feature represents the use of the free market as the main pricing mechanism — that is to allow the demand and supply to dictate the price and allocation of resources within the system — otherwise known as the "invisible hand" of the market [1]. Within this mechanism lies the belief that the most social good can be derived from the rational pursuit of self-interests by all individuals of society [1].

The second feature refers to the notion that property are privately owned by individuals — that is, a "property right entitles its holder to a strong form of authority over an asset, called ownership" [11].

Sternberg's third feature of capitalism — the lack of coercion — is omitted from this definition, as in praxis, few or no forms of capitalism display a lack of coercion within its system [9]. Strictly speaking, this would represent a society without any form of governance, which would be highly improbable in praxis. As this paper is more concerned with the forms of capitalism present in societies, this feature, while vital in helping to defend capitalism, would be irrelevant in the form of capitalism investigated within this paper.

Such a definition would be sufficient to differentiate capitalism from other forms of economic systems, while at the same time, flexible enough to allow for variants of capitalism to be examined, allowing one to look into different forms of the economic system, and to investigate if, out of the many, a form of capitalism that is both socially, economically, and environmentally sustainable, could exist. Given such a definition, the case study chosen for the research, Singapore, could then be determined as adhering to a capitalistic economic system, wherein the main pricing mechanism lies within the market, and the capitalist concept of private property remains present within the society.

3.2. Capitalism with a Human Face

The main objective of the paper is to investigate if a humane form of capitalism could exist in praxis — a term described as capitalism with human face. Before moving on to discuss this in detail, it is necessary to clearly define what constitute a humane form of capitalism. Taking insights from the sustainability discourses, a humane form of capitalism should exhibit evidence of an equal development of not just the economic aspects of society, but the social as well as environmental aspects [4]. Within the social sphere, justice should be observed, which in this paper, would be conceptualized as the meeting of needs of individuals within society [12]. A socially justified system would thus be required to address problems of material inequality, and built on a system which allows for equal access to resources, where basic needs such as food and healthcare are made accessible to the masses [13]. Economic development should also come in tandem with environmental stability — a form of sustainable development that "meets the needs of the present without compromising the ability of future generations to meet their own needs" [14].

4. Methodologies

4.1. Case Study Research

The main research strategy employed in this study is the case study method, which is an "empirical inquiry that investigates a contemporary phenomenon within its real-life context" [15]. What this essentially means is that the case study method takes into account the current social conditions in which the event or object of study is embedded in [15]. This research strategy would take data from a multitude of sources — and thus providing multiple perspectives on the same phenomenon [15,16].

The usage of case studies as is more preferable under three conditions — the first of which has already been discussed in the previous paragraph. The two remaining conditions revolves around the nature of the research question, as well as the amount of control over "behavioural events", or conditions that influence the object or phenomenon of study [15]. In order for the case study method to seen as a viable research strategy within a study, its research question should revolve around why and how questions, or what questions, if the study is an exploratory one, and seek to examine contemporary events where context cannot be divorced from the phenomenon, and where the researcher has little control over the variables and events that lead to the phenomenon [15].

The use of the case study method is thus appropriate within this study, fulfilling the three conditions laid out in the previous section. The main research question of the paper is, is for the most part, an exploratory one, and seeks to understand if capitalism could be a humane form of economic system—thus fulfilling the first condition of case study research. However, other forms of strategies could also be employed if the nature of the study is an exploratory one, and historical, surveys, experiments could similarly be used in the study, if solely based on the first criteria [15]. What then makes the use of the case study the most appropriate method for this study, is the fulfilment of the other two conditions mentioned in the previous section. This research examines capitalism in contemporary societies, thus requiring the examination of the context in which it is in embedded in, while at the same time, the large range of variables and conditions that influence the object of interest does not fall within the control of the researcher—thus requiring the use of the case study method for this research.

According to Yin [15], there are five components to consider during the design phase of the case study research, which are the following: (1) Study's questions, (2) Study propositions; (3) Unit of analysis, (4) Logic linking data to propositions and the (5) Criteria for interpreting the findings. The first component has already been addressed in the earlier section, wherein the question points towards one that is exploratory in nature. The second component examines the specifics of the question—and directs what should be studied in the research [15]. Due to the exploratory nature of this paper, only the purpose of the study would be stated—which is to investigate the nature of capitalism, and to determine if capitalism could be adopted as the main economic framework within a sustainable society.

The unit of analysis in this study would be Singapore. The selection of this case study is intentional—Singapore adhere strictly to a capitalistic model, but seems to exhibit traces of sustainability, and an investigation into its economy could provide greater insight into the nature of capitalism, and determine if capitalism could have a humane side to it. Within the case study, data with regards to the three spheres of sustainability—economic, social and environmental—would be collected, to determine if these economies are truly sustainable in nature. Indicators for the three spheres could take in the form of statistical data, such as the Gross Domestic Product (GDP) of the country, Gini Coefficient, as well as other environmental indicators; but could also be determined through an analysis of economic, social and environmental policies present in the two cases. The two components refer to the data analysis part of the research, which in this case, would be aided by prior discourses on sustainability—and theories such as the Treadmill of Production, and the Ecological Modernisation Theory, would be used to link the data to the proposition, as well as to interpret the findings.

4.2. Discourse Analysis

A discourse analytical framework will be applied in investigation of the nature of capitalism in this study. Discourse refers to "a specific ensemble of ideas, concepts and categorizations that are produced, reproduced and transformed in a particular set of practices and through which meaning is given to physical and social realities" [17]. The analysis of discourse thus involve itself with the examination of the various ideas that surround the concept at hand—which in this case, would be capitalism—and investigate how these various discourses translate to reality and social action, through its influence on institutional arrangements and practices within society [17]. The use of a discourse analytical framework assumes capitalism as a more fluid and dynamic concept—one that is constantly shaped by various discourses—a system that is constantly being constructed and reconstructed, and moulded to variants of capitalism seen in various societies. Such a framework thus recognizes that capitalism could be portrayed in various forms, and a study of its physical manifestations in various economies, should provide an empirical inquiry into the nature of capitalism.

Discourse analysis has been carried out in a two-step process, wherein the various discourses surrounding the concept of capitalism were first identified, before proceeding to discuss the dominant discourses shaping the Singapore's brand of capitalism.

5. Discourses Surrounding Capitalism

Three discourses can be identified in the discussion on the nature of capitalism. The first of which follows a liberalist tradition—taking the perspective that capitalism, and its free market, provides a most socially optimal method of resource allocation, and thus, in its purest form, is humane in nature [1]. The problems that are often attributed to capitalism—those of social inequality and environmental degradation—are then seen as problems arising not from capitalism itself, but from state interventions or market distortions, and could be remedied with increased liberalization of the market [9].

The second discourse takes on the opposing stance—and sees capitalism as an innately exploitative mode of production [2]. Proponents of this discourse views capitalism as an inherently problematic system, in which a humane form of capitalism is seen as unachievable, and paradoxical in this case. The problems observed in contemporary societies—such as extreme social inequality, as well as widespread environmental degradation, could then be attributed to the capitalistic mode of production [3,18,19]. This discourse hence seeks to push for more radical changes in existing institutions, in order to address the many issues, perceived to be a result of the capitalistic economic system [3].

The third discourse takes a more nuanced approach towards capitalism. While acknowledging the problems associated with the current form of capitalism, proponents of this discourse take the stance that solutions to these problems could be attained without radically changing the current mode of production [20]. Capitalism is thus seen as a flexible system, one that is adaptable enough to allow for more sustainable development [20]. Such a discourse would thus shape society to continue to pursue economic development through capitalistic means, in the belief that with slight modification, capitalism could be moulded into a humane system, which promotes economic growth, without having negative impact on the social and environmental aspects of the country [20].

While there exists a large range of literature that could be categorized into the three different discourses identified, the limited space of this paper, and the need to keep brevity, only allows for the discussion of only certain key theories and perspectives—such as those by Adam Smith and Karl Marx. These two theorists essentially adhere to the first and second discourses on capitalism respectively, with Smith advocating for a liberalist approach, and taking on the view that capitalism is humane in nature, while Marx takes on the opposing stance, viewing capitalism as an inherently problematic and inhumane system. Both theorists arguably provided a basic framework of discussion in more contemporary perspectives on capitalism, which would be expounded on in the later section.

5.1. Classical Liberalism and Neoliberalism

5.1.1. Adam Smith & the Invisible Hand of the Market

Classical Liberalism and Neoliberalism perspectives could be subsumed under the first discourse, taking a positive position with respect to the ethicality and morality of the free market. The Classical Liberalism perspectives was first put forth by 18th century philosopher—Adam Smith—in his seminal work of An Inquiry into the Nature and Causes of the Wealth of Nations [1].

Smith puts forth a critical view on the mercantilist economic system that was in practice in the 17th to 18th century, viewing the protectionist measures implemented by different state economies as an impediment to the workings of the market [1]. Instead, Smith introduced the concept of the "Invisible Hand", a metaphor used to describe the self-distributary mechanism of the free market [1]. Such a view is built on the assumption that the self-interests of individuals will result in the most socially optimal outcome—a process that can only occur in a laissez-faire economy [1]. Such an economy is characterized by the lack of intervention from the state, or the absence of severe distortions in the free market, such as the presence of monopoly—and instead made up of small business owners and individuals, who do not have significant influence on the workings of the market individually [1]. The invisible hand of the market will then, in theory, act as an indicator of pricing and production levels, and a result in a situation in which there would be a socially efficient allocation of resources [1].

While Smith advocates for the establishment of a free market—one that is free from state intervention—his conception of capitalism does not push for a completely unfettered market—and the state continues to play certain roles in this system. In Smith's conception of capitalism, the role of the state should be limited to three responsibilities—(1) Defense; (2) Upholding justice; and the (3) Provision of public works [1,21]. The first responsibility lies in the protection of the state and country from physical threats, such as foreign invasion [1]. The second and third responsibilities refer to the provision of public and merits goods—goods that are required, or beneficial to society, but would not be financed through the workings of the free market—such goods include transportation infrastructure, or education [1,21]. Nevertheless, Smith's conception puts forth an optimistic view of capitalism, wherein social justice—a key part of a humane economic system—is seen as a natural product of the free market, thus arguably pushing for the belief that capitalism is essentially humane in nature.

5.1.2. Neoliberalism

The Neoliberal perspective on capitalism is essentially a revival of classical liberalist ideas, but updated to the global economy—and focuses on the worldwide spread of free market and free trade ideologies [22]. Neoliberalism stays true to its liberalist roots, requiring the minimization of government intervention in the economy, with its role limited to the establishment and enforcement of free markets within the world economy [23]. The eventual outcome of neoliberalism was the establishment of a world market—which should, in theory, stimulate economic growth on a global scale [22].

Such a discourse first gained prominence in the 1980s and 1990s, after the failure of Keynesian economics in addressing major economic downturns observed in the 1970s, and could be identified in two distinct waves [23,24]. The first wave occurred in the 1980s, wherein Neoliberal and Monetarist ideas began to replace the previously Keynesian style economy practiced in the United States and the United Kingdom [24]. The second wave was started in the 1990s, continuing the neoliberal ideas of market liberalization and deregulation, but with an added agenda of addressing social and environmental issues previously ignored in the 1980s [24].

Similar to the Classical Liberalist views, Neoliberalism adopts the same positive stance towards capitalism, and such discourse thus aims to shape economies to move towards a more liberal market—a market that has minimal intervention—in the belief that such a structure of the economy could result in a more socially efficient division of resources—in a sense, a more humane type of economic system.

5.1.3. Criticism of Neoliberalism

Neoliberalism has arguably been the dominant discourse for the past three decades, especially in leading economies such as the United States and the United Kingdom, but has been met with a considerable amount of criticism, since its implementation. While neoliberalism continues the classical liberalist stance that capitalism, and its free market, would provide the most socially optimal allocation of resources, and thus result in the well-being of all in society, the implementation of neoliberal principles have been criticized for the creation, and intensification of social inequality on a global scale—resulting in a process of "accumulation by dispossession" [24,25]. According to this perspective, neoliberalism merely redistributes wealth from the poor, or poorer nations, to the global elites—a process facilitating social inequality—and thus is inhumane in nature [25].

Other critiques of neoliberalism has dwelled on the problems of the unfettered markets—and the neoliberal perspective on capitalism has been attributed with an array of environmental crises observed in contemporary times—a perspective that will be discussed in detail in the following section.

5.2. Marxists and Neo-Marxism

5.2.1. Marx & Capitalism

While Smith views capitalism and its free market system as the most socially optimal arrangement, Marx takes the opposite stance—viewing capitalism as an innately exploitative system. The exploitative nature of capitalism is illustrated in his Theory of Surplus Value, which assumes that value is derived from labour—which has the ability to generate surpluses [2]. This surplus is however, appropriated by the capitalists, who own the means of production, and then rechannelled into production [2]. The exploitative nature of this form of economic system, will eventually result in a contention between the two classes—the capitalists owning the means of production—the bourgeoisie—and the labourer—the proletariats—and eventually leading to a the collapse of the system [26].

The class struggle between the proletariats and the bourgeoisie is not the only problem associated with the capitalist system of production—other contradictions embedded within capitalism will also generate problems which would eventually result in its downfall, such as the problem of overproduction observed within capitalism. In order to maintain a constant stream of capital accumulation, there requires a need to constantly expand the markets, either through the establishment of new industries, or creation of new wants [26,27]. When the high rate of production cannot be met, a problem of overproduction occurs, resulting in economic crises—which would exacerbate the class division mentioned in the previous section—where failed business owners would eventually fall into the proletariat class [26].

The form of capitalism that Marx envisioned is a problematic one—one that is innately exploitative, and thus in some sense, inhumane, and one in which is unsustainable, due to the eventual collapse of the system—a result of its internal contradictions. Such a perspective on capitalism differs greatly from that of Smith, standing on opposite spectrum on their conception of capitalism.

5.2.2. Treadmill of Production

The Treadmill of Production follows the Marxist tradition on its views on the exploitative nature of capitalism—but expands its scope to address more contemporary problems such as environmental crises on a global context [3,8]. It begins with the narrative that economic expansion in capitalistic societies, is in direct relation to the widespread environmental damage and social inequality seen in modern societies today [3].

Within this discourse, the capitalistic mode of production is observed to adhere to its own unique set of logic—one that is based on economic principles, instead of ecological ones [3]. Humans, like all other species, are consumers, taking resources from the environment for its own survival [3]. However, what distinguish humans from other species is that humans not only consume, but organize production within our own species—or what Schnaiberg labels as sociocultural production [3]. Such a form of production subverts ecological logic for an economic logic, which eventually leads to the devastation of the environment [3].

In a natural ecosystem, the presence of surpluses would be absorbed by the alpha organism— that is, the species at the top of the food chain [3]. This would lead to an initial growth of the alpha organism, leading to increased consumption, which would then naturally lead to a stagnation, or decline in growth due to the reduced resources in the ecosystem—essentially a self-limiting system [3]. The presence of surplus in an ecosystem embedded with man's production system, especially those of a capitalist nature, is instead rechanneled into production, which in turn leads to more surplus accumulation [3]. Due to this rechanneling of surpluses into production, an increased rate of resource depletion would occur within the ecosystem—but due to man's ability to operate across multiple ecosystems, the natural self-limiting principle present in the environment no longer applies to the production system [3]. The economic logic of capitalistic economic system essentially result in an endless expansion of production, and when the limits of an ecosystem is reached, man simply moves on to the next ecosystem—leading to the devastation of the environment, which is unsustainable in nature [3].

While human societies are embedded within the earth's natural ecosystems, its mode of production differs from that of nature, and do not adhere to the law of ecology [3]. What this essentially means is that while consumption in conventional ecosystems are confined by the natural limits of their environment, the capitalistic mode of production subverts this logic through operating across multiple ecosystems, and adheres not to the laws of nature, but to the laws of economy—that is the accumulation of surpluses [3].

In the modern industrial system, ecological disruptions occur at a heightened rate, due to the increased rate of production and consumption [3]. This modern industrial system is characterized by a system of interchange, where natural resources are withdrawn from the environment, transformed into commodities, before being returned back in to the environment through additions as waste [3]. This constant extraction of resources and addition of waste has led to widespread environmental and ecological degradation, placing society on a path of unsustainability due to Earth's finite resources as well as carrying capacity for waste [3]. This system of addition and extraction is facilitated by technology, and technological advancements is observed to be linked to increased additions, of increasing risk, to the environment [3]. Schnaiberg hence concluded that "(e)very technology, if widely practiced enough, will generate environmental problems"—a view that contrasts strongly against the perspectives put forth by the Ecological Modernisation Theory [3].

The treadmill of production is also characterized by a system of social exploitation, a view that stems from its Marxist traditions. Cheap labour fuels this system of economic production, and various social institutions, such as labour unions, education systems, the state, facilitates this process of social and environmental exploitation—and radical structural changes are required to detach from this system of exploitation [3].

5.2.3. Criticisms of the Treadmill

The main issue of the treadmill of production, is its overly pessimistic view on capitalism— wherein the only solution for the problems associated with the current form of economic system, is a radical reform of socio-political and economic institutions [24]. While such a perspective may be useful in identifying the main source of social and environmental problems in contemporary societies, its pessimistic nature, and its penchant for radical changes, arguably prevents it from entering the elite discourse, and thus would not be as useful in pushing for the change it ironically requires. Such a discourse has arguably failed to gain traction in the different economies—and hence remain embedded in theory, instead of influencing institutional change for the pursuit of social and environmental justice.

5.3. Sustainable Development and Ecological Modernisation

5.3.1. Sustainable Development

The third discourse revolves around the modification of capitalism to address the various social and environmental issues that were associated with liberal and neoliberal capitalism—and stays firm to the belief that sustainable development, under capitalism, is a possibility, albeit with some changes. Proponents of sustainable development thus push for the implementation of policies, in which a balancing of the three goals of sustainability—economic, social and environmental—could be achieved, without radical changes in the current economic system [4].

While this may seem similar to what is proposed under neoliberal principles, sustainable development differs in the sense that some form of intervention is required to modify and restructure the market, to push it towards certain goals [28]. When translated to practice, state interventions and community efforts are often present to uphold social justice, usually through the implementation of redistributive policies, as well as the active promotion of green industries, to push for the cause of sustainable development [4,28].

5.3.2. Ecological Modernisation Theories

Within the various perspectives within sustainable development, the Ecological Modernization Theory (EMT) stands out as one of the few paradigms that offer a practical solution to the problems associated with modern day capitalism. EMT maintains a positive outlook on capitalism, proposing that capitalism itself is flexible enough a system, and could be molded into a sustainable economic system, with the aid of technology or through institutional changes. Such a perspective fits under the third discourse, of the belief that capitalism could be a humane economic system, wherein sustainable development could be achieved under it.

Through a process of super-industrialization, the industrial system could be transformed into one that is ecologically sound, hence limiting the environmental degradation caused to the Earth [29,30]. While initially emphasizing on the role of technology in aiding environmental and economic reforms, Ecological Modernization theorists have expanded to discuss the roles of other social institutions in the facilitating a green transformation of the economy [20]. Several variants of Ecological Modernization exists as well, such as the State-Led Ecological Modernization suggested by Wong [6]. Nevertheless, the commonality of these vairants of EMT is the implicit faith that capitalism could be altered towards a system that is sustainable, and humane in nature.

While EMT and the Sustainable development discourse arguably stands on the same position, with respect to its view on capitalism, and thus could be grouped under the same discourse, some differences do exist between these two perspectives. The main focus of EMT has been traditionally been placed on the environmental aspects, with little emphasis on the social justice aspect—one of the three goals put forth by proponents of the sustainable development discourse [4,5,31]. The similarity of these two perspectives have arguably problematized the structure of sustainable economies in certain countries—which could have profound effects on the sustainability of capitalistic economic system in purported sustainable societies [31]. The problem of a conflation of both perspectives have been highlighted in the Singapore case study—an issue that will be discussed in the later sections of the paper.

Nevertheless, the similar position that EMT and the Sustainable Development discourse takes on, is that of an optimistic perception of capitalism—tending to push for institutional or technological changes in existing capitalistic system, in an attempt to pursue a more humane society—or to put forth a type of capitalism with human face.

5.3.3. Criticisms of EMT

The main criticism of EMT is that too much optimism and too much faith is placed on the role of technological change, in solving the many social and environmental problems observed in society today [24]. Technology that were meant as solutions to certain ecological problems, such as the use of nuclear energy as a clean form of renewable energy, often carry with them their own risks and undesirable effects [24]. As such, more evidence may be required in order to determine if sustainable development and EMT is truly sustainable in nature—which would also help to provide greater insights into the nature of capitalism.

5.3.4. Comparison of Discourses

The various discourses compete for dominance in constructing perceptions on the capitalistic economic system, and take different positions in their views on the role and type of institutional arrangements that should or should not be present, in the pursuit for a humane and sustainable society. The key features and differences between the discourses have been distilled in Table 1 below.

Table 1. Comparison of Discourses.

	Neoliberalism	Neo-Marxism & Treadmill of Production	Sustainable Development and EMT
View on Capitalism	Capitalism is a socially progressive economic system—the presence of social problems is a result of insufficient market liberalization	Innately problematic, requires radical structural changes to capitalistic economic structures to remedy the environmental crisis	Flexible nature allows for the movement towards sustainable development, without altering the capitalistic system significantly
Technology	Technological development is required for greater efficiency and productivity	Accelerates environmental degradation through increased withdrawals and additions and perpetuates social in equality by facilitating worker exploitation	Key institution which allows for the transition towards a sustainable capitalism; through green technology
Social Institutions (State)	Advocates for minimal state intervention, and market should be kept as free as possible	Social institutions facilitate the treadmill of production—state promotes this treadmill for its own causes	Institutional changes can facilitate the transformation to a green economy—state intervention may be required in this process
Modernization	Modernization based on capitalist free market is needed in order to get rid of traditional baggage	Modernization and development brought about the environmental crisis, and only perpetuates it	More modernization is required for transition towards a greener economy/sustainable society
Criticism	Perpetuates inequality on a global scale; and even if proven to be beneficial, free markets could not be attainable in praxis	Economically deterministic approach towards environmental problems; overly pessimistic	Overly optimistic about the power of technology—clean technologies in the past have proven to carry its own environmental risks (nuclear, silicon chip)

Adapted from [1–5].

5.4. Identification of Discourses

Amongst the three discourses, the neoliberal and the sustainable development discourse arguably receives the most reception in the elite discourse, and its ideas often shaped the realpolitik of various economies and societies. While the treadmill of production presents a rather convincing case against capitalism, and offers a deep insight into the origins of the various social and environmental problems observed in society today, its pessimistic outlook, as well as its need for radical changes in various social institutions, arguably prevents it from entering the elite discourse—and hence only useful as a theoretical criticism—one that is unable to incite the change that it promotes.

While neoliberalism and its various principles have been the main discourse guiding various leading economies since the 1980s, such as the United States, and the United Kingdom, its popularity has arguably dwindled in recent years. The advent of the 2008 Financial Crisis has arguably pushed for an ideological shift away from neoliberal principles, with various forms of government intervention—such as fiscal stimulus and deficit spending—implemented during the period of economic downturn [32]. As such, the neoliberal discourse has arguably lost much relevance, and other discourses have replaced it as the dominant perspective on capitalism, and in the process, once again reshaping the economy and the various social and political institutions.

What then remains, and continues to be popular since its inception, especially in Central European and Scandinavian countries, is the sustainable development discourse. Such a discourse arguably portrays a most humane version of capitalism—one that is able to allow for the co-development of the social and economic aspects of society, without compromising on the environment. The ideas perpetuated by this discourse, has been aggressively promoted by the United Nations, the World Bank, and other global institutions, and various state policies have been altered to fit the sustainability agenda [14,33].

The influence of the discourse could be seen in Singapore—a strong advocate for the sustainable development cause. This goal is pursued through an amalgamation of methods—such as the use of technology to facilitate the greening of industries, as well as the implementation of certain socially redistributive policies—but such methods have arguably achieved varying degrees of success. Nevertheless, the case provides the opportunity for an empirical inquiry into the nature of capitalism. The effects of a specific form of sustainable development could be observed and evaluated, therein providing some insights into the possibilities of a form of capitalism that could in fact, be humane in nature.

6. Case Study—Singapore

A detailed of examination of various state policies, as well as economic, social and environmental indicators, have been conducted in this study. The investigation of the various policies implemented in Singapore has provided some empirical evidence of the influence of the sustainable development discourse, in shaping the form of capitalism displayed in Singapore. Through the examination of the various economic, social and environmental indicators, such as the Gini Coefficient, and the various sustainability rankings, some form of evaluation of Singapore's sustainable capitalism model could be carried out—allowing us to determine if the Singapore's model could portray the type of humane capitalism that is of interest to the study.

An examination of the various social and environmental policies implemented in Singapore, seemed to indicate an effort to move towards sustainable development in the country. Various policies point towards an economy that is mostly structured around the sustainable development discourse—and seems to indicate genuine effort towards some form of humane capitalism. Reasons supporting this assumption have been summarized in Table 2.

6.1. Singapore's Capitalism

The examination of various economic, social and environmental policies, seem to indicate that Singapore's economic system is more structured in the form of a sustainable capitalism—a model that is largely influenced by the sustainable development discourse. Specific policies, such as the Singapore Sustainable Blueprint, laid out concrete plans for achieving ecologically sound goals, through energy reduction, or recycling initiatives [37]. Investment in green technology, such as green buildings, have also helped to push for more sustainable living habitats in Singapore, and such goals have seen some form of success, displaying some evidence of a form of humane capitalism, wherein environmental goals are actively pursued.

Other institutional changes have also been undertaken, wherein the state has actively taken on the role of market restructuring—to help establish a niche green market, through increased investment in R&D of green technology [6]. The establishment of the green market, arguably provides an optimistic outlook of sustainable capitalism, indicating the possibilities of the co-development of economic and environmental goals. The movement towards the green market, could also be observed in bi-lateral state-led initiatives—the Sino-Singapore Tianjin Eco-city project—which has created opportunities for the Singapore economy to achieve economic growth through profiting from green business opportunities [6]. These various initiatives indicate that the sustainable development discourse has significantly influenced Singapore's form of capitalism, wherein environmental goals are seriously pursued through a process of state-led ecological modernisation [6].

Table 2. Summary of features of discourses and coherence of policies to these features.

Discourse	Characteristics	Do Policies Cohere to These Characteristics?	Reasons
Neoliberalism	(1) Minimal state intervention, market should ideally be free from intervention and other forms of distortions (2) Market should ideally be made up of small players, as monopolies and oligopolies act as a form of disruption to the market	No	Pursuit of economic goals has been obtained through active state intervention. Examples of such behavior are listed below: (1) Establishment of government-linked companies (GLCS) - GLCS are companies in which partially or directly owned by the state—usually through the government investment company, Temasek Holdings - State directly aids in positioning such companies in a global economy, a form of intervention that goes against Neoliberal principles - GLCS usually hold significant market share—and hence are able to exercise some form of monopoly power—thus acting as some form of disruptions to the market; Neoliberal policies should attempt to resolve this through increased deregulation, but the absence of such policies indicate that the Neoliberal discourse has less influence on Singapore's brand of capitalism (2) Market restructuring through national strategies - Markets have been actively altered and structured, through state policies, since the 1960. - Through various policies, Singapore's economy have been altered through the years, from a labor intensive manufacturing based economy, to a financial and services hub—intentionally altered by the state to position its economy in an increasingly competitive world economy - Such active state involvement in the market seems to indicate that Singapore's brand of capitalism does not adhere fully to Neoliberal principles

Table 2. *Cont.*

Discourse	Characteristics	Do Policies Cohere to These Characteristics?	Reasons
Neo-Marxism	(1) Capitalism is an unsustainable economic system, and is inherently exploitative—therefore requiring radical change in its economic structure (2) Prioritization of social and environmental goals	No	(1) No major structural changes to the economy—Singapore's economic remain strongly embedded in the capitalistic system of free market economics (2) Economic growth remains one of the top priorities, and various institutions have been set up in pursuance of such goal Examples include the Committee of the Future Economy, which aims to identify potential industries and markets, and through various strategies, help to ensure that Singapore's economy remains in a strong position in the global economy of the future. - An examination of its various goals (based on a press release by the Committee of the Future Economy), seems to indicate that an economic imperative is present within the institution; and while it seems to address the need to achieve a "sustainable growth", the institution seems to place more emphasis on economic growth, through the need to "reinforce economic advantage" As such, these various reason indicate that the Neo-Marxist discourse have little influence on Singapore's brand of capitalism
Sustainable Development	(1) Capitalism can be altered to pursue humane and sustainable goals (2) Pursuit of technological advancements and institutional arrangements that could result in the achievement of a sustainable form of capitalism	Yes/Maybe	(1) Policies put forth in the Sustainable Singapore Blueprint seems to indicate a movement towards the pursuit of an ecologically sound society, with a serious attempt at achieving sustainable goals (2) Various attempts at closing the social inequality gap, through the use of redistributive policies such as the Workfare Income Scheme (WIS) (3) Active promotion of sustainable goals through institutional change

Adapted from: [1–4,34–36].

Various social policies have also indicated that the aspect of social justice has not been neglected by the state—and some attempts have been made to address the issue of social injustice within Singapore. Redistributive policies such as the Workfare Income Scheme (WIS), provides low-income wage earners with some monetary supplements, which should allow more income for such groups to meet their basic needs. Other initiatives, such as the progressive wage model system, similarly helps to push for a more sustainable society, through the stipulation of a minimum wage for certain types of jobs in some industries, such as the cleaning industry [38,39].

6.2. Social and Environmental Outcomes

6.2.1. Environmental Outcomes

While the various policies have indicated that Singapore's economy is structured around a sustainable capitalistic model, the effectualness of such policies have been relatively varied. Some success have been seen with Singapore's environmental policies, and sustainability indexes, such as the Asian Green City Index, have placed Singapore the top of its rankings, indicating that ecological sustainability has been somewhat achieved [40]. The positive ranking has arguably provided some evidence of the success of the Singapore's model of sustainability, wherein its investment of green technology and various institutional arrangements, targeted at establishing green markets and achieving green goals, have largely been a success, indicating the possibilities of a co-development of both economic and environmental goals—therein providing a rather optimistic view of sustainable capitalism.

The success of the Singapore's case provides evidence that the profit motive—an aspect of capitalism more associated with exploitation, and seen as the source of various social and environmental problems—could instead provide the necessary impetus for the pursuit of green goals—wherein the sustainability industry provides a potential avenue in which economic benefits could be derived from. As such, the case seems to point towards the possibility that capitalism could be employed to pursue sustainable goals, through some modifications, and hence seems to indicate the possibility of a capitalism with human face.

6.2.2. Social Outcomes

While attempts have been made to address the social aspect of sustainability, the various redistributive policies could be seen as somewhat lacking, with success within this area being relatively limited. The social income gap remains a problem within Singapore, with social indices, such as the GINI Coefficient, indicating that the problem of social inequality remains somewhat prevalent in Singapore. The Gini Coefficient has amounted to a figure of 0.37 in 2015, a relatively high figure that is only 0.03 away from the problematic range indicated by the UN-Habitat [41,42]. However, some evidence of a closing social inequality could be seen, with the GINI coefficient following a decreasing trend in recent years, decreasing from 0.412 in 2014 to 0.37 in 2015, which should indicate that the various redistributive policies have at least some positive effects, thus pushing for a more socially equitable economy [43].

Nevertheless, certain problems remain embedded within Singapore's form of capitalism. While Singapore's economic growth continues to remain vibrant and strong, and its policies seem to indicate a movement towards a more sustainable society, elements of the Treadmill of Production could be observed in its economy. Economic growth has arguably been sustained, at least partially, through the use of lowly paid foreign labours, who work in various industries including the construction and transportation sectors. The absence of a minimal wage system, coupled with the relatively high supply of workers, meant that wages for foreign workers are sometimes depressed to minimal amounts [44].

This brings up a secondary, but yet crucial question on the concept of a sustainable society. In the context of globalization, and in the presence of the increasing flows of people and resources across nation-state borders, what then constitute a society, and therein, a sustainable society? The GINI coefficient tabulated only includes census obtained for Singaporeans and permanent residents living in Singapore, and does not account for the foreign workers, who work here on a short term basis—whose total population could amount to as high as 1.3 million [45]. If such workers are accounted for within the calculation of the Gini Coefficient, it is likely that the actual figure may increase by a significant amount. The various redistributive policies seldom include this group of foreign workers, as they do not fall under the requirements for the supplementary income scheme provided by the WIS. Wages therein remain significantly low for this specific group of workers, and perpetuate the issue of social inequality in Singapore.

Unless this issue is addressed, it is likely that the form of sustainable capitalism practiced in Singapore, may not be as humane in nature—as the pattern of exploitation is merely shifted from within the nation state to other countries—a problem that is illustrated in Wallerstein's World Systems Theory [19]. In highlighting the plight of the foreign workers in Singapore, this paper reiterates that a critical assessment of globalization is required, and global flows must be taken into account in the assessment of sustainability in any society.

The persisting problem of social inequality, as well as the evidence of a more subdued pattern of exploitation within Singapore's sustainable capitalistic model, could point towards some issues with this specific discourse of Sustainable Development. The case study has suggested that there could be evidence of a conflation of the Sustainable development discourse, and its sub-discourse, the Ecological Modernisation Theories. The over-emphasis on ecological and environmental goals could arguably be seen in Singapore, with social goals being somewhat overshadowed. Such a problem would have significant effects in the pursuit of a humane capitalistic system, wherein problems of social injustice may be overlooked by policy makers, in the more pronounced success of the other two sphere of sustainability—hence straying further away from the goal of a truly humane form of capitalism. The rhetoric of sustainable development may in fact problematize the issue at hand, acting as a human mask, possibly disguising the exploitative nature of capitalism that lies within.

7. Conclusions

This study has attempted to investigate the nature of capitalism, through the use of discourse analysis. The case study of Singapore has highlighted the challenge of achieving a humane form of capitalism, showing the difficulties of achieving a tri-development of the social, economic and environmental spheres of society. This study has highlighted some issues with Singapore's brand of sustainable capitalism, noting that while much success have been seen within the environmental and economic aspects of Singapore's form of capitalism, its social justice aspect remains relatively stunted, and hence providing a relatively skewed progress towards a truly sustainable economic system.

The conflation of the sustainable development discourse, with its sub-discourse of EMT, seems to be present in the manner for which Singapore approached the issue of sustainability. This arguably resulted in the dominance of EMT in the structuring of Singapore's various sustainable policies, resulting in limited success in improving the social dimensions required in a sustainable society. However, the case has also provided some optimistic evidence towards the possibilities of a humane type of capitalism. The profit motive of a capitalistic system has been shown to be complementary with green goals, which should provide greater incentive for the achievement of sustainable goals if the profit motive could be properly positioned, such as in the case of Singapore. If the social justice aspect of Singapore's sustainable capitalism could be better addressed, Singapore's capitalism could perhaps provide some hope of a fairer and more humane form of capitalism—a type of capitalism, with a human face.

Author Contributions: Both authors designed and conceptualized the paper. Chua Yuhan prepared the initial draft. Both authors then made revisions and finalized the paper.

Conflicts of Interest: The authors declare no conflict of interest.

References

1. Smith, A. *An Inquiry into the Nature and Causes of the Wealth of Nations*; Soares, S.M., Ed.; MetaLibri Digital Library: Amsterdam, The Netherlands, 2007.
2. Marx, K. *Capital: A Critique of Political Economy. Volume 1*; Translated by E. Mandel; Random House: New York, NY, USA, 1977.
3. Schnaiberg, A. *The Environment: From Surplus to Scarcity*; Oxford University Press: New York, NY, USA, 1980.
4. Holmberg, J. (Ed.) *Making Development Sustainable: Redefining Institutions, Policy, and Economics*; Island Press: Washington, DC, USA, 1992.
5. Mol, P.J.A.; Sonnefeld, A.D.; Spaargaren, G. *The Ecological Modernisation Reader—Environmental Reform in Theory and Practice*; Routledge: New York, NY, USA, 2009.
6. Wong, C.M.L. The developmental state in ecological modernization and the politics of environmental framings: The case of Singapore and implications for East Asia. *Nat. Cult.* **2012**, *7*, 95–119, doi:10.3167/nc.2012.070106.
7. Sonnenfeld, A.D. Contradictions of Ecological Modernisation: Pulp and Paper Manufacturing in South-East Asia. In *Ecological Modernisation around the World: Perspectives and Critical Debates*; Mol, A.P.J., Sonnenfeld, D.A., Eds.; Frank Cass: Portland, OR, USA, London, UK, 2000; pp. 235–256.

8. Gould, A.K.; Pellow, N.D.; Schnaiberg, A. *The Treadmill of Production—Injustice and Unsustainability in the Global Economy*; Paradigm Publishers: London, UK, 2008.
9. Sternberg, E. Defining capitalism. *Econ. Aff.* **2015**, *35*, 380–396.
10. Scott, B.R. *Capitalism—Its Origins and Evolution as a System of Governance*; Springer: New York, NY, USA, 2011; ISBN 978-1-4614-1878-8.
11. Rubin, H.P.; Klumpp, T. *Property Rights and Capitalism*; Oxford University Press: Oxford, UK, 2011.
12. O'Neill, O. Transnational Justice. In *Political Theory Today*; Held, D., Ed.; Polity Press: Cambridge, UK, 1999.
13. Sen, A. *Development as Freedom*; Alfred A. Knopf: New York, NY, USA, 1999.
14. United Nations General Assembly. *Report of the World Commission on Environment and Development: Our Common Future*; United Nations General Assembly, Development and International Co-Operation: Environment: Oslo, Norway, 1987.
15. Yin, K.R. *Case Study Research: Design and Methods*, 2nd ed.; Sage Publications: Newbury Park, CA, USA, 1994.
16. Baxter, P.; Jack, S. Qualitative Case Study Methodology: Study Design and Implementation for Novice Researchers. *Qual. Rep.* **2008**, *13*, 544–559. Available online: http://www.nova.edu/ssss/QR/QR13-4/baxter.pdf (accessed on 2 January 2018).
17. Hajer, A.M. *The Politics of Environmental Discourse: Ecological Modernization and the Policy Process*; Oxford University Press: Oxford, UK, 1995.
18. Frank, G.A. The Development of Undevelopment. *Mon. Rev.* **1966**, *18*, 17–31.
19. Wallerstein, I. The Rise and Future Demise of the Capitalist World Economy: Concepts for Comparative Analysis. *Comp. Stud. Soc. Hist.* **1974**, *16*, 398–415.
20. Mol, P.J.A.; Sonnerfield, A.D. *Ecological Modernisation around the World: Perspectives and Critical Debates*; Psychology Press: Hove, UK, 2000.
21. Eecke, V.W. *Ethical Reflections on the Financial Crisis 2007/2008: Making Use of Smith, Musgrave and Rajan*; Springer: Berlin/Heidelberg, Germany, 2013.
22. Steger, M.B.; Ravi, K.R. *Neoliberalism: A Very Short Introduction*; Oxford University Press: Oxford, UK, 2010.
23. Gamble, A. Neo-Liberalism. *Cap. Class* **2001**, *25*, 127–134.
24. Islam, M.S. *Development, Power and the Environment: Neoliberal Paradox in the Age of Vulnerability*; Routledge: New York, NY, USA, 2013.
25. Harvey, D. *A Brief History of Neoliberalism*; Oxford University Press: Oxford, UK, 2005.
26. Marx, K. The Communist Manifesto. In *Karl Marx Selected Writings*; McLellan, D., Ed.; Oxford Univeristy Press: Oxford, UK, 2000.
27. Marx, K. Economic and Philosophic Manuscripts of 1844. In *Marx's Concept of Man*; Translated by E. Fromm; Continuum: New York, NY, USA, 1995; pp. 90–196. (First published in 1932)
28. Pearce, D. *Is Sustainable Development Compatible with a Free Market?*; CSERGE Working Paper PA 96-02; Centre for Social and Economic Research on the Global Environment: Norwich, UK, 1996.
29. Huber, J. *Die Verlorene Unschuld der Okologie: Neue Technologien und Superindustrielle Entwicklung*; Fischer: Frankfurt, Germany, 1982.
30. Huber, J. *Die Regenbogengesellschaft: Okologie und Sozialpolitik*; Fischer: Frankfurt, Germany, 1985.
31. Langhelle, O. Why ecological modernization and sustainable development should not be conflated. *J. Environ. Policy Plan.* **2000**, *2*, 303–322.
32. Patomäki, H. Neoliberalism and the Global Financial Crisis. *New Political Sci.* **2009**, *31*, 431–442.
33. Soubbotina, P.T. *Beyond Economic Growth—An Introduction to Sustainable Development*; The World Bank: Washington, DC, USA, 2000.
34. Ramírez, D.C.; Tan, L.H. *Singapore, Inc. Versus the Private Sector: Are Government-Linked Companies Different?*; IMF Working Paper. WP/03/156; International Monetary Fund: Washington, DC, USA, 2003.
35. Ministry of Trade and Industry Singapore. Committee on the Future Economy to Review Singapore's Economic Strategies and Position Us for the Future. Press Release. 2015. Available online: https://www.mti.gov.sg/NewsRoom/SiteAssets/Pages/COMMITTEE-ON-THE-FUTURE-ECONOMY-TO-REVIEW-SINGAPORES-ECONOMIC-STRATEGIES-AND-POSITION-US-FOR-THE-FUTURE/Committee%20on%20the%20Future%20Economy%20to%20Review%20Singapores%20Economic%20Strategies%20and%20Position%20Us%20for%20the%20Future.pdf (accessed on 2 January 2018).
36. Yeung, W.C.H. State intervention and neoliberalism in the globalizing world economy: Lessons from Singapore's regionalization programme. *Pac. Rev.* **2000**, *13*, 133–162.

37. Ministry of Environment and Water Resources & Ministry of National Development. *Our Home, Our Environment, Our Future—Sustainable Singapore Blueprint 2015*; Ministry of the Environment and Water Resources: Singapore, 2014.

38. Workfare. Workfare Income Supplement (WIS) Scheme. Available online: https://www.workfare.gov.sg/Pages/WIS.aspx (accessed on 2 January 2018).

39. Ministry of Manpower. Order by Commissioner for Labour under Section 80H(2) Environmental Public Health ACT. Available online: http://www.mom.gov.sg/~/media/mom/documents/employment-practices/order-by-commissioner-for-labour.pdf (accessed on 2 January 2018).

40. Siemens Economist Intelligence Unit. The Green City Index. 2012. Available online: https://www.siemens.com/entry/cc/features/greencityindex_international/all/en/pdf/gci_report_summary.pdf (accessed on 2 January 2018).

41. Ministry of Finance. *Income Growth, Inequality and Mobility Trends in Singapore*; Ministry of Finance Occasional Paper; Ministry of Finance: Singapore, 2015.

42. Smith, J.C. *A Handbook on Inequality, Poverty and Unmet Social Needs in Singapore*; Lien Centre for Social Innovation: Singapore, 2015.

43. Department of Statistics Singapore. Key Household Income Trends, 2014. Press Release. Available online: https://www.singstat.gov.sg/docs/default-source/default-document-library/news/press_releases/press16022015.pdf (accessed on 2 January 2018).

44. Basu, R. $1.50 an hour is just too little for anyone. *The Straits Times*, 12 February 2014. Available online: http://www.straitstimes.com/singapore/150-an-hour-is-just-too-little-for-anyone (accessed on 2 January 2018).

45. Department of Statistics Singapore. Foreign Workforce Numbers. 2015. Available online: http://www.mom.gov.sg/documents-and-publications/foreign-workforce-numbers (accessed on 2 January 2018).

sustainability

MDPI

Editorial

Towards an Environmental Sociology of Sustainability

Md Saidul Islam * and Chua Yuhan

Division of Sociology, Nanyang Technological University Singapore, 14 Nanyang Drive, Singapore 637332;
YCHUA014@e.ntu.edu.sg
* Correspondence: msaidul@ntu.edu.sg

Abstract: The various articles within this book have put forth perspectives that the field of environmental sociology has to offer to the issue of sustainability. Its unique theoretical traditions have been introduced in the first section, with it highlighting some of the more prominent frameworks and methodological innovations available in this field. Articles in the second section focused on the exercise of such frameworks, through the empirical analysis of various case studies regarding issues of sustainability. This particular editorial serves as a concluding chapter to this Special Issue on "Sustainability through the Lens of Environmental Sociology", with it summing up some of the key points discussed in the earlier papers. This editorial will attempt to piece these frameworks into a cohesive whole, and in so, unveil a holistic lens towards an environmental sociology of sustainability.

Keywords: environmental sociology; sustainability; environmentalism; treadmill of production; ecological modernisation; social constructivism

1. Introduction

In the beginning chapter of the book, an introduction into the brief history of environmental sociology has been put forth, with it presenting various ideas that have influenced and continued to influence this field. In the following chapters, within the very same section, contemporary concepts, theories, as well as various methodologies employed, have been systematically laid out. These provided readers with a comprehensive idea of what environmental sociology comprised of, as well as how this field of inquiry contributes and addresses issues of sustainability today.

A second section then presented the readers with case-studies of various contemporary environmental challenges, with them not only providing an exercise of frameworks presented in the preceding section, but also, more importantly, highlighting the need for societies to place more emphasis on sustainability issues. The problems identified demonstrated that issues of sustainability can often be multifaceted in nature, with varying scales that could occur on both regional and global levels. A reflexive evaluation of attempts put forth to curb these environmental problems also revealed the complexity of these issues, with solutions sometimes panning out in unexpected manners. Yet such case studies highlighted the imperative need for a continued focus on providing, or at least attempt to, providing solutions to the environmental challenges that we see today. Throughout this special issue, the various articles demonstrated the strength of environmental sociology as a lens to understand the complexities of the issue of sustainability. Consequently this field of inquiry is seen to present a possible manner to attain a state of sustainability that our society, and our planet severely require.

In this concluding chapter, this editorial will review the various articles presented in this special issue, presenting evaluation of the current state of environmental sustainability, as well as its way forward. In doing so, this editorial will consolidate the various perspectives and frameworks of environmental sociology discussed, therein presenting a cohesive lens that can be used to understand and manage issues of sustainability today.

2. Environmental Sociology—A Lens into Sustainability

Before presenting the various frameworks available in this specific field of inquiry, it is necessary to once again reiterate, what environmental sociology entails. As put forth in the very first chapter in this compilation of articles, environmental sociology, in essence, is the study of the interaction between two systems—the social and the ecological [1]. Environmental sociology places the focal point on how social systems organize themselves, respond to, and affect ecological systems, with the two systems sharing a complex but deeply intertwined relation [2]. This gives us a useful tool in examining issues of sustainability—a tool that has arguably become crucial—given the spate of complex environmental problems that are present in today's world.

While environmental sociology was only officially established relatively recently, in the 1970s, the traditions supporting this specific field have had a relatively long history, with many stemming from the classical frameworks present in sociology. These frameworks were then reintroduced by later scholars, who provide a reinterpretation of the classical traditions, and applied it for the specific use of studying socio-environmental relations. More contemporary theories can be loosely categorized into three perspectives—Neo-Marxist theories, Neoliberalism theories, and Symbolic Interactionism. Each tradition offers a different lens into issues of sustainability, though their commonality lies in how their focus still remains on the relation between the society and the environment.

In looking at the different frameworks, Neo-Marxist theories seem to offer a more critical view on issues of sustainability. Theories within this tradition often looked at how economic systems, specifically the system of capitalism, negatively impact society and the environment. Notable theories include the theory of metabolic rift—a theory first put forth by Karl Marx and later reinterpreted by John B. Foster. The theory discusses capitalism itself introduces an unnatural segmentation, or otherwise termed as rift, between society and ecology—systems which are initially deeply intertwined as a single and whole entity [3]. In another theory put forth by Allan Schnaiberg, the critical view on capitalism is once again reiterated, wherein it was suggested that a system of withdrawals and additions characterize the current economic system present today, creating a never-ending, self-perpetuating, and highly destructive treadmill of production [4]. This critical view on the economy has been reiterated by the proponents of the World System Theory, a framework first put forth by Immanuel Wallerstein. This theory categorizes the world economy into an interplay between three positions of actors, countries who make up the core, those who make up the periphery and others who are in-between—the semi-periphery. In essence, the theory suggests that economic benefits flow from the periphery to the core, with ecological and social destruction in the latter emerging as an ensuing result of these unequal flows [5].

While theories within the Neo-Marxist tradition attributes a significant part of ecological destruction to the role of capitalism, theories and concepts from the Neoliberalism put forth a significantly different perspective. Such theories do not contest the existing political and economic structures, with some perspectives viewing environmental problems as a natural part and extension of the how society functions. The concept of risk society, put forth by Ulrich Beck, frames environmental issues within the confines of a modern society, wherein the process of modernization is seen to have contributed to the creation of hazards, but also, with it, the fostering of a mind-set and methods of organizing and dealing with such hazards. While threats from environmental extremities have always existed, in the form of natural disasters, what is distinct in modern society is that these threats are no longer natural, but in fact are manufactured by human society—derived from the emergence of new technologies. These risks are constantly calculated on a rough balance of benefits versus threats, and in some sense, deliberately controlled [6]. Nevertheless, the risks within a modern society are often ill-managed—resulting in environmental and social catastrophes [6].

One other theory more poised towards that of the Neoliberalist tradition, is the Ecological Modernization Theory. While there exists various perspectives within this broad spectrum of theories, ecological modernization generally proposed that environmental solutions can be resolved through increasing modernization [7]. Capitalism is seen a malleable system, one that can be shaped to address the problem of sustainability, instead of contributing to the destruction of the environment. A more positive view of social and political institutions is also put forth, for which it is suggested that such institutions would naturally—in the presence of environmental disasters—develop a sense of awareness, and therein organize themselves to address the environmental challenges of today [8]. This forms a more nuanced stance within the various discourses surrounding the issue of sustainability, a middle ground between Neo-Marxists who propose that only radical changes can resolve the problem, and fervent capitalists who deny or downplay the presence of environmental problems [1].

Detaching itself from these macro debates, symbolic interactionism puts forth a more micro-meso understanding of individuals' interactions with and perceptions of the environment. In exploring the manners for which meanings are assigned to our surroundings, the concept of naturework discusses how human beings come to their state of relation with the "environment", through constant negotiation and discussion of what constitutes nature. A separation of nature and society emerges from this negotiation of meanings, and inevitably influences what and how we perceive as environmental challenges [8,9].

Debates and mutual critique exists between the different frameworks, most notably between that of the Neoliberalist and the Neo-Marxist traditions. Nevertheless, these various theories provide a greater insight into the crux of environmental problems present today, offering different perspectives and opinions on how current environmental challenges emerged, how they can be understood, and most importantly, how they can be resolved.

3. Contemporary Theories, Concepts and Frameworks

While the frameworks discussed in the preceding section can be seen to provide a strong theoretical understanding of environmental challenges present today, other contemporary concepts have also emerged. These emerging frameworks, put forth in the articles within the first section of the special issue, provided refreshed perspectives towards understanding the relationship between the society and the environment.

In the first article following the introduction, Longo and his co-authors proposed that a critical re-examination of mainstream conceptions of sustainability is required, in order to provide a more comprehensive examination of sustainability issues. It was suggested that current mainstream sustainability concepts are largely entrenched in what the authors termed as a "pre-analytic vision", arguably forming a problematic and narrow view of sustainability issues today. Mainstream discourses of sustainability and sustainable development—perpetuated by political and social institutions—are seen to have been built upon neo-classical economic foundations, which both elevates and normalize the role of a capitalist economic system in a sustainable society [10].

Within the pre-analytic vision, crucial questions and criticisms regarding economic growth are almost non-present, creating an inadequate analysis, and subsequently, the creation environmental solutions and policies that are much less effective than desired. It is hence crucial to shift away from this pre-analytic vision of sustainability, and move towards a more integrative view on socio-eco relation. With this in mind, the article discussed the ever-importance of environmental sociology as a lens to sustainability, proposing that various critical theories within the field, such as the treadmill of production, can be actively employed to and integrated into mainstream concepts, for a more holistic understanding of contemporary and future sustainability challenges [10].

In the following article, Arias-Maldonado explored the ways for which sustainability can be theorized and examined when placed within the context of a post-natural epoch. The article attempts to bridge perspectives from environmental sociology with that of the discipline of Earth-system science, bringing into foreground questions on the nature of socio-nature relations with the emergence of the Anthropocene. Characterized by a hybridized socio-natural system that has become increasingly pronounced—with the social and the nature being almost indistinguishable and indiscernible—it is suggested that sustainability studies today should consider this new perspective in constructing their visions of the nature of sustainability [11].

4. Methodological Innovations

The two articles mentioned earlier provided clear theoretical and conceptual contributions to the field of environmental sociology, in the context of sustainability. Longo and his co-authors presented a compelling need for the inclusion of critical theories of environmental sociology in mainstream conceptions of sustainability [10]. Arias-Maldonado, on the other hand, pushed for the incorporation of a new perspective on socio-natural relations in environmental sociological studies of sustainability. By introducing the geological concept of Anthropocene, the paper arguably bridges two disciplines, therein promoting a more integrative and updated environmental sociological lens on the issue of sustainability [11].

The subsequent two papers in the section puts forth new methodologies that can be employed in the field of environmental sociology. Brown's paper demonstrated the use of large scale textual and discourse analysis as an analytical method in the study of how multinational corporations (MNCs) involved in the extraction of natural resource understand and practice sustainability. The paper revealed that within the selected category of MNCs, firms employed specific vocabulary and grammar in their business reports, in a way which allows them to simplify their management of natural spaces [12]. Terms used often allow firms to quantify their operations in these spaces, therein facilitating the incorporation of sustainability management into the firms' usual operational processes. While this method is arguably still infant in the study of sustainability, given its significant limitations in applicability to other types of organizations and businesses, as well as doubts on its adequacy as a proxy for the level of sustainability, Brown's innovative approach does indeed present a refreshing manner to study sustainability issues. Such an approach could be further refined with a mixed-methods approach, and could perhaps be promising when applied not just to business reporting, but other texts including governmental policies and speeches.

The other paper by Sing Chew and Daniel Sarabia, puts forth a distinct, yet equally unconventional methodology in the study of sustainability issues. Their paper proposed a historical analysis of nature-culture relations, for which they tracked the interaction between world systems and the environment in various epochs and regions for the last 5000 years. It is suggested that the long term tracking of the past would provide greater insights into the current status of development, and present a more holistic overview of socio-nature relations. The paper revealed that changes in society's structure can often be tied to transformations and crises stemming from nature and the climate [13]. Historically, both nature and society have arguably shared a system of feedback. The use of the historical methodology could therein provide hints into the future development of socio-natural relations, presenting a distinct perspective that can be incorporated into the field of environmental sociology.

5. Environmental Sociology in Praxis—Issues, Problems and Case Studies

The methodological innovations mentioned earlier signal the availability of new tools for which environmental sociologists can employ to examine the crucial issue of sustainability. Yet, tools and frameworks are merely conceptual, if not put into praxis. The second large section of this volume has consolidated papers which demonstrated an exercise and application of the various tools available in the arsenal of environmental sociology. These case studies not only highlight the strength of environmental sociology as a lens of inquiry, but also reiterate the severity of the environmental challenges of today.

In the first paper in part 2 of this special issue, Islam and his colleagues applied an environmental sociological lens to the study of transboundary haze pollution in South East Asia [14]. Their paper provided insights into the emergence and persistence of the haze issue in the region, through its application of the theory of the Treadmill of Production and the environmental governance approach. The paper identified the use of the notorious agricultural method of Slash and Burn as a key contributor to the haze pollution in the region. More notably, the paper identified the underlying reasons and factors encouraging the use of such method, highlighting the various parties involved in allowing for the emergence, and persistence of haze pollution in the region. Given the complexity of this transboundary issue, the paper suggested a pluralistic framework of sustainability, putting forth various recommendations that could help to mitigate the problem of transboundary haze [14].

In the following paper by Hui-Ting Tang and Yuh-Ming Ling, a commendable plan to consolidate a framework of sustainable urban development was put forth. This was carried out by drawing upon information available across different disciplines, as well as deriving insights from both historical and contemporary definitions and indexes of sustainability across different regions [15]. In the process, the paper concluded with the successful amalgamation and integration of these different frameworks, therein presenting a common framework of sustainable urban development. This framework—which consists of three themes covering both quantitative and qualitative measures of the state of social and environmental wellbeing, as well as strategies for their management—provided an ease of reference for policy makers with a stake in sustainable urban development. More importantly, this framework arguably paved the way for a more strategically organized sustainability effort—one severely required in the highly discursive field of sustainability.

On a different note, Lisitza and Wolbring examined sustainability within "academic ecohealth" literature through the use of textual analysis. This emerging field of ecohealth is partially influenced by sustainability discourses, and places the focal point on how changes in the social, physical, biological and economic environments affect the well-being of individuals, animals and ecosystems, in a bid to improve their health. The paper concluded that sustainability discourses are poorly represented in majority of the texts analysed, and if represented, are not explored adequately in a conceptual manner [16].

The subsequent articles moved on to discuss sustainability in the context of food and energy systems. McGee and Alvarez applied the theory of metabolic rift in the study of the practice of certified organic farming. In their paper, it is argued that though the implementation of organic farming could and have benefited the environment and consumers, the circumstances for which organic farms emerged and subsequently developed, have prevented the full realization of its potential to positively impact the environment. With its growing reliance on organic fertilizers and pesticides, organic farming merely presented a lesser-of-two-evils scenario in impacting the environment, when compared to conventional agricultural systems. It was suggested that a metabolic rift of organic farming is similarly present, wherein the practice is seen to contribute to increase in water pollution in a manner not dissimilar to that of conventional practices [17]. This meant that these so-called sustainable practices merely sustain a larger socio-economic system that perpetuates environmental degradation, possibly hinting that only fundamental changes could bring about the resolution of current environmental problems. This critical perspective on organic farming therein revealed the complex nature of what could be perceived as sustainable practices, once again bringing about crucial questions on how true sustainability could be achieved.

The following paper by Galli and Fisher presented a theory of a much different nature compared to that of McGee and Alvarez, providing a more Neoliberalist perspective on sustainability issue in the United States through the Ecological Modernization Theory. The paper noted the emergence of hybrid arrangements in the governing of energy efficiencies in various US cities, following the implementation of the Energy Efficiency and Conservation Block Grant (EECBG) program administered by the US Department of Energy. Due to the short term nature of the grants, hybrid arrangements at the sub-regional level, and across different actors, are arguably a necessity in the prolonged sustenance of the energy efficiency programs and projects [18]. This pluralistic, multi-actor model of climate governance, which include public-private and civil partnerships, highlighted the possibilities of alternative interpretations and implementation of sustainability related policies [18]. Importantly, findings from the paper support the Ecological Modernization Theory, providing an affirmation of the possibilities of a form of socio-political modernization required in addressing the environmental challenges of today [18].

In the last article of the section, Mol and Oosterveer conducted an enquiry into the effectiveness of various sustainability certifications currently present in the global agro-food value chains.

Four ideal types of value chain traceability are identified—management traceability, regulatory traceability, consumer traceability and public traceability, differentiated by their purposes, as well as the different target audience requiring or interested in tracing the value chain. In the context of sustainability, the paper puts forth a systematic analysis of the various traceability systems, with a focus on consumer and public traceability, employed in the tracking of sustainability in the agro-food supply chain. Four certification models have been explored in the paper, notably the track and trace model, the segregation model, the mass balance model and a new and emerging traceability system— the book and claim model. Determining the appropriateness of the various models of traceability largely boils down to the combination of at least five factors, factors that are socially, economically, market, and product determined, with different effectiveness and cost-effectiveness in tracing sustainability, based on their different contexts. Through this assessment of the sustainability traceability models currently present in agro-food value chains, it was determined that economic and market driven logics are increasingly influencing traceability models, and therein could have consequences for sustainability in agro-food value chains in the future [19].

6. Environmental Sustainability and the Way Forward

The third and final part of the special issue started off with an article which explored the nature of capitalism, as well as a discussion of some of the existing discourses on contemporary globalization and sustainability. In this article, Chua and Islam once again bring the argument back to the theoretical, looking into the existing discourses that surround the field of sustainability. The discussion and critique of the discourses, notably between those of Neoliberalist and the Neo-Marxist traditions, highlighted the different lenses used to examine issues of sustainability. By presenting a case study of Singapore, a country that has notably achieved a level of standard commendable in global polls of sustainability, the article provided a crucial empirical backdrop to the various contesting frameworks available in the field of environmental sociology. By exploring the case study through a discourse analytical framework, the paper discussed what is perceived as a conflation of discourses and sub-discourses surrounding sustainable development in Singapore—an issue that has problematized and placed limits on the success of sustainability in the region [20].

The article has demonstrated that a certain level of difficulty lies in the pursuit of sustainability, largely due to pre-existing tensions between the three established spheres required of a sustainable society. Such difficulties are amplified when issues of sustainability are examined not only within a closed system, or in this context, a singular case study of Singapore, but on an open system. As suggested, a society that could be sustainable when viewed on its own, could be deemed otherwise, when the flows of globalization are taken into account. In this case study of Singapore, the issue of social justice becomes prominent, with the inclusion of the treatment of foreign workers being placed in the "calculation" of the sustainability of the country. This suggests that there may be a need to re-evaluate how sustainability is traditionally examined by policy makers and institutions, as well as more importantly, the confines and boundaries for which a sustainable society is defined in [20]. The debates on contemporary globalization and sustainability are brought to the foreground, along with an open-ended and arguably long-standing question: Does capitalism have a human face?

7. Concluding Statement

The many articles included in the special issue have highlighted the various frameworks and methodologies available in the field of environmental sociology. Their application on both local and global case studies has also revealed the strength of the discipline, demonstrating its aptitude in providing critical perspectives on issues of sustainability. Supported by a wealth of established theories, loosely categorized into the Neo-Marxist, Neoliberal and Symbolic Interactionist traditions, environmental sociology as a field of inquiry provides a crucial method to understand the complex socio-nature relations of today. More importantly, environmental sociology presents a possible manner to understand the persistence and increasing scale of contemporary environmental problems, allowing for a critical reassessment of the dire environmental challenges facing contemporary societies.

These various papers have cemented the position that environmental sociology holds in the current study of sustainability. Yet, it is important to postulate the future status of this field of inquiry, whether in studying sustainability, or things of other nature. The emerging frameworks and methodological innovations within this field—as highlighted in the first section of the special issue— have already signalled that this specific field of inquiry is one that is ever-evolving, with respect to the ever-changing environmental challenges and issues of sustainability. It is especially crucial for the field to remain updated and relevant, for many of the pressing environmental problems are complex issues which are arguably increasing in both scale and intensity.

It should be noted that environmental problems are not exclusive to the current period. History has highlighted that environmental challenges are not just a contemporary phenomenon, but are in fact present in various epochs of the past. A longstanding and complicated relation can be seen to exist between society and nature, with changes in respective spheres resulting in mutual systemic crises [13]. This suggests that environmental challenges, at least with respect to human existence and interaction, has always persisted, and may continue to persist, albeit in different manners in the future. As such, environmental sociology as a discipline, and as a lens to understand socio-nature relations, will most probably remain relevant and perhaps increasingly crucial, in not just issues of sustainability, but perhaps those of Mother Nature, in the near future.

Author Contributions: Both authors conceptualized and designed the paper. Chua Yuhan wrote the initial draft, and Md Saidul Islam finalized it. Both authors claim equal authorship.

Conflicts of Interest: The authors declare no conflict of interest.

References

1. Islam: M.S. Sustainability through the Lens of Environmental Sociology: An Introduction. *Sustainability* **2017**, *9*, 474.
2. Gould, K.A.; Lewis, T.L. (Eds.) *Twenty Lessons in Environmental Sociology*; Oxford University Press: New York, NY, USA; Oxford, UK, 2009.
3. Foster, J.B. Marx's Theory of Metabolic Rift: Classical Foundations for Environmental Sociology. *Am. J. Sociol.* **1999**, *105*, 366–405.
4. Allan, S. *The Environment: From Surplus to Scarcity*; Oxford University Press: Oxford, UK, 1980.
5. Wallerstein, I. *The Modern World System I: Capitalist Agriculture and the Origins of the European World-Economy in the Sixteenth Century*; Academic Press; New York, NY, USA, 1974.
6. Beck, U. *Risk Society: Towards a New Modernity*; Sage Publications: London, UK, 1992.
7. Hannigan, J. *Environmental Sociology: A Social Constructivist Perspective*, 2nd ed.; Routledge: New York, NY, USA, 2006.
8. Islam, M.S. *Development, Power and the Environment: Neoliberal Paradox in the Age of Vulnerability*; Routledge: New York, NY, USA; London, UK, 2013.
9. Fine, G.A. *Morel Tales: The Culture of Mushrooming*; Harvard University Press: Cambridge, MA, USA, 1998.
10. Longo, S.B.; Clark, B.; Shriver, T.E.; Clausen, R. Sustainability and Environmental Sociology: Putting the Economy in its Place and Moving toward an Integrative Socio-Ecology. *Sustainability* **2016**, *8*, 437.
11. Arias-Maldonado, M. The Anthropocenic Turn: Theorizing Sustainability in a Postnatural Age. *Sustainability* **2016**, *8*, 10.
12. Brown, M. Managing Nature-Business as Usual: Resource Extraction Companies and Their Representations of Natural Landscapes. *Sustainability* **2015**, *7*, 15900–15922.
13. Chew, S.C.; Sarabia, D. Nature-Culture Relations: Early Globalization, Climate Changes, and System Crisis. *Sustainability* **2016**, *8*, 78.
14. Islam, M.S.; Hui Pei, Y.; Mangharam, S. Trans-Boundary Haze Pollution in Southeast Asia: Sustainability through Plural Environmental Governance. *Sustainability* **2015**, *8*, 499.
15. Tang, H.-T.; Lee, Y.-M. The Making of Sustainable Urban Development: A Synthesis Framework. *Sustainability* **2016**, *8*, 492.
16. Lisitza, A.; Wolbring, G. Sustainability within the Academic EcoHealth Literature: Existing Engagement and Future Prospects. *Sustainability* **2016**, *8*, 202.

17. McGee, J.A.; Alvarez, C. Sustaining without Changing: The Metabolic Rift of Certified Organic Farming. *Sustainability* **2016**, *8*, 115.

18. Galli, A.M.; Fisher, D.R. Hybrid Arrangements as a Form of Ecological Modernization: The Case of the US Energy Efficiency Conservation Block Grants. *Sustainability* **2016**, *8*, 88.

19. Mol, A.P. J.; Oosterveer, P. Certification of Markets, Markets of Certificates: Tracing Sustainability in Global Agro-Food Value Chains. *Sustainability* **2015**, *7*, 12258–12278.

20. Chua, Y.H.; Islam, M.S. Capitalism with a Human face: Debates on Contemporary Globalization and Sustainability. In *Environmental Sustainability through the Lens of Environmental Sociology*; MDPI AG: Basel, Switzerland, 2018, in press.

MDPI AG

St. Alban-Anlage 66

4052 Basel, Switzerland

Tel. +41 61 683 77 34

Fax +41 61 302 89 18

http://www.mdpi.com

Sustainability Editorial Office

E-mail: sustainability@mdpi.com

http://www.mdpi.com/journal/sustainability